PRAISE FOR DEEP TIME

"It would require encyclopedic knowledge, penetrating insight, and critical assessment of extant scientific knowledge to condense approximately five billion years of Earth's history into just about five hundred pages. **'Deep Time: Earth's Geologic Journey from Eternity to *Homo sapiens*'** is an extraordinary, data-driven, and evidence-based work. The author's erudition makes lucid the seemingly mysterious. Kudos to Dr. Santosh K. Ghosh."

Dr. Abhijit Basu
Herman B Wells Professor Emeritus,
Indiana University

"***Deep Time*** brings 3 rare strands together: an encyclopedic and rigorous knowledge of the subject matter relating to geology; a highly developed ability to bring in cross-disciplinary perspectives from the arts, religion, and other areas of human inquiry; and finally a warmth and humanity and a passion for the human story that conveys itself on every page."

Andrew Cross
Educator

"***Deep Time*** is a deeply educational book with an easy, engaging style that motivates the reader to keep turning pages and enjoy the journey. An awesome effort."

Upendra Pai
Chemical Engineer

"***Deep Time*** will be a permanent reference book to be enjoyed and revisited over time. It is written with the rigor of a textbook, yet remains accessible to anyone interested in the history and development of the Earth, humanity, and its potential future."

Dr. Carlos Molinares
Exploration Geologist, Petrobras
Bogota, Colombia

DEEP TIME

Earth's Geologic Journey *from* Eternity *to* Homo Sapiens

Santosh Ghosh

EARTH'S JOURNEY STORYWORKS

Deep Time: Earth's Geologic Journey from Eternity to *Homo Sapiens*

Published by Earth's Journey Storyworks
www.santoshghosh.com | sghoshgeology@gmail.com
First paperback edition June 2025

Layout and cover design by Robin Ghosh for Ghosh Creative Endeavors LLC, Nashville, TN, USA
Proofreader: Cheryl Antao-Xavier
Indexing by Ritu Amrita

First Edition
ISBN 978-1-0691169-0-1

Catalog
A copy of the book stored in the Library and Archives Canada, Ottawa

Disclaimer
The narrative, views and opinions in the book are the author's own. The stated facts, events, stories, etc., have been verified to the extent possible with numerous chapter-wise, cited references at the end of the book.

To you, dear readers,
for rewarding my journey

ABOUT THE BOOK

Geology, the Natural World, and the Human Species

The author's love of nature started during his geological journey in the far corners of India like Meghalaya, Arunachal Pradesh, in the Northeast, in the late 1960s. Later, his extensive work and travel in Asia, Europe, the Americas, especially many decades in Venezuela, Mexico, Belize, Colombia, Peru, the Caribbean islands, etc., deepened his understanding of the uniqueness of Planet Earth. Geology runs deep under superficial lines on the surface of Earth. He remembers seeing in 1977, a 3-faced concrete cairn marking boundaries of Brazil, Guyana, and Venezuela, on top of the inaccessible, uninhabited, 2,800m-high, wind-swept Mount Roraima Table Mountain, made famous by Sir Walter Raleigh's 1595 El Dorado Expedition.

Earth, always on the move, is like a living being with its invisible core, mantle, crust, enveloped by the benign atmosphere vital for all life. The lighthouses at the easternmost point of North America at Cape Spear, Newfoundland, Canada, and on the westernmost point of Europe at the Cabo da Roca promontory, Portugal, tell us of the drifting plates at the Mid-Atlantic ridge passing right through Iceland. The lofty Mt. Kilimanjaro volcano and the elongated Lake Tanganyika are proofs that the East African Rift System is splitting apart Africa before our eyes. The kaleidoscopic *Aurora borealis* and *A. australis* remind us how Earth's geomagnetic field, created by the liquid outer core, is fiercely combating the harmful solar ultraviolet radiation.

Wandering geologist Santosh Ghosh unravels the layered secrets of the evolving Earth in a simple, engaging way. The story starts at the Big Bang zero hour and ends at the last visible milepost of the Present when the *Homo sapiens* reign supreme. Ghosh stresses the oneness of nature and mankind. He tellingly explains how the future of Earth hinges on the untamed forces of nature, as well as the prudence of our species.

Foreword I

I have worked with Santosh Ghosh on numerous oil exploration projects over a span of twenty years. While an expert exploration geologist, this book demonstrates a thoughtful side of him that I was less acquainted with. *Deep Time* refers to the concept of cosmic time in billions of years. This work attempts to decipher the history of the universe and our solar system on a deep time scale, and so by definition is incredibly ambitious!

Geologists are routinely challenged by incomplete records, one analogy being trying to understand a complete mystery novel of which the reader only has a few non-sequential, torn out pages. But the processes recorded in rocks can be directly related to and understood by examining present-day processes, although perhaps on different scales. Our concepts of the history of the universe are much more theoretical than our knowledge of Earth's geology. This work looks at the results of recent space-based telescopes to examine the dimensions of the universe and discuss the possible origins of the universe itself. Ultimately, this is a philosophic discussion which shows connections among the creation stories of great religions of the world.

Part One of *Deep Time* deals with the origin of the universe and our solar system. The book uses isotopic studies of moon rocks to demonstrate that the moon and Earth had a similar origin, which was most likely due to collision of a Mars-sized planet with Earth early in its history. For Earth, this resulted in the formation of a dense iron-nickel core, an intermediate mantle composed of iron and magnesium silicates, and a crust composed of aluminum and silica-rich rocks.

Deep Time goes on to outline general processes of rock formation. These processes include intense meteorite impact on the early Earth, the production of sediments through the action of water, waves, and wind, and the formation of sedimentary rocks by evaporation. The book continues with explanations of the formation of carbonate reefs and of laterites (the source of aluminum ore) by intense tropical weathering. *Deep Time* next covers the formation of metamorphic rocks by the application of heat and pressure to sedimentary and igneous rocks.

Following on from explaining how rocks are formed, the book explores how early geologists made sense of the field through studying fossil assemblages and discovering the concept of superposition. Part One concludes with the discovery that lithospheric plates are mobile and a discussion of the formulation of the theory of plate tectonics and an exposition of our current understanding of the process. This includes explanations of the formation of linear island chains and mid-oceanic ridges, and discussion of the puzzle of cratonic basins.

Part Two of *Deep Time* deals with the evolution of Earth, from the lengthy period when minimal life was present, to a sudden explosion of lifeforms, followed by successive eras characterized by different dominant lifeforms. Throughout this period, the book covers several mass extinctions, caused variously by temperature changes, volcanic activity, and meteorite impact.

Starting with the Precambrian eons, *Deep Time* describes eras of intense meteorite bombardment, the development of primitive one-celled organisms, and the formation of cyanobacterial stromatolite structures.

Next the work discusses the dramatic change that occurred at about five hundred and thirty nine million years ago (539 Ma) with the rapid explosion in number and variety of complex hard shelled organisms. *Deep Time* moves on to the post-Cambrian Paleozoic Era, characterized by high sea levels and the continents of Gondwana and Laurentia, which ultimately collided to form Pangaea. This section also covers the Carboniferous Period, when major coal reserves developed worldwide, and the Permian period, when a major extinction occurred.

The following section covers the Mesozoic Era, from the Triassic period when dinosaurs appeared, through their extinction at the end of the Cretaceous period, likely due to meteorite impact. This discussion of the Mesozoic Era concludes with an examination of the evidence for this extinction in the rock record.

From here, *Deep Time* covers the Paleogene Period, when mammals thrived, followed by the Neogene Period, when the geography of modern Earth began to take shape. The book outlines the changing boundaries of bodies of water and the birth of modern mountain chains. It also discusses the development of various humanoid species during the Neogene and Part Two concludes with an exploration of glaciation and of the advent of *Homo sapiens*.

Finally, in **Part Three**, *Deep Time* speculates on the future of our planet and galaxy. Predicting the geologic future is even more difficult than interpreting the past. The book deals with this challenge by exploring various possible future cosmic events, offering multiple potential outcomes for each scenario.

Deep Time accomplishes the ambitious task that its author sets out to do by highlighting the geologic forces and events at work on our planet over an impossibly long time scale and provides an understanding of Earth's geology that is accessible for even the curious layman. Because this book describes many locations of geologic interest worldwide, it has much to offer armchair travelers with the vivid pictures it paints, and to real world travelers with its potential as a travel companion. In addition to providing a solid foundation for understanding Earth's geology, Santosh Ghosh has produced a work that will encourage curiosity and thoughtfulness in regards to the nature of our planet, solar system, and universe.

Robert G. Hickman

Formerly Coordinator of Structural Geology and Remote Sensing, Unocal Corp.

President, Wabash Exploration, Inc., Houston

Foreword II

My first encounter with Santosh Ghosh occurred in the late summer of 1997, in Caracas, Venezuela, where I worked as an English teacher at an international school. My enduring memory of that first meeting is of a soft-spoken, courteous man with a kindly expression and a highly-developed ability to listen attentively—not a common attribute in today's noisy world.

We talked about our respective professions, and Santosh explained his work as a petroleum geologist. As improbable as I know this will sound, given the very few apparent touchpoints connecting our career paths, that early conversation led to many more in the years to come. I gradually learned how to ask better questions, which would inevitably elicit illuminating responses. As we became better acquainted, Santosh's passion for his work and his ability to translate incredibly complex topics into accessible language pulled back the curtain on the volatile world that lies beneath our feet. Little did I know on that long-ago summer day that I was not the only one living and working in a field laden with poetry and imagination.

It is difficult for me to conceive of any writer setting themselves a more ambitious task than to chart the incomprehensibly vast 4.5-billion-year history of our planet and the solar system of which it is a part. It is not a project for the faint-hearted. In painting his vivid portrait of planet Earth, its origin story and its place in the universe, Santosh Ghosh has produced a work of extraordinary scientific rigour that brings to bear five decades spent working within the field of geology.

While I am hopeful that a scholarly readership will find its way to **Deep Time: Earth's Geologic Journey from Eternity to *Homo Sapiens***, it is not a textbook in the traditional sense. This is a work which, like the Earth itself, is filled with hidden gems. Its author narrates, often in unexpectedly luminous language, a story that this non-specialist reader found unfailingly engaging. Its pages are filled with the joy and excitement of a lifetime of discovery on the leading edge of geological exploration. As he notes in the introductory pages, "The book is for people curious and willing to admire the captivating story of our Blue Planet."

The narrative which unfolds, layer by layer, takes us on a journey through time with a deftness and lightness of touch that verges at times on poetry. Spiritual traditions and their various ways of comprehending our planet and the universe are described alongside references to scientific discoveries dating back to the ancient Greeks and on to Copernicus and Galileo. Santosh Ghosh's own unique life experiences as a global citizen give a texture and colour to the themes he unravels. This is a work of deeply felt compassion for all living beings which inhabit our planet. As his story draws to a close, he turns his focus to the urgent environmental crises which have arisen from climate change and explains their impact on the natural environment: the lungs and beating heart of the planet. His closing sentence is one which I hope all of us who live on what Carl Sagan famously described as this "pale blue dot" will heed: "Let us strive so that the Earth-human relationship does not turn into a zero-sum game."

Deep Time: Earth's Geologic Journey from Eternity to *Homo Sapiens* is a work of remarkable scholarship and profound humanity. It deserves to reach a wide and diverse audience, for it tells a story that must be heard if life on this planet is to have a future as bright as its past.

Andrew Cross
Head of School, *Lauremont School*
Richmond Hill, ON, Canada

CONTENTS

BACKSTORY

deep

/diːp/

1. vast, immeasurable, or profound.

2. extending far into Earth's geological past.

time

/taim/

1. The immense chronological span of Earth's 4.6 billion-year history.

Geology touches many aspects of life on Earth – land, water, atmosphere, living beings, natural resources, climate change, natural hazards, to name a few. Understanding our planet is vital, it highlights the intertwined nature of Earth's many facets, where humans increasingly play a crucial role. It is always rewarding to share fascinating, yet intriguing stories and secrets of the multihued planet we call home. After my retirement, the thought of writing a book slowly crystallized in my mind as I got increasingly involved with a group of retirees of varied backgrounds in my hometown of Toronto, Canada. While acting as a catalyst for the group's quarterly social program, we organized talks on subjects such as rocks, minerals, the cosmos, climate change, and myths around geological wonders etc. Most people enjoy tales of the evolving, ever-changing Earth story in simple, lively language. The big picture is what matters when the canvas is the whole Earth.

I grew up in the heartland of the Ganges River Delta, India, where no rocks are visible for nearly 200km from Calcutta, where I lived. Yet, I had the good fortune to be able to spend close to 50 years of my life as a wandering geologist studying rocks in many countries, wearing different hats as the occasions demanded - field geologist, petroleum and coal geoscientist, research specialist, academia, consultant, and public sector advisor. As I look back, I marvel at the common thread of geological continuity of rock formations across national boundaries everywhere. Names of geological formations may change, but everywhere rocks of a certain age display equivalence, based on intrinsic geological-paleontological criteria. Geology runs deep under superficial lines on the skin of Earth.

The title of the book reveals my goal – a discussion of deep time starting with the nascent, incandescent phase of the Gaia. The chapters follow a predictable, chronological step by step evolution of Earth, pin-pointing the geological uniqueness of each time slice. *Chapter 1* starts with an introductory framework of the larger picture – the cosmos, the big bang theory, galaxies, solar systems, moons, and asteroids. All planets, including Earth, inherited everything from cosmic objects. *Chapter 2* deals with the history of the Moon, which is an essential element that influences Earth, whilst at the same time helps us understand our planet's evolution better. *Chapter 3* explains the basic concepts in geology – Earth's internal make-up, the hierarchical time scale, exciting rock types, amazing fossil flora and fauna. It also explains the occasional topsy-turvy physical appearance of rock layers, related to tectonic structures like folds, faults, etc. *Chapter 4* dwells on the wonders of plate tectonics - how it works, its effect on climate, seafloor spreading, mountain building, and the birth or demise of supercontinents. *Chapter 5* is an account of the Precambrian age, lasting 4 billion years of the entire 4.57-billion-year-old planet history. The *Chapters 6 – 12* examine the broad geologic succession of the planet's last 539 Ma i.e., 539 million years, revealing the secrets of the Paleozoic, Mesozoic, and Cenozoic Eras. These describe characteristic rocks, the rise and demise of fauna and flora, the icehouse-greenhouse cycle, and the history of the last supercontinent. *Chapter 13* tries to divine the future of our planet, equipped with the knowledge from the pages of the long travelogue. We muse – where we are likely to be heading under different future scenarios – very short term to very long-term geological.

An identical format consisting of three principal headings marks the layer-by-layer Earth history in the core *Chapters 5-10* – Paleographic Highlights, Stony Testimony, and Miracle of Life. 'Paleogeographic Highlights' refers to tectonic movements affecting the contour and distribution of landmasses and oceans in each geological period. Discussions under the other two headings highlight distinctive rocks, and extant life that flourished in each period respectively. *Chapters 11 and 12* complement each other. The former deals with the paleogeographic highlights and characteristic rocks during the latest 2.58Ma-old Quaternary Period, while the *Chapter 12* deals with the fauna and flora of the period, especially the emergence of *Homo sapiens*.

Occasional overlap of ideas between chapters happens because many geological events are recurring in nature. What happened in a lower rung of Earth's time-ladder might recur in a higher rung. It is a way to refresh one's memory of earlier discussions, and make every chapter stand-alone, complete, and coherent.

I have a personal take on the throbbing Earth story, though the idea of a living planet is as old as humanity. Animism, predating religions, holds that all beings and inanimate objects possess spirits, i.e., even rivers, oceans, mountains, trees, etc., are all sentient. Ancient Greeks visualized Earth as goddess Gaia. Hindu and Buddhist goddesses - Bhumi and Vasundhara, are personification of Earth. Many folkloric references center around geological monuments worldwide. The Paleoproterozoic Vishnu schist, Brahma schist and others, in the Grand Canyon, USA, got their mythical names from India and elsewhere. I have often stressed on Earth's ever-changing nature, and complex milieux of inner workings. Earth's ceaseless heartbeat continues creating all the physical changes manifest everywhere. Its dynamic hydrosphere, atmosphere and geosphere provide a goldilocks ecosystem for the thriving biosphere. Nature amazingly seems to temper many apparently chaotic internal and external forces to make

Earth a cozy place. We experience daily the intrinsic order, and pulsating seasonality of all natural phenomena.

Unfortunate events like earthquake, magma eruption, tsunami, hurricane, tornado, wildfire, abnormal flood, or drought, etc., sometimes inflict sufferings on humanity, yet these are only but manifestations of nature's routine events, amenable to scientific explanation. The same severity of nature that had troubled our ancestor *Homo erectus* from time to time, still visit *Homo sapiens* today. It will be evident how climate change is baked into the earth system. In our heliocentric solar system, there is value in being a bit 'geocentric', rather than totally anthropocentric.

The book is for people curious and willing to admire the captivating story of our Blue Planet. Geoscientists, paleobiologists, and others interested in the evolution of Earth's flora/fauna, paleoclimate, and paleoceanography will benefit from the broad overview of our planet presented in the book. Hopefully, the headings and subheadings in each chapter will make it easier to enjoy. Especially, the subheadings offer non-technical cues to the subject that follows. I have added a summary – "Chapter Highlights", narrating the salient geologic happenings, at the end of all 13 chapters. Numerous chapter-wise references, cited as superscripts in the text, provide additional resource for readers.

Let us embark on our exciting rocky journey up to the last visible, transient finish line of the 'Present'.

*"The world is my country;
science is my religion"*

Christiaan Huygens, mathematician
(1629-1695)

PART ONE

ORIGINS *and* FOUNDATIONS

1

Origin of the Universe and Solar System

EARTH'S COSMIC BACKGROUND

The exploration of the cosmos is a voyage of self-discovery because we inherited everything from it.[1] Any story about Earth, its intricacies and clockwork inner workings may seem like far removed from the cosmos. However, our blue planet is closely linked to other heavenly bodies, especially the Sun and filial planets. Our solar system forms only a miniscule part of the Milky Way Galaxy—one of the more than two hundred billion galaxies strewn across the almost empty universe. The same primal building blocks of matter, energy and space permeate all space bodies as well as everything on Earth. All we treasure on our planet, is a gift of the cosmos. Everything we feel with our senses is made up of elements coming from space objects, big and small.

An understanding of the heavenly big picture is helpful to delve into the entrails of Earth. Where does it all fit into the larger context of the universe, galaxies, stars, planets, our solar system, and our satellite Moon? We will not discuss black holes and black energy here. Where did all the elements—the ingredients of mineral, rock, mountain, plant, and animal kingdom, come from? How did land, ocean, and atmosphere evolve? Our goal is to sketch the main components of the cosmos, especially their bearing on earthly features.

Earth happens to be the only habitable planet we know in the whole universe. A few exoplanets of stars, light-years-away, may exist in far corners of the universe with a similar potentially favorable scenario. Some astrobiologists postulate that primitive biological matter on Earth could have come from other planets, via impacting asteroids and comets. We will discuss later a real possibility of rare goldilocks habitable conditions elsewhere. Indeed, a few exoplanets eerily mimic an earthly living condition. The same stellar ash is the raw ingredient of the whole arc of animate and inanimate matter, everywhere in the cosmos. To know us better, we need to probe where we came from.

How and When Did it All Start?

Humans have been asking this question from time immemorial—Where did it all start? In the Vedic philosophy, the mantra 'SOHAM' (*So Aham*) means 'I am that'—'the universe exists within me, as much as I exist within the universe.'

The brief discussion will start at the very beginning of the cosmos, as we comprehend it today. It's amazing how the big bang would have unleashed the creation of cosmic interstellar clouds of particles, gas, and dust—raw ingredients of matter and energy for every cosmic object. We will discover how atoms, molecules and nascent elements were nurtured in the womb of supernovae and stars, and then expelled during the violent end of the celestial objects. Elements of the Periodic Table are building blocks of all matter on Earth and elsewhere in the cosmos.

THE UNIVERSE—BIG PICTURE

How Old? What Do Myths & Religion Reveal About the Universe?

The early astronomers, philosophers, scientists—all speculated about the creation event. In many religions, it is God who created the universe out of nothing or *ex-nihilo*. Aristotle (384–322 BCE), like Ptolemy before, adopted a model in which four basic elements—earth, water, fire, air, are the prime constituents of all matter. Ancient Greeks had a fifth element in addition—ether. Ancient Hindus mentioned similar five basic elements in Sanskrit—Kshiti (earth), Apas (water), Tejas (fire), Vayu (wind), Akasha (sky). Aristotle thought that the universe is eternal, but not infinite. He put Earth at the center of the cosmos in his then popular 'geocentric model.' Aristarchus (310–230 BCE) was the first astronomer who visualized the Sun at the center of the solar system or the universe.[2] Of course, Aristarchus was right about the Sun, but he was a lone voice born long before his time.[3] Polish astronomer Copernicus (1473–1543 CE) first formally proposed the heliocentric model, and Galileo Galilei (1564–1642) later established it.

Earlier scriptures and theologians, scientists, offered a historical perspective on the predicted creation date. We are told that Creation is less than 6,000 years old. Indeed, the biblical creationists invoked the literal reading of the holy book—six days of creation in Genesis, the tale of Adam and Eve and the forbidden fruit, their fall from heavenly grace, the historical biblical Noachian flood and so on. St. Augustine (Book of Genesis) posited the universe as approximately 6,000 years old.

Ancient cultures defined the age of the cosmos differently. All except Hinduism, fell far short of the modern scientific consensus on the age of the universe. The Hindu sages debated the origin of the universe in the 'Song of Creation' in Rigveda 10.129, as far back as 3,500 ka.[4] The famous German Indologist, Max Muller, translated this hymn many years ago. To paraphrase him—Who knows the secret? ... Whence is this manifold creation?... The God came later after the creation of the universe...Perhaps He knows, or perhaps not even he knows the answer...[4] One wonders if the sage writer(s) of Nasadiya Sukta (or Hymn of Non-Eternity) had an agnostic viewpoint about the origin of the universe.

Carl Sagan noted that only the Hindu philosophy echoes a view somewhat in line with modern cosmological interpretation.[1] The Vedic sages posited cosmos itself underwent cycles of birth and demise, i.e., expansion and contraction, spanning 8.64 billion years, that is, one complete day and night of Brahma. The cited age is longer than the antiquity of our solar system, and about half the time since the big bang happened. Each Vedic cycle likely represents a 'big bang,' followed by a big crunch. Many western scientists like Niels Bohr and others, saw this similarity between Hindu philosophy and modern cosmology.[5] The life cycle of the infinite universe is supposed to be endless. The Vedas tell us the universe is as old as Brahma's lifespan—100 years or 36,000 days. By multiplying Brahma's life span by the duration of one complete Brahma day, the age of the universe becomes 36,000x8.64 billion, or ~311 trillion years. However, it is unclear how the Vedic seers arrived at the above numbers.[6]

The Universe—Flat or Round?

NASA estimates more than 200 billion galaxies in the universe, each one of them having many more stars. The stars have their own sphere of influence with a collection of obedient planets, satellites, asteroids, and comets. Add to that, all the black holes spread across the vast, almost empty cosmic expanse. The visible matters make up less than 5% of the space of the universe (4% gas, 0.5% planets and stars). The rest is made of enigmatic dark matter (24%) and mysterious dark energy (71%) pervading most of the empty space.

The universe is a big place. How big is it? Scientists have calculated the distance to far-off galaxies floating at the edge of the ever-expanding universe. Astronomers are now observing light from these galaxies, hitting their telescopes after travelling for about 13.8 billion years. Data shows the distant galaxies are 46.5 billion light years away. In other words, radius of the observable universe is approximately 46.5 billion light years. That makes the diameter of the universe ~ 93 billion light years.[9] Light travels 300,000 km per second. We may possibly grasp the dimension even better when we realize one light year means the distance light travels in a year— ~9.46 trillion kms. The size of the universe is 93 billion light years X 9.46 trillion kms.

What is the shape of the universe? Experts have long debated about it. Is it upwardly convex like a spherical body, or a flat surface like any sheet of paper, or a downwardly convex saddle-like body? According to NASA 2013 WMAP results, the universe is flat with a miniscule 0.4% margin of error.[10] To note, dissenting voices claim the universe is anything but flat.

Albert Einstein once commented about the comprehensibility of the universe.[11] Nature is now fathomable to a point. The universe is all we see and beyond—existing space, matter, and energy, englobing all the galaxies, stars, planets, and inter-stellar space.

The Cosmos—Several Theories Explain its Origin

Astronomers proposed several alternative theories including the out-of-favor 'steady-state model.' James Jeans in the 1920s first proposed it, later Fred Hoyle, Thomas Gold, and Hermann Bondi reformulated it in 1948. They argued the universe is ever-existing, eternal, and ever-expanding, but keeping its general characteristics of homogeneity and unchanged density. However, astronomical evidence contradicted the group's assertion. The fact that galaxies are moving away from each other means that the universe had once been denser, contrary to what the steady-state model predicted. A later discovery (we will explain soon) drove the death knell into the steady-state model, as it proved the universe had a beginning.

Two other important theories are worth mentioning, namely, cosmic inflation and cyclic universe theory. However, both are considered like refinement and outgrowth of the big bang theory. About 40 years back, the 'cosmic inflation theory' tried to explain our flat, smooth universe. The theory, spearheaded by several theoretical physicists, explains the exponential expansion of space in the early universe. Now it is part of the Inflationary Epoch, the period 10^{-36} to 10^{-33} second after the Big Bang.

The alternative model to the big bang is the cyclic contracting and expanding universe of Steinhardt.[7] Like the ancient Vedas, he argued the universe went through unending cycles in the pre-Big Bang past. Each cycle began with a big bang (expansion) and ended in a big crunch (contraction). The universe is trillions of years old, and cycles may recur every 10^{12} years.

Many newer hypotheses also exist about the origin of the universe.[8] All are possible, including the cyclic universe concept; however, scientists cannot probe beyond the latest 13.77-Ga-old big bang event.

In the book, we will express geological ages as Ga (giga-annum), Ma (mega-annum) and ka (kilo-annum) for a billion (10^9), million (10^6) and thousand (10^3) years respectively.

BIG BANG, A SINGULAR MOMENT—QUEST FOR THE '0' HOUR

Georges Lemaître: Father of the Big Bang, Priest, Astronomer, and Mathematician

The Russian astronomer Alexander Friedman (1922) showed how the universe could expand (or contract) with time. One can visualize Friedman's concept with 2 painted dots on a partially inflated balloon; the dots will move further apart as the balloon inflates more. The converse happens when the balloon is deflated.[3]

The seed of the expansion theory started with Friedman and earlier works of Einstein. But it was Georges Lemaître (1894–1966) who formally advanced the idea in 1927. Lemaitre was a polymath Belgian priest, astronomer, mathematician, and a professor with a Ph.D. from MIT, USA. He conceived of an expanding universe model to explain the 'red shift' of cosmic spiral nebulae. This theory accorded well with contemporaneous observations, namely, the phenomenon of redshift of galaxies. What is redshift? Redshift is simply what happens when light wavelength stretches as space expands—everything space holds also expands in consequence. In a spectrum of light, the wavelength at the red end is longer. The converse is true for the blue end of the spectrum. When astronomers see light emitted from a particular space object gradually getting redder, they know the object is receding away as space is expanding. Lemaître realized a 'creation-like event' occurring at some point in the distant past.

Soon thereafter, he jointly calculated the Hubble-Lemaître constant that explains the relative rate of cosmic expansion. In a 1931 conference in London, he formally proposed the universe expanded

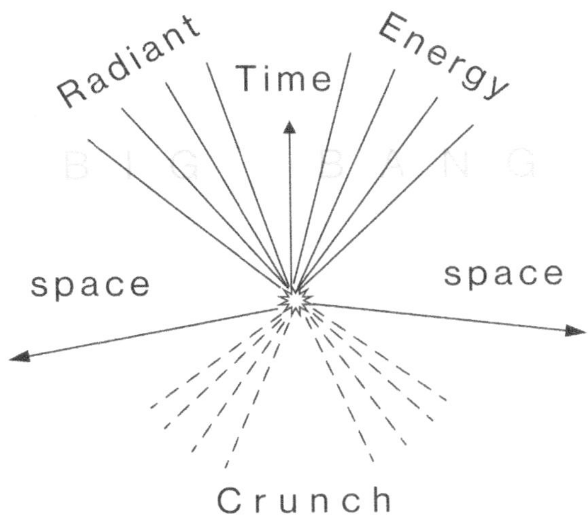

1-1 Traces of a particle path (worldline) in 4-dimensional space-time, at the post-big bang moment. The conjectural bottom half suggests a prior contraction phase of the universe.

from an initial concentrated mass of nuclear matter (Fig. 1.1). The matter later dispersed in the form of gas, dust, 'ice'—the raw ingredient of all later cosmic bodies like galaxies, stars etc. He wrote in a brief paper in the journal Nature the same year, that the beginning of the universe coincided with a unique quantum or cosmic 'atom.' The dense 'atom' carried the entire mass of the universe.[12, 13] Astronomer Fred Hoyle dissented to Lemaître's idea, as he championed a 'steady-state universe.'[3] Hoyle later playfully called Lemaître's proposal as the 'big bang' theory, in a 1949 BBC interview.[14] Little did Hoyle realize that his moniker would endure in serious astronomical literature and discussions ever since. The big bang marked the singularity of the momentous cosmic expansion phase (Fig. 1.1). Confirming Lemaître's theory, some contemporary astronomers, most notably Edwin Hubble in the U.S., provided concrete proof in 1929. Hubble's outstanding work showed that all galaxies are indeed moving away in all directions due to the progressive expansion of the universe.[15]

Stephen Hawking noted the expanding universe model was a great intellectual revolution of the 20th century. The big bang stuck. The astronomical evidence by Hubble and colleagues clearly meant that if the universe is expanding now, it would have had started from a concentrated quantum at the beginning of the expansion. Hubble's idea later came to be known as the 'expanding universe theory.' All the raw material for galaxies, stars and planets were fused together into a superhot and dense diffuse mass of matter and energy. It was what Lemaître earlier described as a 'cosmic egg.' Lemaître's idea was based solely on the current expansion rate of the universe extrapolated backward in time. The singular cosmic event triggered the expansion phase of the universe. Scientists calculated that the big bang happened about 13.77 Ga ago. In 1933, after hearing Lemaître describing his theory, a hitherto skeptic Einstein embraced the expanding universe theory of Lamaître and Hubble.

Jesuit priest Lemaître believed in God, yet he based his ideas solely on scientific theory and reasoning. He made it clear that there was neither a connection nor a contradiction between his religion and science." [17] Interestingly, much to his surprise, some theologians and scientists started to cite the big bang theory as evidence of a divine creation of the universe. Pope Pius XII commented in his address to the Pontifical Academy of Sciences, Nov 22, 1951, that he accepted this scientific discovery about the big bang happening eons ago. He reasoned a Creator was responsible for this Creation if it took place out of nothing at a fixed point in time.[18] "Therefore, God exists." In a 1964 television interview, Lemaître commented that the philosophical problem of creation disappears in some way if his expansion theory is found valid.[54] According to Stephen Hawking, it is possible that a creator was behind the expanding universe event, but it does restrict him on the timing of his creation job.[19]

Astronomical observations with optical and radio telescopes became very useful after the big bang theory of the late 1920's. Newer telescope designs permitted recording lights in the visible, infrared, microwave, and radio wave spectrum. Important contributions also came from space observations using other specialized telescopes. These recorded x-ray and gamma ray emissions peered deep into the universe without atmospheric interference. The input of valuable work done by theoretical physicists was also noteworthy. The James

Webb Space Telescope, NASA, launched in December 2021 (to replace NASA's retiring Hubble Space Telescope), is proving to be an out-of-the-world boon for astronomers who are peering even deeper into the cosmos.

In summary, most scientific evidence supports the basic premise of the big bang. It may not be completely satisfactory today, but it is the only one that has stood the test of scientific scrutiny and analysis. A latest study posits the universe may be 27-Ga-old.[49] Some astronomers think there may be a multiverse scenario, where parallel universes form like bubbles side by side; cosmic development may not follow the same natural laws that we know in a parallel universe.

EARLY UNIVERSE

Radiation and Matter Era—
Two Cosmic Stages from the Big Bang to the Present

An inconceivably hot, and infinitely dense plasma point source existed at the beginning. About 13.77 Ga+/-40 Ma ago, the unique supercharged source exploded—the big bang occurred. It created energy, matter, space, and time. Progressive development of the present universe, starting at the big bang, happened in two successive eras—Radiation era, followed by the Matter era.[20] Each era is again subdivided into epochs—8 in the Radiation era, and 3 in Matter era.

Radiation era—It started from the moment of the big bang and lasted for about 50 ka. The Radiation era was a time of all-pervasive darkness shrouding the universe. At this stage, it was just radiant energy, no trace of matter anywhere, in the early expanding universe. Only one unified super-force, the ancestor of four fundamental forces of nature existed at the unthinkable Planck temperature (10^{32} Kelvin). An unspecified energy field soon inflated the original dense point, filling it with hot dense plasma of subatomic quark particles. In the later '*quark epoch*' when temperature cooled a bit, all four fundamental forces—gravity, electromagnetic, weak, and strong nuclear forces, went their own ways. The gravity and electromagnetic forces have an infinite range, whereas nuclear forces act at short range, at the atomic level. In the *'lepton epoch'*, new leptons included electrons (mass 1/2000th of a proton), neutrino (mass 500,000 times lighter than an electron), and the nascent subatomic particles. The end of the Radiation era witnessed protons and neutrons fused into the atomic nuclei due to strong nuclear force. It heralded the birth of the first post-big bang element—*helium*.

Matter era—This era of matter dominating radiation continues to this day. The gravity forces on matter became more dominant in the era. The formation of early matter started in the *'atomic epoch'* when electrons and protons combined to form first tangible matter, the *hydrogen* atom. "Let there be light" of *Genesis 1:3*, became true. Transparent light dawned on the hitherto dark universe, about 380 ka post-big bang, thanks to the cosmic microwave background (CMB), we describe below (Fig. 1.2). In the succeeding *'galactic epoch*,' the collapsing helium and hydrogen gas clouds apparently initiated the large-scale structures of most galaxies. The current 'stellar epoch' of the *Matter era* started about ~3 Ga post-big bang, when nuclear fusion made the jewel box of stars sparkle within galaxies. The stellar nuclear fusion slowly started producing elements lighter than iron (Fe), of Mendeleev's Periodic Table—the raw material of matters and objects, celestial or mundane. Heavier elements are formed by 'neutron capture.' The expansion of the universe accelerated during the Steller epoch that continues today.[21, 22]

COSMIC MICROWAVE BACKGROUND (CMB) RADIATION: A VITAL CLUE TO ANTIQUITY

Two independent lines of evidence fully supported Lemaître's expanding universe theory—telescopic detection of the cosmic microwave background (CMB) radiation from the nascent universe, and the correct prediction and observation of the chemical composition of the universe.[16] In 1948, cosmologists R. Alpher, G. Gamow and R. Herman correctly predicted the existence of cosmic microwave background (CMB) radiation. Later, the accidental discovery of CMB in 1964 by A. Penzias and R. Wilson (Nobel laureates, 1978), proved to be a game changer in astronomy, at par in significance as the 'red shift' of galaxies, as one expert commented.

NASA's advanced satellite—Wilkinson Microwave Anisotropy Probe (WMAP) has clearly documented the CMB over a 9-year period (Fig. 1.2). The cosmic microwave background signature is virtually identical all over space, exactly as the big bang model had predicted. The super-cold (near absolute zero) radiation in the microwave segment of the electromagnetic spectrum, is telltale evidence of the primordial big bang. Its widespread, uniform presence everywhere

1-2 NASA/WMAP Cosmic microwave background (CMB) radiation map of the universe about 380,000 ka after the big bang; it is the remnant heat after the big bang. The colors (artificial) show incredibly small temperature variations (± 200 microkelvin).

confirms that it is the remnant heat from the big bang.[23] Astronomers learnt from CMB about the primeval conditions of the nascent universe, long before galaxies, stars and planets debuted. How did they do that? What astronomers observe today as the CMB, has taken nearly 13.8 Ga to reach them via powerful, modern earth-bound telescopes. The countless twinkling stars in the night sky are many light years away from us. Because speed of light is finite, light from distant stars takes a long time to reach Earth. We are seeing today only a snapshot of how they really appeared 100's to 1000's of years ago. Just to have an idea, photons from sunlight take about 8 minutes to reach us, and the Sun is only about 150 million kms away.

In a nutshell, the post-big bang universe was dark, and an unimaginably dense opaque plasma of undefined energy. It was so hot that even atoms and smaller quarks could not exist; quarks with fractional energy are internal constituents of neutrons and protons. Soon, ~380 ka after the big bang ('Recombination era'), simple hydrogen atoms appeared when temperature cooled down a bit. The released photons after hydrogen formation, now appear as the CMB radiation.[24]

The Universe, a Delicate Balancing Act—Expansion vs. Gravity

One may rightly ask—How can two seemingly opposing forces such as 'expansion and gravity' coexist in the universe? Undoubtedly, there must be a fine balance between expansion and gravity, such that all matter will not be scattered farther apart due to excessive expansion, nor a very sluggish expansion triggering gravity to collect and bunch together all matter, preventing the formation of cosmic objects. The expansion rate had to be just optimum to maintain all the heavenly bodies in the universe. Einstein, though not a believer, marveled if God could have created the cosmos in any other way than what we perceive today.[11]

A VIOLENT BEGINNING—CHRONOLOGY OF SPACE BODIES

Vital Building Blocks of Matter in the Nascent Universe

The powerful telescope of Edwin Hubble, the brilliant stargazer astronomer, revealed the universe extended far beyond our galaxy Milky Way. He documented innumerable other galaxies, each containing many more stars and planets. Our nearest galaxy Andromeda is 2.48 light-years away from us. Star formation began about 200 Ma post-big bang. By about 1 Ga post-big bang, more than 100 billion galaxies had already formed. Our solar system formed ~9.1 Ga after the big bang. Radiometric data reveal that the solar system, including Earth and other planets, is 4.57 Ga old.

Scientists posit that less than a mere second after the big bang, electrons started forming, and afterwards, the protons. When the universe cooled below ~100 million Kelvin ($10^{8\circ}$C), protons fuse with other protons and neutrons, forming the atomic nuclei. Only low mass atoms dominated in this primordial universe. 90% of the nuclei were hydrogen (H) and 10% helium (He), with traces of deuterium, tritium, lithium, and beryllium. Below 3000° Kelvin (2,726.85°C, i.e., half of the Sun's surface temperature), all free electrons combine with atomic nuclei completing the creation of particles and atoms in the nascent universe.[25]

Supernova nucleosynthesis—If hydrogen and helium came from the big bang, what about the myriad other elements? A supernova is a stellar nursery generating almost all atoms heavier than hydrogen and helium. Scientists argue that 'sequential hydrostatic burning process' creates heavier elements from the fusion of lighter elements. The ashes of one nuclear fuel became the fuel for subsequent burning stage, after compressional heating. Examples: hydrogen burning into helium, helium (4 protons) burning to carbon (6 protons). Likewise, carbon burns to oxygen with 8 protons, which in turn produces silicon (14 protons) atoms and so on. The supernova explosion even creates elements heavier than nickel.[26]

A New Creation Story—
The Post-Big Bang Era: Galaxies—Stars—Planets—Satellites

Once big bang started the ball rolling, the task of creation of galaxies, stars, and planets went into full swing. Gravity played an important part in the early stages of star and planet formation. It produced a horizontal separation of disperse exploded material into a discoidal shape. Nascent protostars like our Sun, stayed at the center of the disk where fusion reaction soon started to happen. The denser, refractory, stony material stayed nearer to the proto-star, and lighter gas-rich, fine dust and debris migrated to the outer part of the disk. All floating debris and dust in the proto-planetary disk coalesced and began to form planetesimals i.e., smaller cosmic bodies. Planets arise gradually from growth of planetesimals by continuous collision and accretion. The Proto-Earth begins to form as gravity and centripetal force give the planet a spherical shape. The core and mantle begin to differentiate by density. Satellite moons are born out of the ring of debris, probably when a wayward space body shatters into an earth-like planet. Eventually atmosphere develops from volcanic gases and impacting cosmic bodies including comets. The spewed moisture creates the oceans.

Sagan's Unique Cosmic Calendar—
A Novel Way to Read Cosmic Milestones

Astronomer Carl Sagan[1a] earlier created an easy, visual cosmic calendar that compresses the nearly 14 Ga year history of the universe into 'one cosmic calendar year.' Each cosmic year represents 1.17 Ga of time. On this scale, our galaxy, the Milky Way, formed in early February. Between February and September, many stars lived and died contributing to the reservoir of both heavier and lighter atoms in the universe. The solar system came about in early September, about 4.57 Ga ago. The late September saw the first signs of life on Earth. The Cambrian life explosion happened in mid-December. The dinosaur demise took place by the End-December. The entire human history unfolded in the last two minutes of Sagan's cosmic calendar.[16]

Galaxies, Stars, Planets—
The Spiraling Mystery: There are Billions of Galaxies

Galaxies are the nurseries of stars, planets, and all other smaller cosmic bodies. Most galaxies occur in clusters and are made up of gas, dust, and billions of stars of all sizes and vintage. There are about 100 billion or more stars in each galaxy. According to some estimates, the total number of stars is about 200 billion trillion, and likely an equal number of planets.[35] A 'black hole' occurs at the center of most galaxies.

Our 'Local Group of Galaxy' cluster consists of twenty constituent galaxies, including our neighbor Andromeda, spread several billion light-years across. The Hubble Space Telescope of NASA discovered a new spiral galaxy NGC2985, in the Ursa Major Constellation, 70 million light-years away. Its outer boundary merges and disappears in the vast darkness of the intergalactic space.[27, 28]

A curious thing about the universe is that it is mostly empty space, though we may think that galaxies with countless stars, planets, and satellites, occupy most cosmic space. The amazing galactic structures make up only ~5% of the mass of the universe, meaning the universe is mostly dark energy and dark matter.

How do we know the age of a galaxy? Scientists recently described a way to determine a galaxy age by first studying the 'pulses' of thousands of stars in the distant galaxy Messier 87 or M87.[29] Another neat way to estimate the age is to know the quantity of beryllium in a star within a galaxy. Beryllium acts like a cosmic clock as its percentage in a star increases with time. Lithium is also a useful age marker, but it works the reverse way, i.e., its content decreases with time. It behaves like radioactive uranium and thorium. Stars consume lithium in its core, so an older star's lithium content is minimal. A high-lithium-star is a juvenile.[30]

Our Galaxy—'Milky Way'—A Name from Greek Mythology

The universe created more than 100 billion galaxies during the first billion years. We recall astronomer Edwin Hubble proved in 1925 that the universe extended far beyond our spiraling Milky Way Galaxy.[31] Innumerable other galaxies exist, each with countless cosmic objects. The Milky Way stretches 105,700 light years. Where does the exotic 'Milky Way' name come from? Upon first seeing our galaxy, the ancient Greeks likened it to the squirted milk from the breast of Hera, wife and sister of Zeus, and the goddess of birth and marriage. Later, Romans similarly called our galaxy 'via lactea,' a literal translation from the Greek, and the Hindus called it 'Akash Ganga'—'the heavenly Ganga River.'

The Milky Way Galaxy is 13.6 Ga+/- 0.8 Ga old. The universe is slightly older—13.77 Ga +/-0.2 Ga, based on the cosmic microwave background radiation data.[32] The Milky Way would have formed only ~100 million years after the "dark ages" of the post-big bang Radiation era. Some stars are really very old—13.4 Ga, i.e., only ~0.3 Ga after the big bang.[33] Another thing, galaxies change. Scientists estimate that the Milky Way and its neighbor Andromeda are on a collision course, they will merge in about 4 Ga from now.

The powerful NASA Hubble Space Telescope could peer back at galaxies as old as 0.45 Ga after the big bang. But the new NASA James Webb Space Telescope (JWST) is truly a game changer today. It is observing galaxies formed only 100-200 Ma post-Big Bang.[34] NASA JWST is 100 times more powerful than the retiring NASA Hubble Space Telescope. NASA released five amazingly clear infra-red images of cosmic bodies on July 12, 2022: The 0723-galaxy cluster is the deepest image of thousands of galaxies in the universe so far. However, it shows only a small slice of universe, comparable to a grain of sand held at arm's length by someone on Earth. The Carina Nebula region displays the birth of countless stars and planets in 'cosmic cliffs and valleys' (Fig. 1.3). The Southern Ring Nebula, about 2,500 light-

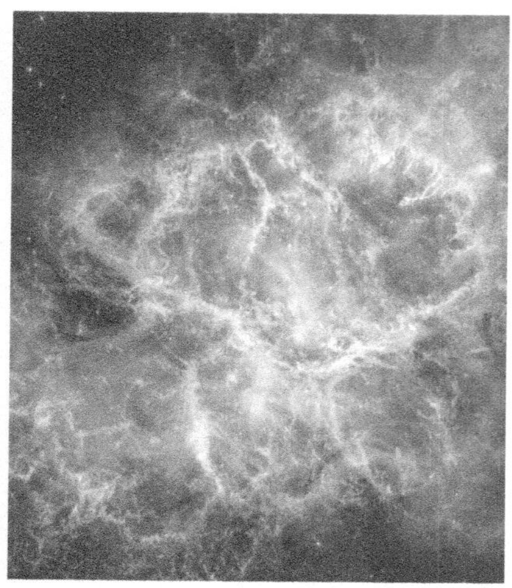

1-3 Gifts of NASA—JWST. **left** The area, called the Cosmic Cliff, is a gaseous cavity within the cluster NGC 3324 in the Carina Nebula, ~7,600 light-years away. It reveals for the first time numerous hidden stars, and many background galaxies with amazing clarity. **right** Crab Nebula image using James Web Space Telescope in infrared via its specialized camera and instrument. It may allow researchers to know the composition of the nebula and gain a better understanding of the supernova that created it.

years away, documents in detail the demise of a star and the striking ring of gas and dust it has created. NASA's Stephan's Quintet is a captivating image of five galaxies showing millions of nascent stars and planets. The Crab Nebula image gives scientists clues regarding the supernova that created it (Fig. 1.3). A NASA-JWST big discovery is the presence of water in the atmosphere of a giant hot gas planet—WASP-96 b, orbiting a star ~100 light-years away from Earth.[53]

Specks of Silver Dust in the Ink-Blue Sky—How are Stars Born?

Stars have mesmerized poets since ancient times. From the childhood rhyme of 'Twinkle, Twinkle, Little Star' to the musings of poet Walt Whitman, examples are galore:

When I sitting heard the astronomer where he lectured with
Much applause in the lecture-room,
How soon unaccountable I became tired and sick,
Till rising and gliding out I wander'd off by myself,
In the mystical moist night-air, and from time to time,
*Look'd up in perfect silence at the stars.**

WALT WHITMAN (1819–1892)

*Walt Whitman, 1819-1892, "When I Heard the Learn'd Astronomer": https://poets.org., Public domain.

Indeed, we have never stopped wondering what those twinkling stars really are. Simply put, nature needs two ingredients to produce a star—voluminous clouds of gas, dust and debris, and the pull of gravity working on them. The gravitational force pulls the disperse gas and particulate matter inward toward a central core. The concentrated inter-stellar gas and dust becomes so much hotter and denser that it collapses to form a nascent star under its own weight and gravity. A budding star can claim its stellar status, when it can effectively burn hydrogen into helium in its core. The temperature rises to about ~10 million Kelvin or ~ 7.3 million°C when hydrogen eventually burns, certifying the birth of a real star.[36] Depending on its size, it may take at least 1 million years to form a star.[37] Stars shine because of the energy released in the process of nuclear fusion of 4 hydrogen nuclei into a helium nucleus. In its essence, Einstein's formula *e=mc2* means mass (m) and energy (e) are interchangeable. It is a 2-way street—conversion of mass and energy. Because of the law of conservation, energy is released when 4 hydrogen nuclei produce one helium nucleus of lesser mass.

Stars come in all vintage and size—from small (may be one tenth or so in size of our Sun), to intermediate or average, a moniker that fits our daystar. The Sun is a typical, not too large not too small a star; it burns by nuclear fusion of hydrogen atoms into helium. A large star may have 100 times more mass. Our Sun will likely burn up all its hydrogen fuel in the next 4–5 Ga. Curiously, larger stars with high luminosity exhaust their fuel sooner than smaller stars, though they have more fuel to start with. The life of a star may last from several billions to a few hundred million years.

The oldest stars, born soon after the big bang, inherited hydrogen, and helium from the cataclysmic event, along with traces of lithium, boron, and beryllium. A 2013 study narrowed down the age of the oldest known star Methuselah or HD140283, by using NASA's Hubble Space Telescope. It established that Methuselah in the constellation Libra, known for over a century, could be as old as 14.5 Ga+/- 0.8 Ga. Its younger possible age of 13.7 Ga, tells us that it would have had formed very soon after the big bang,[38] The star shows high deficiency in heavier elements, a testament to its very early origin.[39]

An important word of caution here about all the cited ages of big bang, galaxies, stars, and planets throughout our discussion now and hereafter. The relative vintage of each may be perhaps more telling than the literal numbers cited.

WHEN STARS DEBUT—PLANETS CANNOT BE FAR BEHIND

Our Sun—The Only Star That Really Matters to Earthlings

The nebular hypothesis of Swedenborg, Kant, and Laplace [40] in the 18th century, posited that the solar system arose from a giant molecular cloud, probably because of explosion of a giant supernova, and ensuing gravitational collapse of the same. With the collapse, the solar nebula would have gradually heated up, and spun, resulting into a disk. The new star, our Sun, formed near the hottest central part, an area of greatest mass. All dust particles and debris surrounding it, eventually coalesced into the solar planets. The Sun stayed at the center of the massive inter-stellar cloud as a new star in the spiral Milky Way galaxy. The planets, including Earth, formed, and stayed on orbit around the Sun in a proto-planetary disk. NASA-Spitzer Space Telescope 2003 data on nearly 500 new stars in the Orion Nebula confirm that the disk absorbs some of the star's angular momentum. It prevents the star from disintegrating due to excessive spin.[41] The planets have a solid central 'core' of heavy metals that formed early from planetesimals attaching to each other and enlarging over time.

Our solar planets are not unique in the Milky Way Galaxy, due to some unusual cosmic event. Planet formation is a natural by-product of the birth of a star. As of 2024, NASA has confirmed the existence of more than 5,000 exoplanets, i.e., planets of other stars in our galaxy.

The Sun has a diameter of 1.39 million kms; its surface temperature is 4,310°C.[42] Sun's chemical composition is H 70%, He 28%, carbon + nitrogen + oxygen 1.5%, neon + iron + silicon + magnesium + sulfur 0.5%.[43] It shines safely 24/7 about 151 million kms away from Earth. Our life-giver star will die when it finally exhausts its precious hydrogen fuel supply in about 5 Ga.

Earth pulsates with life and activity only because of the Sun. It gives us precious light and heat. Heating of the air and oceans creates wind pattern and ameliorates global climate, otherwise the blue planet would have become a cold inhospitable place. There will be no liquid water for us without solar heat, so essential for all forms of life. The solar machine runs the whole food-chain enterprise of the animal world. Sunlight provides the vital energy needed for photosynthesis that sustains plants and forests. Photosynthesis ceaselessly creates pure oxygen for us to breathe at the expense of the toxic carbon dioxide we exhale and that is generated by nature and humans. The critical ozone layer is also an indirect gift of the Sun as its ultraviolet radiation chemically breaks down oxygen molecules and helps to produce ozone (O_3) in the stratosphere. The Sun is potentially an inexhaustible source of clean, non-polluting, safe, renewable energy for the entire mankind.

The Sun's Obedient Planets—Where Are They in the Firmament?

Earth and the neighboring six planets are orbiting our solar system. The ones nearest to the Sun are the four stony planets—Mercury (the closest), Venus, Earth, and Mars. The first two are naturally the hottest and heaviest, while Earth is in a 'goldilocks position,' being not too hot, not too cold, and blessed with liquid water.[25] Mars is quite a bit cooler than Earth. The outer, less dense planets are larger, and mainly gaseous in nature. For example, Jupiter, a largely gas-rich planet, is a distant outwardly neighbor of Mars in a separate orbit. Jupiter is 1000 times larger than Earth, and many times larger than all the solar planets combined (Fig 1.4). It has 95 moons, each orbiting it separately. The Sun gave birth to Mars at an early age and our blue planet arrived soon after, probably ~0.1 Ga later. It appears that Mars in its early formation remained as a puny cosmic body, while Earth continued growing into a bigger planet.[44] A belt of countless asteroids exists in a separate solar orbit between Mars and Jupiter.

Earth orbits the Sun in an elliptical path—from the shortest 147 million kms, to longest 152 million kms. Our nearest star Alfa Cen-

1-4 A montage of images of solar planets, taken by different spacecrafts at different times, managed by the Jet Propulsion Laboratory in Pasadena, CA. From the top: Mercury (closest to Sun), Venus, our home Earth with Moon behind, Mars, Jupiter, Saturn, Uranus and Neptune. *(not to scale)*

tauri shines over 4 light-years away, it is ~40 trillion kilometers away from us. We are seeing light emitted by the star 4 years ago.[45] Imagine the distance between the nearest stars.

The Element Soup—Mendeleev's Periodic Table Secrets

Do we know where the gold in your ring came from, and how it formed? Do we know why the ruby displays such a crimson red hue, or the emerald the color of a green parrot? Why the sapphire reminds us of the blue waters of the Caribbean? Everything we are, and anything we touch and feel with our senses, are chemical compounds with atoms of different elements. For example, water—a liquid essential for our life, consists of 2 simple elements—hydrogen and oxygen. Hydrogen is highly combustible and super reactive. Oxygen is also a very reactive, non-flammable element that feeds fire in a big way. The two gases chemically combine to produce water that quenches fire—a liquid essential for all living things. I still remember the excitement I felt as a college freshman when our chemistry professor produced a drop of water in the lab from hydrogen and oxygen (in air), accompanied by a rather loud sound.

Another example of elemental miracle is sodium chloride or common salt, Salt is composed of sodium, an alkali metal, a soft reactive element, highly explosive in water, a poison by itself. Chlorine is a halogen gas with a choking smell and causes constriction in the chest when inhaled. A concentration of one part in thousand in air can cause death in a few minutes. However, NaCl is an ingredient essential for all mammals. Jesus told his disciples "Ye are the salt of the Earth" highlighting salt's value.

The same element can have very different chemical and physical properties when they appear as different allotropes (with different chemical bonds), such as carbon in its various incarnations as coal, graphite, and diamond. Graphite—a metamorphic mineral, forms as a crystal sheet, each carbon atom bound to three neighbors in a plane; diamond, on the other hand, forming very deep inside the mantle of Earth, displays a tighter tetrahedral bond, meaning each atom is bound to four neighbors in a tetrahedral pattern.

Dimitri Mendeleev in 1869 formulated the Periodic Table of Elements. Later the continuous update of the table with additions of newly discovered elements is a testament to scientific progress over the years.[46] One can predict the chemical property and characteristics of any element just by observing its position in the table (Fig.1.5). The horizontal rows are periods, and the vertical columns are known as 'group' or 'family.' It tells us the atomic number (i.e., number of protons in the nucleus) and atomic weight (mass of protons and neutrons in an atom) of all 94 elements. Both atomic number and atomic weight increase as one moves across a period or down a group. Atomic number starts with 1 (hydrogen) at top left and ends with 94 (plutonium). The synthetic elements with atomic number 95-118, follows plutonium.

The elements of the same horizontal period show a progressive increase of atomic number and atomic weight from left to right. For example, the second 'period' from top shows lithium, beryllium, boron, carbon, nitrogen, oxygen, fluorine, and neon with increasing atomic number from 3 to 10; atomic weight also increases from left to right, i.e., from lithium 6.941 atomic-mass-unit (amu) to neon 20.179 amu.

All elements of the same vertical 'family' have the same number of electrons in the outer shell, having similar chemical properties: The first 'family' on the far-left side of the table shows similar chemical

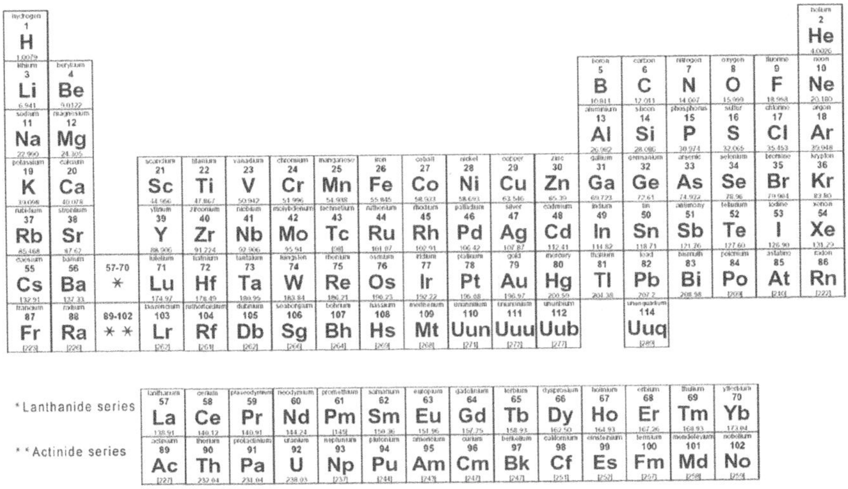

1-5 Cosmic source of elements in the Periodic Table. Row = period. Column = group or family.[48]

properties. Alkali metals lithium, sodium, potassium, rubidium, cesium, and francium, all have 1 electron in the outermost shell. The last 'family' of inert gases (neon, argon, krypton, xenon, radon), on the far-right column of the table shows similar chemical properties. The elements in the second group or 'family' from left, i.e., beryllium, magnesium, calcium, strontium, barium, and radium, also show similar physical and chemical properties.

Heavenly Bodies—Elements They Cook Up

The big bang started it all, but its direct contribution to the cornucopia of elements in the periodic table is insignificant with only hydrogen and helium coming directly from it. After the big bang, hydrogen (75%) and helium (25%) essentially dominated the dark early universe. When cosmic rays hit hydrogen or helium in the interstellar space, lighter elements such as lithium, beryllium, and boron form in trace amount. Most of the lighter elements and some heavier elements are produced by stars—small, average, and large. A giant star, 100–1,000-million-kms-across, may be 100–1,000 times larger than an average star like our Sun.[50]

Elements heavier than nickel likely formed from merging neutron stars and supernova. Exploding massive stars produce supernovae.[50] The neutron stars are the smallest and densest of all. Scientists in the USA detected an amazing collision of two neutron stars for the first time in 2017. The large stars release enough energy and neutrons to produce elements heavier than iron, such as gold and uranium.[47] The heaviest elements known—osmium and iridium, formed out of merging binary neutron star collisions. All the expensive gold, silver, and platinum jewelries we cherish, or the metals of daily use in autos, aircrafts and electronics were made in the heavens, billions of years back.

Nature's Gift—Only a Few Elements Make Up Life on Earth

It is amazing how a few essential elements create all living beings on Earth. The human body consists of six principal elements—oxygen (65%), carbon (18%), hydrogen (9.5%), nitrogen (3.2%), calcium (1.5%) and phosphorus (1.2%). The rest are traces of a few other elements—potassium (0.4%), sodium (0.2%), magnesium (0.1%), sulfur (0.2%), chlorine (0.1%), others >1%. If we look at trees and plants, we see the same elements in them, though the proportion may vary from tree to tree or among plants to some degree. Typically, wood in a tree is carbon (50%), oxygen (42%), hydrogen (6%), nitrogen (1%), and traces of other elements (1%)—calcium, potassium, sodium, magnesium, iron, and manganese. The same four major elements oxygen, carbon, hydrogen, and nitrogen, largely make up both humans and trees. Animals and plants breathe the same air with 79% nitrogen and 21% oxygen—the two main components. It appears that the elemental difference between *Homo sapiens* and trees is a matter of degree. A few major elements (and many minor elements) make up the space, the atmosphere, and the organic world. Every element is nature's gift in all animate and inanimate objects.

Mystery of Coveted Gemstones—Magical Spectrum of Colors

A curious aspect of some elements is that even a trace amount of it may impart a certain desirable tint or color to naturally occurring compounds; these make minerals more appealing and attractive to the human eye. For example, the gemstone ruby is so prized for its vibrant red color. Ruby is corundum, an aluminum oxide, with trace amount of chromium impurities replacing aluminum atoms (Fig. 1.6). Why does chromium bestow a magical red tint to ruby? The chromium atoms absorb certain wavelengths of light such as yellow-green, and only allow the red part of the visible spectrum to reflect, making the ruby enticingly red. Sapphire, also a corundum, gets its sea blue color from the traces of titanium and iron impurities. Likewise, the mineral beryl (a beryllium aluminum silicate), may show different colors in its two gem varieties —emerald and aquamarine. Trace elements of vanadium, chromium, and iron add a lovely green color to emerald. The blue color of aquamarine is due to trace impurities of iron. The rare earth elements or REE (a misnomer, as neither are rare, nor earths) play an important role in crystal structure of some minerals. An example, monazite is a primary ore, an oxide of phosphorus and a few rare earth metallic elements—thorium, cerium, lanthanum. Thorium in monazite is abundant in placer (beach) deposits in India. The highly radioactive thorium is strategically important as it can replace uranium in nuclear power generation. The rare earth elements freely substitute one another in the crystal structure of monazite.

1-6 A Ruby crystal, about a centimeter cube (corundum—Al_2O_3) from the Proterozoic marble (the white matrix below the ruby), Jegdalek Ruby Mine, Kabul Province, Afghanistan. The ~30-Ma-old ruby's genesis is related to the Oligocene Himalayan orogeny-related granite and pegmatite intrusions, producing rubies in several countries other than Afghanistan.[55]

Precious Metals—Gift of Meteorites to Humanity

From the dawn of humanity, our love, admiration and longing for beautiful stones and metals has never abated. Where and how did they form? Geology and astrogeology are useful tools to know the origin of precious metals and minerals. Molten denser elements like iron sank due to gravity to form the core during Earth's nascent stage, taking along the coveted gold, platinum, etc., deep inside Earth. Actual estimates reveal that enough precious metals may exist in Earth's core to cover its entire surface with a 4-m-thick layer.[51] But the early formed metals are eternally locked in the vault of the unreachable inner core, while the mantle and crust remain almost depleted of precious metals. Where is the gold, platinum and other metals stored, so humans can tap them? The ultra-high precision Tungsten isotope 182W analysis revealed poor precious metal content of some of the oldest samples (4 Ga) from Greenland, compared to younger rocks. Researchers offered important clues to their origin in the shallow crust.[52] They argue that our planet's accessible precious metal reserve in the silicate mantle came from cataclysmic bombardment of meteorites ('the early bombardment phase'). It happened during the first 200 Ma of Earth's existence, when the inner core had already formed. Asteroids delivered on the early Earth the desired metals now available for our economies and key industrial processes.[52] Tectonic and magmatic processes along deep fractures brought up the coveted gold, platinum, etc. from mantle to shallow crust. Some ores are exposed on the crust for exploitation and multiple societal use. Rivers also wash down some grains of precious ore mixed with sands, easily recovered by the traditional panning method.

Colombia is known for amazing gold artifacts of pre-Hispanic fame, a replica seen in Fig. 1.7. The central character is a naked cacique covered in gold dust for his initiation ceremony as Muisca chief. He made offerings of gold pieces to gods, and then submerged himself and the gold raft in the Guatavita Lake (eastern Cordillera of Colombia). The source of the metal was from the nearby Magdalena river where workers still pan gold from the riverbed and channel bars.

1-7 A gold-plated replica of the *Muisca Raft*, the 'Golden Raft of El Dorado.' The priceless original Muisca Raft is now part of the 34,000 pre-Hispanic gold pieces in the world-famous Gold Museum of Bogota, Colombia.

CHAPTER HIGHLIGHTS

- The broader view of the cosmos—its age, origin, size, shape, composition, and other factors.
- The main Big Bang protagonists—such as Priest Georges Lemaître and astronomer Edwin Hubble—and the theories on the origin of the universe and its constituents.
- The significance of the Big Bang's "Radiation Era" and "Matter Era," along with the "Cosmic Microwave Background" as testimony of the Big Bang.
- A discussion of Earth's larger cosmic context—surrounding galaxies, stars, planets, our solar system, and our satellite, the Moon—all gifts from the 13.77-billion-year-old Big Bang.
- The significance of the Sun and its orbiting filial planets within the solar system.
- Description of Mendeleev's 'Periodic Table' of elements, derived from the remnants of disintegrating stars.
- Ingredients of the mineral, rock, mountain, plant, and animal kingdoms—gifts from cosmic bodies. Primarily oxygen, carbon, hydrogen, nitrogen, phosphorus, and calcium form the building blocks of the animate world, including *Homo sapiens*.
- The origin, distribution, and availability of gemstones and precious metals within Earth's crust.

WHAT'S NEXT?

Now that we have traveled through the limitless universe of big bang, galaxies, stars, and planets, we will make a brief stop at our only Moon close by, to know some of its alluring secrets in the next chapter.

2

The Moon: Earth's Life-Long Companion

THE MOON AND HUMANITY

The Moon is Earth's closest cosmic companion, revolving steadily around us. It alternately bathes our planet in a mellow unearthly light and merges it with total darkness periodically. The softly glowing sphere in the night sky intrigued our ancestors from prehistoric times. They would have wondered about the changing phases of the Moon. They would have stared at the grey patches and pondered if those were mountains or oceans. The Moon helped weave an endless rich tapestry of mythology across cultures. Perhaps they could not explain why the Moon caused the clockwork appearance of tides that routinely affect many mundane phenomena. They would have seen beautifully designed, rippled tidal flats that we observe today in the intertidal zone but did not know why and how they formed. Our ancestors knew that fishing is better closer to a full Moon phase. The Moon was a living power to them—a deity, enriching our planet and life in multiple ways. All cultures including the prehistoric ones, considered the Moon to be a god. It has always played an important role on the nocturnal habits of animals in the air, land, and ocean.

We will briefly review some facts about the Moon, and its importance for the living world. The origin of the Moon is intertwined with Earth as it tangibly contributes to a more stable climate pattern on Earth. The 20th century lunar missions including the U.S.-manned Apollo series, produced a trove of valuable data about the lunar geology, resources and environment. The data is helping scientists plan future midway lunar bases for space exploration.

DAUGHTER OF THE BLUE PLANET

Moon Deities—Revered Across Cultures

The Moon plays a starring role in mythological stories of all cultures as it had mesmerized our ancestors since the dawn of humanity. The lunar emblem holds a position of pride on flags of many nations, especially Islamic ones, after the example of the Ottomans in Turkey. The proof of Moon's eminence is everywhere: Artemis and Selene in the Greek mythology, Luna, and Diana in the Roman empire, while Mani is the personification of the Moon in the Norse culture. Khonsu (child of Amun and Mut), is the ancient Egyptian Moon goddess. Sin or Nanna was venerated in ancient Sumeria, so was Chang'e in ancient China, and Soma or Chandra in India. In Mexico, the tongue-twisting Aztec Moon goddess, Coyolxuahqui, is the daughter of Earth goddess Coatlicue. The Aztec story of filial bond between the Earth-Moon pair is a telling myth, now supported by space scientists.[7]

Earth's Age Constrains the Moon's Antiquity

We need to first assess the origin of Earth to know the Moon's age, because it is tied to Earth since its birth. Ancient cultures have delved into the antiquity of Earth since prehistoric times. Based on Genesis 1:1–5, Old Testament, the estimate of the precise date of Creation was ~4,050 BCE (~6,000 years ago). Another estimate for the creation date is 4,045 BCE.[1] Famous scientists of the 15th–16th centuries—Johannes Kepler and Sir Isaac Newton, were also in line with the teachings and doctrines of the church of the time. Kepler (1571–1630), who formulated the historic law of planetary motion, gave a creation date of 3,993 BCE.[2] Soon after, Sir Isaac Newton (1643–1727), the brilliant physicist/mathematician, suggested an age of 4,000 BCE. It is rather amazing today that Kepler and Newton—two of the greatest scientists in the post-Genesis history of the world, both believed the world was so young. Newton supported Archbishop James Ussher's 1650s chronology of the universe—Oct 23, 4004 BCE. He emphasized that cosmic history and human existence did not extend beyond 6,000 years.[3] The more recent religious biblical scholars also stick to a creation age of 6,000 years.

The above estimates contrast with modern empirical scientific data on the age of Earth. Earth's 4.57 Ga radiometric age is relevant to understand the vintage and origin of the Moon.

How Old is Our Nearest Neighbor?

Apollo lunar samples yielded a range of radiometric ages for the Moon rocks.[4] The basaltic samples from the lunar dark maria region are 3.16 Ga old, whereas light color rocks from the lunar highlands are much older. The pyroxenes from plagioclase-rich anorthosites gave an age of 4.46 Ga, based on samarium-neodymium radiometric data.[5] Zircon crystals from Apollo 14 samples yielded a U/Pb radiometric date of 4.51 Ga.[6] The maria basin lava flows are distinctly younger than the original Moon rocks. The youngest lunar basalt extrusion occurred much later, ~1.2 Ga ago. The oldest measured ages on Earth and the Moon are practically the same, between 3.8 to 4.55 Ga, and they originated in quick succession.

The Moon—Anatomy and Vital Statistics

The information from NASA's many exploratory missions over the years, has been invaluable to our understanding of the Moon.[8] Here are some salient facts: diameter—~3,480 km, and distance from Earth 384,400km; elliptical orbit distance—apogee (farthest) 405,500 km, and perigee (closest) 363,300 km; orbit time around Earth—27.32

Earth days, axial tilt—1.5^0, no seasons on the Moon. Its mass and gravity are ~1%, and ~1/6th respectively of Earth.

Other salient data include daytime temperature of about 120°C at equator (Lunar Reconnaissance Orbiter [LRO] Diviner provided temperature data)[9] and nighttime temperature—minus 183°C at equator. It is colder at the Poles, almost as cold as liquid oxygen. Temperature may be near absolute zero (minus 273°C) in areas with no direct sunlight. NASA study of the dark polar regions confirms the presence of abundant frozen water at the poles. Earth appears brighter, with reflectivity near 30%, compared to the lunar reflectivity of only 12%.

Nothing Reveals Earth's History Like the Moon

Earth and its closest neighbor are very similar in many ways. The Moon is an accessible laboratory to better understand the origin and evolution of Earth. Noah Petro, a NASA scientist, echoed that the Moon is like the Rosetta Stone that lets us understand the rest of the solar system.[10] Meteors and comets impacted the Moon, Earth, and other planets in their infancy, equally and as often. We clearly see many impact scars on the lunar surface. The Moon presents a snapshot of the very juvenile Earth. Lunar history is revealingly intact because it lacks atmosphere (erosion) that would have erased the first chapter of the Moon saga. Absence of plate tectonics and mountain building events also contributes to the pristine nature of the lunar surface.

Lunar Exploration—A 20th-Century Undertaking

The Moon intrigued people from all walks of life—astronomers, scientists, philosophers, poets, and others. They wanted to know the secrets of the not-too-distant, photogenic, romantic companion of our blue planet, fortnightly bathing us in its soft borrowed glow. Galileo Galilei (1609) had the first telescopic view of the Moon and proved that it has mountainous terrains, contrary to conventional wisdom at the time. Ancient Greeks and Romans thought that the dark areas are oceans (or 'mare,' 'maria') and light areas were land ('Terre').

The United States spearheaded lunar exploration in the 1960s, later Russia, China, India, Israel, and others joined in. Early attempts by the U.S. were the NASA Rangers impact probes (1961–65). Other NASA missions like the Lunar Orbiters (1966–77) mapped the lunar surface for future landing sites, and almost simultaneously, the NASA Surveyors probed as soft landers. NASA's six Apollo missions (11, 12, 14, 15, 16, 17) successfully landed 12 American astronauts on the Moon between July 24, 1969–Dec 11, 1972. Exploration continued with numerous remote missions in the nineties and beyond. The United States plans to send a manned lunar mission again in 2026. NASA started the new endeavor with the Artemis-Orion Mission launch in November 2022.

The Israeli spacecraft 'Beresheet' crashed on the lunar surface on April 11, 2019, in a failed landing attempt, spilling among other things thousands of tough, almost indestructible 'moss piglets', i.e., microscopic, dehydrated tardigrades. These cute critters can withstand extreme temperatures on the Moon. They can weather heat of 149°C and cold of minus 200°C.[12] Hopefully, a future mission will find them in a resuscitated state, nourished by water molecules hidden in the lunar regolith.

The Moon is so small compared to the Sun. Why does it appear so big? It has 1/400th the diameter of the Sun but is also only 1/400th as far from us compared to the Sun, making it appear so big to earthlings.[11]

Lunar Internal Onion-Like Layering

Scientific input from seismometer equipment of the Apollo mission, and data of Moon's gravity field and rotation, proved that the Moon has a concentric layered structure as Earth. It has an outer crust, an intermediate solid to plastic mantle, and an iron core at the center (Fig. 2.1).

The Core—A Miniature Version

Seismometer data reveals a much smaller core than a 1,000-km-

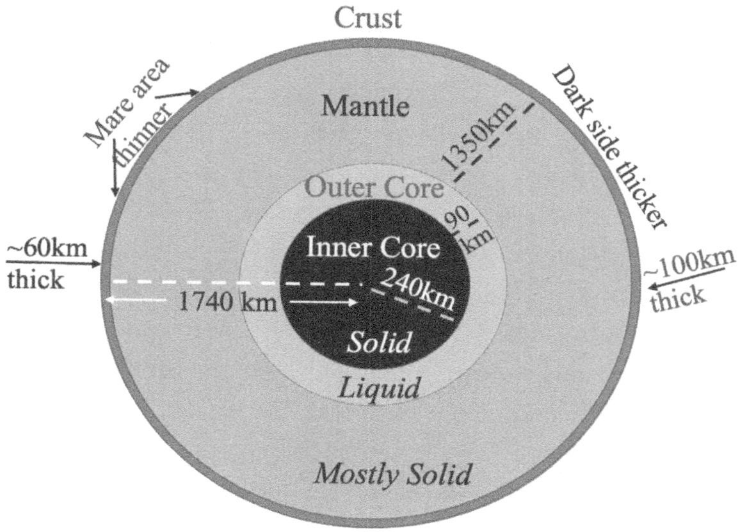

2-1 The Moon's near side with thinner crust and more of Mare (Maria) lavas than dark side *(not to scale)*.

across suggested by gravity data, probably because the Moon's gravity is uneven between the near and far side. The small, 240-km-radius, solid inner core is enveloped by a liquid, outer core 90-km-thick. The total core amounts to less than ~20% of the ~3,460-km-across Moon. Earth's core (~7,000-km-across) measures >50% of our planet (~12,800-km-across). The Moon's lack of magnetic field may be either due to a small liquid outer core, or it is devoid of iron.[19, 20]

NASA's seismometer had detected moonquakes 700-km-deep because of tidal effects. Earth's gravitational pull causes 'tugging and stretching internal structure of the Moon.' Space body impacts may create occasional shallow moonquakes. Extreme temperature-related expansion and contraction may also cause minor moonquakes.

Lunar Mantle and 'Mascons'

The mantle beneath the lunar crust is roughly 1,350 m thick, accounting for 72 percent of lunar mass. New data suggests that dense pyroxene-rich norite and olivine-rich troctolite began to form 3.85 Ga ago in the partly molten lower mantle. Some authors have divided the mantle into an upper solid basalt, and a lower partly liquid layer of mostly silicates and some metal oxides. The mantle contains in some cases the enigmatic 'mascons,' i.e., mass concentrations, under deeply buried lunar 'maria.' The 'Mascons' are either buried meteorites or denser magma bodies.[18]

Lunar Crust—Pockmarked Surface, Lava Sea, and Towering Mountains

The heaviest elements gravitated to the core, while the lightest stayed near the surface of the crust.[13] The overall lower lunar density is partly because the Moon's original molten material came from the lighter, silicate-rich, metal-poor crust of Earth. The heavier elements i.e., most metals, had already sunk to Earth's core by the time of the Moon's birth. The 60–100-km-thick lunar crust makes up only six percent of total lunar mass. It is asymmetric—thinner on the Earth-side, and thicker on the far dark-side. The lunar highland areas consist of less dense rocks like anorthosite compared to the mare (maria) regions with dark basalt lavas. The Mons Huygens (5.5 km), Mons Hadley (4.5 km) and Mons Rumker (1.1 km) peaks adorn the highlands.

What are the typical rocks in the crust and mountains? The basalt 'magma ocean' of the Moon crystallized within 100 Ma and 'fractional crystallization' had gradually set in within the magma oceans (depth ~500 km). What is fractional crystallization? In liquid magma, dense minerals like pyroxene and olivine formed first and started sinking and settling to the bottom in the mantle. The relatively light anorthite feldspar crystals literally floated to the top in the still-liquid crust when crystallization was 75% complete. Low density minerals and metals are typical of the lunar crust.[14]

The special calcium-rich feldspar—anorthite, produced abundant anorthosite in the crust, granite being rare. Other crustal rocks are olivine-rich pyroxene gabbro, and 'KREEP' basalts— sourced from the mantle as igneous bodies. The KREEP acronym means rocks rich in potassium (K), rare earth elements (REE) and phosphorus (P). A KREEP-rich lunar meteorite once impacted a site in Antarctica.[15]

Craters on the Lunar Crust—Too Numerous to Count

Most impact craters formed very early (~ 4–3.85 Ga ago) in the highlands. The age coincides with the 'heavy bombardment phase,' affecting all solar planets during the infant solar system. The 'Maria' impact-basins formed much later. Moon's circular craters are numerous and of all sizes, though showing more craters of smaller size. Also, crater size and frequency decrease with time. About 10,000 craters on the near-side Moon implies a rate of meteorite bombardment of 100,000 years/crater. Planet Earth also went through a similar history of meteorite impact, but the evidence is largely gone. Geological processes of plate tectonics, weathering, erosion, uplift, and sedimentation mostly erased the vestiges of the earlier earthly craters. The lunar highlands show many more craters than the maria regions. NASA documented the circular ~4-Ga-old, 'Clavius' crater in the southern highlands.[16] It is the second largest on the Moon's near side—diameter 231 km, depth 3.5 km.

2-2 Sample of a vesicular volcanic rock from NASA's Lunar Surface Collection at Johnson Space Center's vault in Houston, Texas.

Craters and Craters—Three Special Ones Stand Out

Three of the most famous craters are the Copernicus, Tycho and Aitken basins near the South Pole. The Copernicus—radius of 93 km, and depth 3.76 km, is about 0.9 Ga old. The younger, 108-Ma-old Tycho is also a relatively small crater—diameter 86 km, depth 4.8 km. NASA Apollo 17 samples from the crater "rays" gave Tycho's radiometric age. Rays are long radial streaks on crater wall made by fine ejecta thrown up when a meteorite hits lunar surface. Tycho impact happened when dinosaurs were still roaming around on Earth. The Aitken basin near the South Pole, is the largest, deepest, and oldest (~3.9 Ga) crater on the Moon, and one of the largest in the entire solar system—diameter of 2,500 km and 13 km deep. NASA thinks the Aitken basin formed during the 'heavy bombardment phase,' affecting all solar planets. The highest point of >8 km on its northeast rim, and the lowest >6-km-deep point at the basin center, are records for the Moon.[14] The basin floor is pyroxene-rich, and the Leibnitz Mountain samples on the northeast rim are typical pale highland anorthosite. Scientists think that a 200-km-across meteorite could have produced the deep Aitken crater. A new lunar mineral 'Armalcolite' honors the first 3 NASA astronauts of Apollo 11 mission—Neil Armstrong, Buzz Aldrin, and Michael Collins. We are not lucky enough to see the real, circular Moon craters. However, an enticing earthly version is accessible in the Labrador region of Canada. A meteorite punched a large 28-km-across hole into the ground 36 Ma ago. Weathering and erosion pared down the size of the original crater to a small one about 16-km-across with a 50-m-high crater wall. The central 3-km-wide Horseshoe Island in the Mistastin impact crater lake represents the original peak ring. The impact would have created a temperature of 2370^{0}C, the highest ever on Earth surface, producing rare cubic zirconia crystals. Crater wall of anorthosite rock of deep crust-mantle origin, mimics the Moon's highland anorthosites.[35]

The Secrets of Maria—Young Crater Oceans Filled with Basalt

Basalt lava filled the depressions of the dark mare or maria regions of meteorite impact origin. A maria may be 1,000-km-across and 5-km-deep. The lavas are less viscous and contain iron and titanium-rich ilmenite. Commonly maria volcanism occurs with pyroclastic eruptions. The explosive bursts of liquid effusive and shattering of solidified magma produce pyroclastic ejecta. Maria basalts contain abundant vesicles, volcanic glass, and rills in lava channels. In contrast, the mountains of lunar highlands are mainly of ejecta origin.

Dark Side of the Moon—Little-Known Far-Side Reconnaissance

The Chinese Chang'e 4 lander touched ground on the dark side of the Moon, the first time ever by any space vehicle. It landed in the small "Von Karman" crater, inside the Aitken basin. Lunar rover Yutu-2 identified olivine and orthopyroxene of likely mantle origin in the crater.[17] These samples differ from the near side lunar rocks. China's Chang'e 6 probe landed on the Moon again and collected and brought back samples from the dark side on June 2, 2024.

Surface Regolith—Scattered Impact Debris, Similar to Earth's Soil

The superficial layer on the lunar crust is regolith, a loose dark grey to reddish meteoroid impact-related dust and debris; it covers most of the lunar surface. It varies in thickness from 5 m on the low, flat maria surface, to about 10 m on the highlands. It is somewhat like hardened soils mixed with lunar ejecta rocks. The Apollo Mission astronauts had brought back drill core of regolith samples, including a titanium-rich vesicular basalt. Escape of gas produces vesicles.

Lunar Tectonics or the Lack Thereof

One may wonder if Earth-like plate tectonic activities affect the Moon. The simple answer is no, because the Moon is smaller, and cooler inside. The mantle is not hot enough to flow easily and create convection currents, the driving force of plate movements. Also, the Moon has only about 1% of Earth's mass. That may explain why the Moon's small core has such a weak magnetic field.[19] However, the retrieved Moon samples by Apollo Missions showed remnant magnetic property, suggesting it had lost its earlier magnetism through time.

Meteorite Bombardment and Moon Ejecta—What Can It Reveal?

Antarctica, Oman, and north African deserts are the sites of 340 recovered lunar meteorite fragments. The 190 kg rocks vary from 100 kg to 0.788 gm each. These rocks were ejected when wayward asteroids had smashed onto the lunar surface with great force. The impact made some of the lunar ejecta fly far enough to enter Earth's orbit and eventually landed in various places on Earth. Scientists used characteristic mineralogy, isotopic and chemical composition to establish lunar origin of the meteorites. They also compared them with the 382 kg of rocks collected by NASA's Apollo missions from 1969 to 1972.

Two definitive proofs of lunar origin are the fusion crusts (thin millimetric glassy surface layer) of meteorites, and 'cosmogenic' isotopes because of penetrating cosmic rays.[29] Breccia rock fragments in earth-bound meteorites originated in the lunar highlands and other places. The meteorites are reliable representatives of the moon from the poles to equator. In contrast, Apollo samples are not representative, as they came from a very small area near the 'mare imbrium' containing anomalously high radiogenic thorium (up to 30 ppm), potassium, and uranium content. The lunar meteorites show very low thorium content—< 1ppm.

EARTH-MOON RELATIONSHIP

How the Moon Affects Earth—Beneficial Gifts Like Tides

Tides were well known to all cultures as tidal cycles influence us in a big way. The Moon rotates synchronously with Earth (i.e., tidally locked); its axial spin and orbital speed is the same around Earth. That explains why we always see the same face of the Moon. Both the Sun and Moon exert gravitational pull on the near side of Earth, but the Sun's pull is much weaker (about half of the Moon). The Moon's tug on Earth's oceans has a positive effect; its proximity keeps Earth's tilt on its axis relatively stable, making our Earth a more livable planet. It moderates the wobble in Earth's axial tilt and contributes to a more stable climate pattern.[21]

Sir Isaac Newton was the first to explain in 1687 that ocean tides are due to gravitational pull of the Moon (also Sun) on Earth (*NOAA's National Ocean Service*). The clockwork tides create the beautiful ripple marks on shallow areas like beaches and intertidal flats (Fig. 2.3).

The highest tidal range in the world occurs in the elongated, tapered Bay of Fundy region of eastern Canada (Fig.2.4). The 16-m-high tides occur at the tip of the bay near Burnt Coat Head Park, Nova Scotia. Up to two-meter-high tidal bores visit the distant Petitcodiac River near Moncton, New Brunswick, about 30 km inland from the Chignecto Bay. Tides are important agents of erosion, transport and local circulation in coastal inlets, estuaries, and tidal deltas. The asymmetric nature of ebb and flood tides may trigger local net erosion and mass transport of water, sediment, and nutrients. Effect of tides has a strong bearing on Earth's sedimentary processes, creating a spe-

2-3 Ripple marks on the intertidal flat. Lucaya beach, Grand Bahama Island.
top The water-covered, ripple-flat shows the shoreward-moving asymmetric ripple train, and the general wind and current directions. Note sharp beach/tidal flat contact.
bottom White beach sands (1) border the tidal flat (2). Berm is the high area behind the beach. The exposed ripple-flat (2) shows a transition from symmetrical single-crest ripples (2a) to double-crest ripples (2b) to wide ripples (2c). The oblique contact of ripple train and beach face is noteworthy. The submerged back bar low (3) separates the ripple-flat from the exposed elliptical sand bar (4). The subtidal area (5) lies beyond the marine bar.

2-4 Bay of Fundy, Hopewell Rocks, New Brunswick, Canada. **left** Ebb tide exposes green algal growth on sea-stacks (arrow). **right** Same area under flood tide.

cial kind of sedimentary structure— 'tidal bedding.' Rise of global temperature and mean sea level, a sequel to climate change, may lead to changes in tidal reach, strength, and surges, drastically affecting local ecosystems and livelihood.

Horizontal flow of water ('Tidal streams') makes it possible to predict tidal activities, important for coastal navigation. Fishing is closely affected by tides—it is best when tide is running during the mid-point of ebb and flood stages. Fish tend to bite more when water flow is continuous. Fish movements and feeding is more active during the full moon phase.

A U.S. study documented increased wobble in Moon's orbit around Earth.[22] It is causing, and will cause significant high tide flooding of North American coasts in the future. The wobble in Moon's orbit runs in 18.6-year cycle, each half-cycle alternately suppresses and amplifies high tide levels. NASA predicts the next cycle of 2030s will have a major effect on coastal areas including coastal populations and infrastructures.

The Moon Has a Hand in the Increase of Day Length on Earth

Tidal friction of the Moon slows down Earth's rotation over time. Our present 24-hour day is a product of a gradually slowing rotation of Earth through geological time (~4.6 Ga) due to 'tidal braking.' The nascent Moon used to orbit Earth every five hours, because it was much closer to us. At that time our earthly day lasted only two to three hours. The gravitational pull of the Moon likely contributed to slow-down of Earth's rotation and consequent increase of day-length. Tidal records of ancient estuaries with seasonal cycles of sand/silt layers suggest that a day had only 21 hours in the 620-Ma-old Precambrian Earth. Compare that with a Dinosaur's Day of 23 hours in the Cretaceous.[23]

Lunar Phases & Moonlight—Unearthly Glow in the Night Sky

Moonlight has been a constant source of inspiration and romantic feelings among poets, song writers, artistes, etc. Indeed, it is soothing, has a contemplative quality to it. Moon's waxing and waning phases seem magical, especially in places like the timeless Taj Mahal, a story of love etched in the uniquely white, 1.5-billion-year-old, Makrana marble, Rajasthan, India (Fig. 2.5). No wonder, it has both scientific and mythological explanations. Moon enjoys 2 weeks of borrowed daylight from Sun every lunar month (28 days) and we see the reflected light as nightly moonlight. The waxing-waning happens because sunlight progressively increases or decreases as Moon is orbiting Earth. In Indian mythology, Chandra (Moon) waxes and wanes because of King Daksha's wrath. Chandra had married Daksha's 27 daughters but loved one of them more than the others. The king cursed the Moon willing him to lose his perennial glow and instead alternately wax and wane.

The Moon and Moods—Affecting Even the Lowly Corals

Scientists know how the Moon affects biorhythm and moods of peoples and animals.[24a] I remember growing up in India, where our elders used to relate health with the phases of the Moon. Women may have a likely biological connection, mirroring lunar cycles. In the animal world, lions normally hunt at night, but they break the rule on full moon nights. They hunt the morning after a full moon. Australian corals feel the primal urge and release massive quantity of eggs and sperms for a unique mass reproduction event during a special, particularly moonlit night.[24] According to researchers, more than 130 species of corals in the Great Barrier Reef, Australia, display the world's largest annual spawning show, when they release eggs and sperm cells within a short window of 30–60 minutes.[25] Environmental controls such as lunar phase, ocean temperature and the daily light cycle, are factors triggering the unique reproduction show.

The Birth of the Moon—Still a Mystery?

Now that we have some idea of the ages of the Earth-Moon pair, how did the Moon become a satellite of Earth? To resolve this issue, scientists proposed several hypotheses—the proto-planet Thea impact hypothesis, fission from Earth, a wayward stellar moon from outer space captured by Earth, simultaneous origin of Earth-Moon pair, etc. We will discuss only the Thea model, as the scientific majority favors it.

Moon's Origin—The Impact Hypothesis of Giant Theia

The Mars-sized proto-planet Theia had directly impacted the nascent, somewhat molten Earth about ~4.5 Ga ago.[26, 27] The collision likely caused a portion of Thea to generate voluminous sheared off debris. The large ring of detritus arose from the combined flotsam of the

2-5. The Taj Mahal, Agra, India—one of the Seven Wonders of the World—captured at dusk.

disintegrated Thea and the flying fragments of the impacted Earth's crust/mantle. The debris field eventually would have coalesced to form Earth's satellite Moon. This widely accepted hypothesis has several advantages. It explains the size, orbit, and tidal locking of the Moon's rotation resulting in one of its sides continuously facing Earth. The Apollo samples revealed that oxygen isotope ratio, a crucial fingerprint, is identical in the Earth-Moon pair, implying their co-genetic origin. However, the impact hypothesis cannot explain the virtual absence of volatiles in the Moon, so abundant on Earth. Some argue that collision energy dissipated all the volatiles from the lunar rocks. To circumvent the problem, a NASA scientist suggested that two celestial bodies (five times the size of Mars) in early solar system collided, and then again recollided. The collision gave rise to a large disc of mixed detritus from which ultimately the Earth-Moon pair was born.[30]

No wonder the ancient Mexican Aztec culture sagely personified their moon-goddess Coyolxauhqui as the daughter of earth-goddess Coatlicue.[7] However, our planet surely will not wish the same fate as the Mexican earth-goddess. Apparently, the Aztec moon-goddess had her mother Coatlicue killed to save her family honor.

The Birth of the Moon and the Origin of Water on Earth

We take water for granted in our day-to-day life because it is essential to sustain life. How did it form, especially when Earth was a partially molten ball at its birth about 4.57 Ga ago? We recall water comes from the chemical reaction between two essential elements—hydrogen and oxygen. The very word 'hydrogen' signifies it generates water, and 'oxygen' implies that it generates acids (mistaken idea). Simple molecules of H_2 and O_2 will not be able to form water, because the process needs energy input to break the covalent bonds of the gases. A spark will add heat to supply the activation energy to start the reaction, water formation and release of heat.

Molybdenum (Mo) isotopes have likely solved the mystery of the source of water.[26] The solar system formed 4.57 Ga ago. Early Molybdenum isotope came from dry non-carbonaceous meteorites from the inner solar system and gravitated to the core because of its affinity for iron. The late Molybdenum in the mantle arrived from the outer solar system via protoplanet 'Thea,' after the core had already solidified. Analysis of Mo isotopic composition of Earth's primitive mantle reveals that it falls between carbonaceous and non-carbonaceous sources. That may mean the Mo of carbonaceous meteorite origin in the mantle is rather late in origin, probably when the moon-forming nascent Thea impacted Earth violently. German scientists argued that Thea was the one that delivered water to the dry inner planet Earth. Only carbonaceous, water-rich meteorites from the outer solar system could have endowed Earth with life-giving water and highly volatile materials.[27] Smectite clay (a swelling clay with water molecules in crystal lattice) in carbonaceous meteorites is further proof of the presence of water in asteroids. According to U.S. researchers 'protoplanet Thea' collided head-on with our planet 4.5 Ga ago. All planets and satellites show a signature oxygen isotope fingerprint; we know the Moon and Earth show identical oxygen isotope content. Thea's head-on collision could explain the cogenetic origin of the Earth-Moon pair.[28] More on the Thea-Moon connection later.

The Moon's Untapped Resources—A Base in Space?

Analysis of NASA Apollo 17 samples suggested stable isotope Helium 3 as a potential source of energy for future space exploration.[31] In 2006, Apollo 17 astronaut-geologist Harrison Schmitt discussed the exploitable energy sources of the Moon. Scientists are closely looking at the promise of *In Situ Resource Utilization* (ISRU) techniques. Also, the discovery of ice and water in the polar regions of the Moon, and Mars, opens the possibility of midway bases in space for distant space exploration. Dissociation of water into hydrogen and oxygen will be a paradigm shift as it will open new vistas. In 2013, geoscientists compiled a book, describing possible resources in the solar system for human settlement.[32] The study reported that lunar regolith near NASA's Apollo 11 landing site in *Mare Tranquillatis*, covers an area of 2 km^2, to a depth of 3m. The small area may contain nearly 100 kg of Helium-3, enough to power a 1 GW fusion power plant for a year.[33] The work reveals the Moon is replete with energy minerals (helium-3, thorium, and uranium), some volatiles (N), metals (titanium, iron, and magnesium), and polar water and ice. These are all usable off-world resources for future space exploration. Also, the Sun may be a handy lunar energy source in the polar areas with steady illumination.[34]

CHAPTER HIGHLIGHTS

- Knowledge of the Moon is useful as it partially mimics Earth's history.
- The Moon's age, size, and characteristics of its visible and dark sides.
- Myths surrounding the Moon across different cultures.
- The beneficial effects of the Moon's proximity on Earth's climatic stability. Moon-induced tides affecting Earth.
- The internal make-up and characterization of the Moon's core, mantle, crust, and lava oceans.
- Meteorite bombardment shaped the Moon's surface, forming craters, highland ejecta, and the lunar regolith (soil).
- The Earth–Moon interdependence relates to moonlight and moods; the Moon affects all animals, including triggering the largest coral spawning on Earth.
- The origin of the Moon is explained by the protoplanet 'Theia' impact hypothesis, including Theia's role in bringing water to Earth.
- History of lunar exploration, its untapped resources, and the potential for a future base in space.

WHAT'S NEXT?

The brief, scintillating space journey ends here. Chapters 3 to 13 will center around our palpable blue planet, narrating remarkable changes through the evolving Eons and Eras. Chapter 3 explains the basics of geoscience—Earth's layered internal structure, geologic time scale, myriad rock types, utility of paleontology, i.e., the study of ancient life, tectonic fold, and fault structures. These core concepts will reappear often in rest of the book.

Cross-stratified, color-banded Proterozoic sandstone pillars in one of the cave temples in the ~1500-year-old Badami Fort, Karnataka, India. The foreground pillar on right shows apparent large scale herringbone tidal bedding.

3

The Blue Planet: Time, Rocks, and Fossils

THE BIG PICTURE

We have all seen NASA's blue marble image of our very old living planet. How old is very old? Unlike someone on the street, a geologist may feel at home thinking in terms of a few billion years. Earth's checkered history is like a multi-tome compendium. It is long, super-complicated with many earth-shattering, literal ups and downs, spanning 4.57 Ga. To make matters more challenging for geoscientists, countless pages of this tome have been erased by cataclysmic processes in many areas. Long, painstaking detective work and observations by countless geoscientists, paleontologists and specialists have produced a reasonably complete and convincing biography of the blue planet, our home.

Discussion in the chapter will center around the universal concepts in geology, stressing on features like Earth's concentric layers—core, mantle and crust, and the nature of rock types—igneous, sedimentary, and metamorphic. Rock structures like faults and folds, resulting from rock deformation, will follow. One will learn how geoscientists determine the age of strata and fossils. Finally, we discuss Earth's stratigraphy, i.e., geological succession of rock strata, and the essential pillar of Earth history—the nested time scale. Rocks, radiometric ages, fossils, seismic images, are tools we use to solve the mystery of Earth's well-guarded secrets.

THE STRUCTURE OF EARTH

Our Onion-Ring Earth—The Core and More

To start with, why Earth and other planets are spherical, not angular like all asteroids, meteorites, and comets. Gravity makes mass evenly distributed in a perfect sphere, while it is not strong enough to make a smaller asteroid spherical.

Earth shows internally four differentiated, concentric layers—inner and outer core, mantle, and crust.[3] Mineral composition and relative solidity define the 4 layers (Fig. 3.1).

Earth is a typical sphere (radius ~6,400 km) having layers of varying density. The solid inner core (radius 1,300 km) is the heaviest, enveloped successively by a liquid outer core (~2,200-km-thick), and a mantle in a solid to semisolid state (~2,900-km-thick). The crust on top, a thin veneer of solid rock, wraps around the mantle. Continental crust is much thicker (up to 80–100 km) than oceanic crust (~7 km). Crust is what appears as mountains, valleys, rivers, oceans, deserts and what lies below.

The Core Question—What Is Known of Something So Deep?

The inner core is a sphere of solid iron-nickel alloy. It is in a solid state even at 5,200°C (Sun's temperature), because of the incredible 3.6 million atmospheric pressure inside. The outer core is also an iron-nickel alloy, but in a liquid state because of lesser pressure and a high temperature of more than 4,500°C.[33] Flowage of liquid iron in the outer core generates Earth's magnetic field—an essential property for life to survive. The *Bullen Discontinuity* separates the two core layers.

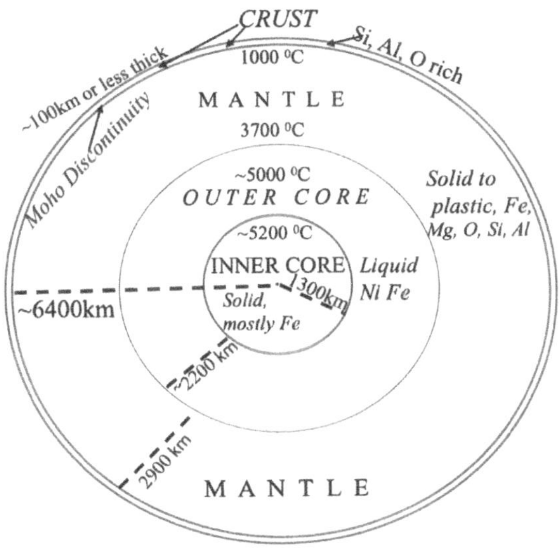

3-1 Earth's four main onion-ring layers—crust, mantle, outer core, and inner core—illustrating the planet's complex internal structure (not to scale).

The Mantle—Center of Most of the Action

The thick mantle is a spherical layer that envelops the outer core and forms more than 80% of the volume of Earth. Beneath a thin crust, upper mantle depth is 100 km–660 km, and the lower mantle (660–2,900 km) reaches the outer core boundary.[3a] Mantle is semisolid, denser than the overlying silicic lighter crust. It is so hot that it is malleable and flows plastically, though very slowly. Mantle pressure is 1.4 million times atmospheric pressure. The unique vertical gradient of temperature, pressure, and density within the mantle, likely cause convection current. It sets up very slow eddy currents, about 10–15 cm/year. It may take nearly a hundred million years for a piece of basal mantle rock to travel 2,900 km to the top of the mantle (below the crust).[1] Mantle rock is olivine-rich peridotite.

Earth's rotation may also contribute to mantle convection that was likely more active during Earth's infancy. Early formed denser metals gravitated toward the core. Thus iron, nickel, and metal sulfides settled deeper, whereas less heavy, floating olivine and pyroxene stayed in the mantle. Convection current in the mantle is probably the key mechanism of 'plate tectonics,' a subject of Chapter 4. It triggers movement of the crustal plates, deeply affecting the crustal layer of Earth. Plate movements cause many geological phenomena of far-reaching impact—volcanic eruption, continental drift, earthquake, mountain building, biogeography, climate pattern variation, among others.

The Crust—Frosting on the Cake

Crust is the solid, outermost layer of Earth, both continental and oceanic type depending on its location. It is an oxygen-rich, ultra-thin layer, forming only ~1% of Earth's radius (6,400 km). It is like the crust of an apple pie, where the underlying semisolid mass may be roughly akin to the mantle and core. In 1909, geophysicist A. Mohorovicic discovered the boundary of crust and mantle—the *'Moho Discontinuity'* (Fig. 3.1). He detected a marked velocity increase of earthquake waves across this boundary due to sudden change in rock properties. The crust, markedly different from mantle rocks, may vary in composition, though is generally much lighter and more silicic.

The continental crust, also known as 'sial' (silicate and aluminum-rich), is much thicker and underlies the continents. It is about 75–80 km thick under the Himalayas, the highest mountain range in the world. The India-Eurasian plate collision has led to excessive compression, buckling and enormous crustal thickening in the Himalayas.[2] It is still going on.

The oceanic crust, known as 'sima' (silicate-mafic composition) underlying the ocean floor, is much thinner than continental crust. It consists of denser extrusive basalt and intrusive gabbro, both rich in mafic (ferromagnesian) and ultramafic minerals such as pyroxene and olivine. The ~7 km thick layered oceanic crust shows an average ~0.5-km top layer of recent deep-sea sediments; underneath is a <2 km layer of basalt lava, breccia, etc. A <5 km layer of crystalline gabbro and peridotite forms the bottom layer of oceanic crust.[3]

Oceanic crust, way younger than continental, is being continuously consumed at plate subduction zones. Dense oceanic crust dives under lighter continental crust till it reaches the mantle. The oldest oceanic crust, still present in the northwest Atlantic, is 200 Ma old (Jurassic). In comparison, the western Australian continental shield is 4.4 Ga old, almost as old as the 4.57-Ga-old Earth itself.

Rocks Are Not All Created Equal: Varied Textures and Origins

Familiarity with some geological terms, definitions and facts will help us to follow the rest of the chapters with ease. The nature and origin of different rock types, mechanics of deformation and consequent rock structures—are important elements of rock history. We may see rocks of different origins if we have a chance to visit different geographic locations. The mode of origin implies that rocks are primarily of three types—igneous, sedimentary, and metamorphic. Igneous rocks make up the primitive, solidified Earth, with abundant granites and widespread lavas. Both sedimentary and metamorphic rocks appeared soon after. All rocks consist of several mineral constituents.

Mineral hardness is another useful concept in understanding nature of rocks. Softer minerals weather and erode more easily. Geologists use the Mohs Scale of Hardness (1 to 10, i.e., talc to diamond) for all minerals (Fig. 3.2).

Mohrs Scale	Mineral	Common Object Hardness
1	Talc	
2	Gypsum	Fingernail (2.5)
3	Calcite	Copper Penny (3.5)
4	Fluorite	
5	Apatite	Glass Plate/Pen Knife (5.5)
6	K-Felspar	Steel file/Streak Plate (6.5)
7	Quartz	
8	Topaz	Masonry Drill (8.5)
9	Corundrum	
10	Diamond	

3-2 Mohs scale of hardness uses 10 common minerals showing increasing hardness from 1 (Talc) to 10 (diamond). Fingernails scratch talc and gypsum, while diamond scratches all minerals.

Igneous Rocks: Chiseled from Fire and Fury

The word 'igneous' comes from Latin 'ignis' meaning fire, like 'Agni' in Sanskrit. Cooling and solidification of fire-like, molten, or semi-molten deep-seated magma melt creates these rocks. Igneous rocks are either intrusive or extrusive. The former intruded into the crust, but always remained below Earth's surface. It slowly cooled down and crystallized over geological time. In contrast, extrusive lava erupts on the surface through deep volcanic chimney or fissures and cools rapidly.

Geologists use two simple criteria to classify igneous rocks—texture, and the ratio of felsic/mafic, i.e., pale to dark minerals (Fig 3.3). Texture basically means crystal size of a rock, i.e., small or large. Volcanic rocks show small crystals, whereas intrusive rocks contain large crystals. A rule of thumb—crystal size is small when magma cools very rapidly. Spilled lava on the surface solidifies very quickly, whereas intrusive rocks slowly congeal below Earth's surface over millions of years. Losing heat slowly, crystals will have more time to grow larger.

Pale and dark minerals define igneous rock texture and classifica-

MAFIC MINERALS →

INTRUSIVE	GRANITE	DIORITE	GABBRO	PERIDOTITE ultramafic
VOLCANIC	RHYOLITE	ANDESITE	BASALT	KOMATITE, PICRITIC BASALT

FELSIC MINERALS ←

3-3 Classification of igneous rocks—Minerals are felsic (pale, silica-rich), and mafic (dark, magnesium-iron-rich). Arrows show increasing content.

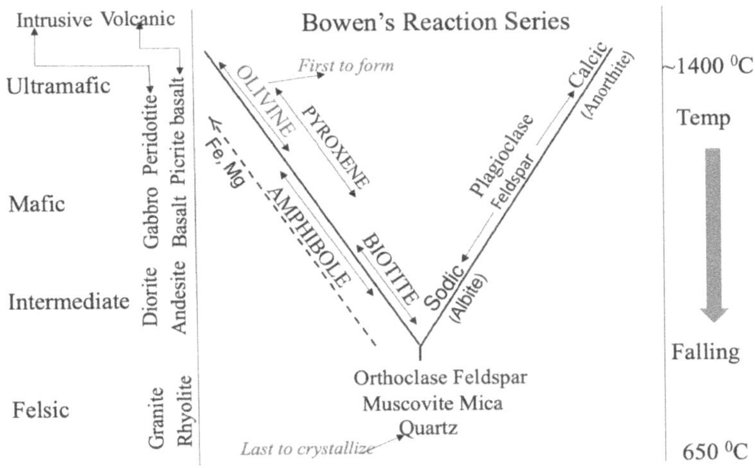

3-4 The 'Bowen's Reaction Series' shows the sequence of mineral growth from cooling magma. High temperature minerals crystallize first.

tion. Silica-rich, felsic minerals are feldspars and quartz (SiO_2). Mafic dark minerals—pyroxenes and amphiboles, are magnesium (Mg) and iron-rich (Fe). The ultramafic mineral olivine is darker, even richer in Mg and Fe, and has very little silica. Intermediate rocks—diorite and andesite, show subequal felsic and mafic minerals.

Dark minerals crystallize first at higher temperatures, and pale silicate minerals crystallize last at lower temperatures. Ultramafic olivine crystallizes from molten magma at the highest temperature (1200–1400°C) and at a much greater depth within Earth. Mafic amphibole and pyroxene crystallize at a lower temperature than olivine, though still high. Whitish quartz and feldspars (orthoclase and albite) crystallize at lowest temperatures (650°C–750°C) from magma at shallow depth.

'Bowen's Reaction Series' shows the path of crystallization of different ferromagnesian and silicate minerals (Fig. 3.4). For us, suffice it to know that olivine-bearing peridotite forms at highest temperature, pyroxene-rich gabbro and basalt form next at high temperature. Intermediate diorite, and andesite form at progressively lower temperature, and finally quartz, feldspar, and mica-bearing granite and rhyolite form last.

All Igneous Rocks Are Not Granite

Granite is such a commonplace name. Who has not heard about a granite countertop in the kitchen? Or the many other items we encounter in our daily life—floor tile, tomb stone, building veneer, paving stone, etc. It seems that any polished, crystalline rock of any texture, color, and composition can easily pass off as a 'granite.' What is it really? Granite is an intrusive rock that crystallizes at ~650°C. It mainly consists of quartz and feldspars (orthoclase and/or albite). Accessory minerals are muscovite, biotite (both micas), hornblende (an amphibole), epidote, and traces of other heavy minerals—rutile, zircon, tourmaline, sphene, etc.

Pegmatite—Treasure Trove of Gemstones

A pegmatite is a granite with crystals larger than 2.5 cm. The word comes from Greek 'pegmat' meaning things joined together. It refers to the interlocking texture of large to abnormally large quartz, feldspar, and other crystals. Typically, pegmatites occur as veins and dikes shooting off from intrusive granitic rocks, often hosting highly prized, large, perfect crystals. Hot, less viscous reacting fluids, and slow rate of crystallization promote large crystal size. Most high temperature minerals exclude water from their crystal structure. At lower temperature, however, plenty of ambient water allows rapid mobility of ions toward growing crystals in pegmatite. It is common in cratons and in large granite bodies i.e., batholiths. Many meter-long or larger, museum-quality crystal pieces come from pegmatites from all continents. Malaikilina, Malagasy, provided the world's largest—a giant, 18-m-long, 3.5-m-across beryl crystal. A spodumene pyroxene from the United States measured 12.8 m.[4] Indeed, gemstones are common in pegmatite—amazonite, apatite, chrysoberyl, garnet, kunzite, lepidolite, topaz, tourmaline, zircon, beryl's like aquamarine and emerald, etc. Pegmatite also hosts many precious metals such as gold, silver, etc.

Lava Rock-Basalt, and Intrusive Gabbro

Basalt is a common dark, very finely crystalline, iron-magnesium bearing, pyroxene-rich, extrusive rock containing calcic plagioclase. It is the most abundant rock that covered the Indian plate in the Cretaceous Period. The modern mid-oceanic rifts and the Pacific Ring of Fire typically produce basaltic and basaltic-andesitic lava rocks respectively of mantle origin. Basalt's coarse variety is gabbro of intrusive origin (Fig. 3.5).

3-5 Images of some common igneous rocks. **1** A pink granite with a probable andesite xenolith (a different rock fragment) with feldspar crystals within it, Giza, Egypt. **2** The history of local fishermen etched on granite by sculptor W. E. deGarthe, near the famous Peggy's Cove Lighthouse, Nova Scotia, Canada. **3** A pegmatite with large crystals of orthoclase feldspar, Canada. **4** An outcrop of columnar basalt on Iceland seashore (Photo: Omar Colmenares). **5** A dark crystalline anorthosite with large iridescent calcic plagioclase crystals (labradorite-bytownite), and black pyroxene and amphibole, Quebec, Canada. **6** The barren tablelands of olivine-bearing rocks in Newfoundland. Leaching of high amount of Ni, Mg, Cr, Fe, Co, makes the soil too toxic for trees and forests, Gros Morne National Park, Canada. **7** A rhyolite porphyry, i.e., large pink orthoclase crystals in a fine silicic groundmass, Patmos Island, Greece.

Sedimentary Geo-History: Layered Rocks Formed by Rivers, Lakes, Oceans, and Wind

The Grand Canyon, USA, is a great monument of stratified sedimentary rocks (Fig. 3.6). Geologists routinely use terms like bed, bedding, bedding plane, while describing stratified rocks. A 'bed' is a sedimentary layer, separated by a minor discontinuity from beds above and below. Each bed has a thickness, width, and length, just like a common bed one sleeps in. A 'bed' is enclosed by bedding planes above and

3-6 Grand Canyon, Arizona, USA, shows multi-km-thick, pinstripe, horizontal sedimentary strata.

below. Beds are the fundamental building blocks of stratigraphy—a branch of geology dealing with vertical order of strata. Individual beds range from 1–100 cm or thicker. A sediment layer <1-cm-thick is a 'lamina.' A bed may differ from adjacent beds by its distinctive physical, chemical, mineralogical, and biological characteristics. A 'bedding plane' lies between beds. It may be gradational, sharp or irregular. The 'bedding' denotes layering in sedimentary rocks. It may be horizontal or tilted, i.e., inclined. Sometimes beds are vertical because of tectonic deformation. Ripple marks form by waves and/or currents on horizontal sandy beds (Fig. 3.7). A 'cross-bedding' means an inclined set of beds between two horizontal beds. It indicates the direction of flow in ancient rivers, oceans, etc.

The Grand Canyon walls and towers are great examples of layered sedimentary rocks. However, vertical rock sequences may not always be continuous. An erosional, irregular bedding plane may separate two contiguous strata of dissimilar age, contrasting fossil content or bedding style.

The 3.8-Ga-old Isua sequence of deep-water volcanic-sedimentary origin, and the Banded Iron Formation, Greenland, are some of the oldest sedimentary rocks. Western Australia[5] documents a similar 3.45-Ga-old rock.

We see loose sediments everywhere around us. Source of sediments is always older rocks. Ancient sediments of diverse origin eventually became rock through cementation and compaction. The sedimentary rocks are simpler, they are lithified sediments of river, lake, ocean, and other environments. These rocks are of four types: clastic, chemical, biochemical, and organic. The clastic rocks are shale, siltstone, sandstone, and conglomerate, with increase of grain size. Fig. 3.8 shows some common sedimentary rocks from different places in the world.

3-7 top Unusual ripple marks on a tidal flat, Sinamaica, West Venezuela (*credit: A. Isea*). **bottom** Cross-bedding with angular middle strata between sub-horizontal bedding planes above and below (Socorro Fm., Falcon State, Venezuela).

3-8 Some common sedimentary rocks. **1** Folded sandstone-shale strata by the ocean, Nfld., Canada. **2** 15-cm-wide slab of the 'picture sandstone' of Jurassic desert origin, southwestern United States. **3** The colorful oxidized Ordovician Queenston Shale, Ontario, Canada (photo: Omar Colmenares). **4** Cambrian-Ordovician deep-water allochthonous rocks in Newfoundland, Canada. **5** The abandoned underground mine of bedded Cretaceous salt, Zipaquirá mine, Colombia. **6** A breccia of limestone pebble in a red clay matrix, Patmos Island, Greece. **7** Well bedded Paleogene sandstone, shale, and coal, Táchira State, Venezuela. **8** A view of horizontal limestone beds and beach, Algarve region, Portugal. **9** Red fossiliferous limestone, Monument of the Discoveries, Lisbon, Portugal.

Shale—A Mud Rock of Clay Minerals

Shale consists of mainly clay minerals, finer than 0.002 mm crystal size; it may show red, brown, green, grey, and black colors. Shale is fissile, whereas a 'claystone' is non-fissile. Fissility means a shale can part along bedding planes as thin flakes. Many shales are important reservoirs of gas, often tapped by horizontal drilling and hydraulic fracturing. It is an important and useful rock in oil exploration—as a seal, a reservoir, or a source rock. A seal helps to trap hydrocarbons in a sandy reservoir. A carbon-rich shaly source rock generates hydrocarbon after thermal maturation. A siltstone of silt-size grains (0.06–0.002 mm), is intermediate between a shale and a sandstone.

Sandstone—A Rock of Cemented Sand Grains

What is sand? It is a grain, 2 mm–0.06 mm in size, a product of weathering and erosion of preexisting rocks. Sands are generally of silica, silicate, and carbonate, and rarely of gypsum, glauconite, olivine, black iron minerals, etc. Sand grains become rounded and sorted by current and wave action during transport.

Three principal components of a sandstone are grain (quartz, feldspar, rock fragment), matrix (fine pasty mud), and precipitated cement. The cement holds grains together in a sandstone, while the squashed clay matrix between grains may bind a sandstone.

Sandstone's main constituent is hard quartz grains that concentrate in rivers, beaches and offshore as residue. The white quartz-rich sandstone differs from feldspar-rich arkose sandstone. In special cases, 'beach sands' of distinctive mineral source display striking colors—black (magnetite and basalt), white (pure quartz and comminuted carbonate shells), yellow (limonite-coated sand), red (hematite); green olivine sand occurs in many volcanic island beaches, including Hawaii. Green sands of New Jersey, U. S., contain marine glauconite.

A spectacular sandstone mountain is the Uluru or Ayer's Rock inselberg in Central Australia. The German word 'inselberg' means an 'island mountain.' Indeed, the Uluru looks like an island rising from the surrounding plains. It is the most emblematic natural monument in Australia drawing hordes of tourists. Geologically, it is a ~500-Ma-old arkose of alluvial fan origin. The 345-m-high, monolith of vertical beds, has a perimeter of nearly 10 km. The iron content of Ayer's Rock imparts a distinct reddish color, especially during sunrise and sunset hours (see image at the end of Chapter 5).

The Precambrian table mountains of South America and Africa are impressive structures where sandstone erosional features often mimic eye-catching animal, bird, and human forms (Fig. 3.9).

3-9 Roraima Table Mountain, Venezuela, (1.8-Ga-old). Weathering and erosion sculpted interesting human-like and animal forms.

The Difference Between Sands of Silica and Carbonate Origin

We have discussed sands that only include silica-rich sand grains—quartz, feldspar, etc. Another important variety of sand is plentiful in many beaches. The white tropical sandy beaches of the Caribbean or the Mediterranean consist of pulverized sand-sized marine shells—mostly corals, caused by wave action. 'Sand' refers only to grain size of 2 mm–0.62 mm.

Conglomerate—Made of Particles >2 mm Size Like Granules, Pebbles and Cobbles

It represents the coarsest variety of sedimentary rocks in fluvial, alluvial fan, beach, glacial, and deep-sea submarine channel (Fig. 3.10). Conglomerate pebbles and cobbles may preserve vestiges of many older rocks, as they do not suffer long transport. A breccia is a conglomerate with angular clasts of landslide, volcanic ejecta, or meteorite impact origin.

Chemical Sedimentary Rocks—Minerals Precipitating from Hydrous Solution

The chemical rocks consist of precipitated minerals from super-saturated aqueous solution in oceans, lakes, rivers, and evaporative 'sabkha' (supratidal mudflats on the coast with saline evaporite precipitation). Some important deposits include chert/flint, tufa and travertine, rock salt, gypsum, iron ore, etc. Chert (also chalcedony) is a microcrystalline variety of quartz common in many rocks; flint is cryptocrystalline chert that typically occurs as nodules in limestone,

3-10 Beds of conglomerate (1, 3, 5), and cross-bedded sandstone (2, 4). The base of conglomerate (1, 3) shows erosional scour; Quaternary Bettijoque Formation, Venezuelan Andes.

including chalks. Chert is often associated with black shale of anoxic origin—a common petroleum source rock. The Precambrian 2–3-Ga-old Banded Iron Formation (BIF) is a rhythmic alternation of mm/cm-scale laminae of chert/chalcedony/jasper, and very thin bands of iron oxide (hematite/magnetite). Silica forms shells of radiolarian, some foraminifers, glass sponges, and diatom—a single-celled-alga with shell of opaline silica.

'Rock Salts' were common in the geologic past in many places. Rock salt is typically halite, a sodium chloride. The Permian Zechstein salts of Europe contain halite, and minor potassium chloride (sylvite), and magnesium chloride (bischofite). Halite crystals precipitate when seawater concentration is 90% or more.

Dissolved salts in the ocean can precipitate through evaporation and supersaturation. The Mediterranean Sea evaporated during the End-Miocene, precipitating a 3-km-thick salt column. It now lies only 100–200 m below the modern ocean floor. The Himalayan Salt Range of Ediacaran-Cambrian age is another example of evaporites. The Paleozoic salt beds under Lake Michigan is an important source of commercial salt in the USA. The 11,000 km^2 Salar de Uyuni in the Bolivian altiplano is the largest salt desert in the world.

Gypsum, a calcium sulfate mineral is also an evaporite, widely distributed in the world. Seawater precipitates gypsum when 80% concentrated, through evaporation. The White Sands National Park, New Mexico, USA, has an enormous ~712 km^2 field of white gypsum sands. It is the world's largest gypsum dune field.[7] Gypsum crystals can grow very large. An amazingly large, 11-m-long gypsum crystal occurs in a mine in Chihuahua state, Mexico. NASA discovered gypsum in the north pole region of Mars.

Travertine and Tufa are special cases of calcium carbonate precipitation in a freshwater environment. Many countries trade travertine—Italy, Turkey, Mexico, China, etc. Hydrothermal activity of circulating groundwater, and the process of degassing, produce travertine. CO_2-rich groundwater is acidic, so keeps calcium carbonate in solution. As soon as groundwater surfaces, trapped CO_2 escapes, and calcium carbonate precipitates in springs and caves. The travertine is a limestone of hot springs origin. The Yellowstone Park hot springs and geysers show abundant travertine deposits.

Travertine on a Farmland—
The Fly Geyser, Black Rock Desert, Nevada, USA

It is an unusual site, geysers appeared on a dry bed of an ancient lake in Nevada, USA. Looking for underground water in the desert region, farmers drilled a well in 1916, and accidentally tapped into a hydrothermal reservoir (93^{0}C) that flooded the area. The geyser erected a 4-m-high cone of travertine by the 1960s. It was still spouting water and minerals in 1964, when a geothermal power company drilled a second hole about 100 m away from the first one. The second well produced hot water, but robbed the water from the first one, causing it to dry.[9] However, the company decided to reseal the second well, as the water temperature was not adequate for energy generation. The story does not end there. Soon, three new natural geysers sprouted, sitting side by side, releasing hot hydrothermal waters, and gradually building up eye-catching, colorful, travertine cones. The red and green color of the cones is due to thermophile algae residing in the hot spring water. Nature never stops surprising us.

Cold-water tufa can form in lakes, rivers, and as speleothems in caves (stalactite and stalagmite). Biological activities, and degassing, mediate in travertine and tufa formation, as abundant microbes reside in springs, even in hot ones.

Dolostone and Limestone—
Carbonate Rocks with or without Magnesium

Biological processes give rise to most limestones. Only in particular cases lime mud or 'micrite' precipitate inorganically, like on the Trucial Coast, Dead Sea, the Great Bahama Banks, and the Red Sea.[8] The frequently observed 'whiting' are large milky patches of crystals of aragonite needle ($CaCO_3$) floating in clear ocean water. It is abundant in the warm tropical waters of the Bahamas, as a reaction product of bicarbonate (HCO_3) and calcium (Ca) ions. However, some argue the whiting is due to microbial activity. Like calcium carbonate mud, dolomite mud may form by addition of magnesium atom in the calcite crystal lattice. Some Paleozoic dolostones are good oil reservoirs.

Limestone Ingredients—
Shell Detritus, Lime Sand, Ooids and Reef Fauna

Limestone types are bioclastic, oolite, reef, lime mudstone, and chalk, consisting of carbonate grain, lime mud, and calcite cement. Calcite and aragonite are the principal life builders in the ocean.

Bioclastic limestones contain mostly fossil fragments. Coral, sponge, bryozoan, algae, clam, oyster, and gastropod are important producers of calcium carbonate in the modern environment. Many extinct organisms—trilobite, brachiopod, stromatoporoid (earlier sponge), some crinoid (sea lily), ammonite, rudist bivalve etc., also produced thick, widespread deposits. Waves and currents break down marine shells into grains to create bioclastic limestone.

Interesting to note that parrotfish is also a prolific carbonate sand producing agent.[6] A giant >1-m-long parrotfish chomping on dead corals and reef debris to scrape off algae, may produce >100 kg of sands (Fig. 3.11). They grind up huge amounts of reef material in the process and later excrete the undigested carbonate grains, etc., add-

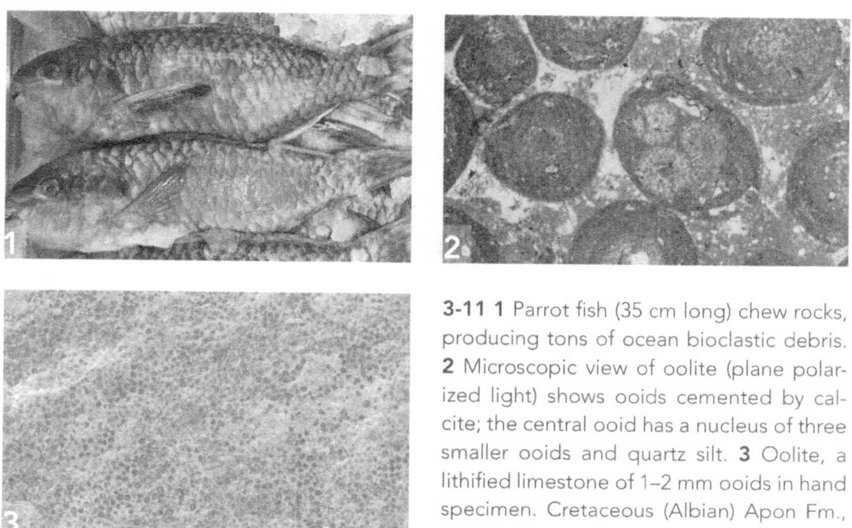

3-11 1 Parrot fish (35 cm long) chew rocks, producing tons of ocean bioclastic debris. **2** Microscopic view of oolite (plane polarized light) shows ooids cemented by calcite; the central ooid has a nucleus of three smaller ooids and quartz silt. **3** Oolite, a lithified limestone of 1–2 mm ooids in hand specimen. Cretaceous (Albian) Apon Fm., Mérida, Venezuela.

ing to the budget of marine sands. *Oolite* was widespread in the past (Fig. 3.11). Today it is found only in the clear, tropical waters around the Bahamas, Great Salt Lake, Baffin Bay (Texas), Persian Gulf, eastern Mediterranean, and Shark Bay, Australia. It consists of spherical ooids (Greek— 'egg'), up to 2-mm-diameter, of shallow marine origin. The Great Bahama Bank is a natural laboratory of modern ooid banks. Each ooid starts with a nucleus of a tiny, microscopic speck of silt, or shell fragments in agitated waters. The ooids grow as concentric rings, as calcium carbonate precipitates around the nucleus. Rolling constantly on the sea floor, they become perfectly spherical and glossy on the surface. Ooids become oolite limestone, when cemented by carbonate cement and/or mud (Fig. 3.11). The rock is used as building material in the construction industry. It is interesting to note that many famous buildings—Buckingham Palace, London, and the Empire State Building, New York, etc., are made of Jurassic and Mississippian oolites. Oolite is attractive as a building stone because it is lightweight, easy to fashion, and strong.

Carbonate reefs flourish by in-place growth and preservation of marine life. Shallow marine coral, alga, sponge, etc., actively build reefs where organisms occur bound together during growth. Think of the Great Barrier Reef, Australia, or the Trans-Central American Reef. Ancient reefs include Devonian reef of Australia, Permian Reef Complex of USA, Jurassic reefs of Europe and Middle East, famous rudist bivalve reefs of USA, Mexico, etc. Many geological reefs of the Middle East and Mexico are important oil reservoirs.

Chalks consist of *coccoliths*—calcareous marine, planktonic, unicellular algae, and subordinate unicellular foraminifers. *Coccoliths* show highly ornamented shell structure and platelets. Average size is 4–5 microns (1,000 microns = 1 millimeter). The majestic chalk cliffs of Dover and Normandy coasts of the English Channel are geological wonders created by submicroscopic coccoliths. Extensive chalk deposits occur in the United States, all the way from South Dakota to Texas. High fluid flow capacity makes chalks attractive oil reservoirs.

The Hierve El Agua Spring—
Open-Air Stalactites Mimic a Foamy Waterfall

The Oaxacan "Hierve el Agua" ('The Water boils') spring, and its cascading waterfall-like white calcified open-air stalactites, is a singular eye-catching sight (Fig. 3.12). The name belies the water temperature as it is cold water that is gushing out. It resembles boiling water as escaping carbon dioxide bubbles out of crevices and cracks of the karst-filled Cretaceous carbonate rocks. In 2018, I visited the remote site in the cordillera region of Mexico; it is about 70 km from the

3-12 Illusion of a waterfall, created by open-air stalactites, Hierve-el-Agua spring, Oaxaca, Mexico.

Oaxaca City, reachable only by a dirt road. It mimics a foamy waterfall cascading down from two rock shelves about 90m and 50m high respectively, above the valley floor. The locals call them 'Cascada del Sal' or 'Cascada Petrificada.' Indeed, it gives the illusion of torrents of gushing, foaming water tumbling down the vertical scarp face.

How does it happen? Rainwater percolates down karst-filled crevices of the calcareous hill country, into underground bed rock. The acidic, under-saturated rainwater dissolves calcium carbonate, and finally emerges as a spring near the ledge of the precipice. The spring forms several small pools on the flat shelf surface, a popular tourist spot for bathing and recreation. The pool edges show ample evidence of hard tufa formation. The water (27°C) is supersaturated with dissolved bicarbonate. As the trickle of water makes its way to the edge of the shelf, it quickly loses dissolved CO_2 in open air. As it plunges down the ledge, it precipitates white calcite, forming long, tapering, vertical stalactites. It is a spectacular calcified escarpment at the end of a projecting ledge. $CaCO_3$ forms 95% of the stalactite with traces of barium, silver, and other elements.[10] The discovery of 550-BCE-old, irrigation canals tapping spring water, reveals a unique feat of the Mesoamerican Zapotec civilization.

Iron Ore—The Backbone of Modern Society

Metallic iron is the basic ingredient for steel manufacture, and many items we use—from a tiny paper clip to a Jumbo airplane. Two different sources yield sedimentary iron ore deposits. The well-known 1.8–2.2-Ga-old, Banded Iron Formation (BIF) is an important source. The other source is Paleozoic ironstone ore of local importance. The BIF is widely distributed—United States, Canada, Brazil, Venezuela, Sweden, Russia, Australia, India, South Africa, etc. Copious weathering of iron-rich rocks went on in the Precambrian Eons. The dissolved iron ended up in oceans, lakes, and accumulated there for a long time before the arrival of photosynthesizing cyanobacteria. Once photosynthesis started, the microbe released oxygen as a byproduct. The oxygen chemically would have combined with dissolved iron to produce hematite (Fe_2O_3) of the banded iron ore. The other constituent of the banded ore is microcrystalline silica polymorphs (quartz, chert, chalcedony, jasper). The alternating hematite and silica laminae would imply a seasonal variation of oxygen supply related to the fluctuation of microbial population. The dissolved silica probably precipitated in oxygen-poor season. The Banded Iron Formation supplies 60% of the world's iron ore demand.

The Paleozoic ironstone is a much less productive ore, especially nowadays. It used to be the mainstay for British source of iron ore centuries back. Laterite and oolite ironstone are of replacement origin with minerals of limonite, hematite, magnetite, chamosite and siderite.

Ironstone oolites are common in England, Saudi Arabia, USSR, Colombia, Venezuela, Alberta in Canada, Malaysia, etc.[12] The oolite formed in very shallow bays where spherical ooids developed by mechanical accretion of kaolinite clay and iron oxide. Soon the kaolinite converted into iron-rich chamosite.[13] Earlier, Colombia had used the Paleogene Paz del Rio ironstone oolite for steel production. In the 1970s, the author had evaluated Paleogene oolite ironstone deposits in Venezuela (Fig. 3.13a).

Laterite is a hardened clayey soil enriched in leached iron and aluminum oxide, typical of hot tropical areas. English surgeon F. Buchanan coined the name 'Laterite' in 1807, when traveling along the coast of southern India. The lateritic soil developed on the thick Deccan Trap lava flows that literally inundated the Indian plate in the Cretaceous (Fig. 3.13b).

The iron-enriched deep red laterite can develop on several weathered rocks—basalt, peridotite, granite, and shale. It is widespread over one-third of the world's continental areas, especially abundant in New Caledonia, Australia, Philippines, and Indonesia.[11] Earlier, a 30-m-basalt-laterite strata used to be a source of iron ore in North Ireland.

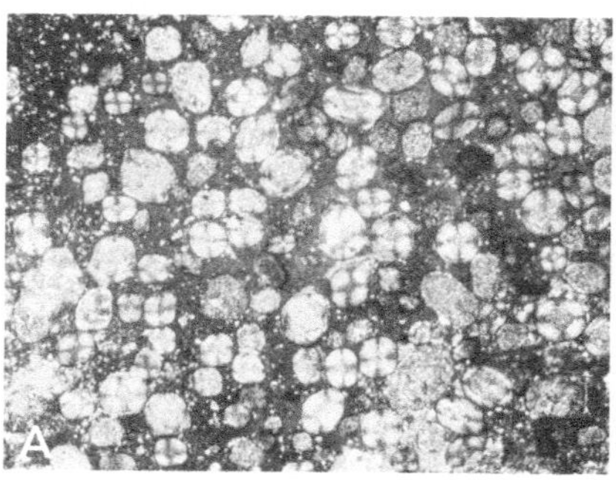

3-13. **A** Polarized microscopic view of sub-millimetric chamosite oolite showing a starry extinction cross texture, due to tangential orientation of platelets. Quebrada Arriba Fm., Humocaro, Venezuela.
B A laterite sculpture (petroglyph) of a dancing musician in Goa, India. Along the 1,600-km-long Western Ghat range of India from Trivandrum, Kerala to Mumbai, extensive laterite tablelands have formed from intense tropical weathering of Deccan basalts on the slopes of the Western Ghat range toward the Arabian Sea.

Metamorphic Rocks—
Older Rocks Get Cooked Up, Not in A Melting Pot

Metamorphic rocks are simply variably cooked older rocks in the gut of Earth's crust at high temperature and pressure. The older rocks may be igneous, sedimentary, or even metamorphic. Metamorphism also happens at tectonic plate junctions. The rocks were rarely in a molten state, rather it is mostly a solid-state chemical reaction where old minerals give rise to new ones.

Ample pressure and high temperature are prerequisites of metamorphism, aided by mineral-rich hot solutions. What happens in a metamorphic rock? The texture of the original rock changes—preexisting minerals may become elongated and aligned; also, entirely new minerals may appear, reacting with hot circulating fluids. Examples of metamorphic reincarnations are many—sandstone to quartzite, shale to slate/schist, limestone to marble, granite to granite gneiss, and basalt to amphibolite.

The Precambrian rocks are metamorphosed preexisting rocks. Migmatite is a common yet special Precambrian rock due to partial fusion of some rocks, not the adjacent others. For example, a granite will crystallize or melt at a lower temperature than a basic rock like basalt. Under high grade metamorphic regime, a granite will melt and mobilize at a lower temperature as dykes and sills, traversing the unaltered basalts nearby. The dykes, stringers and lenses may be coarse-grained pegmatite, or a very fine-grained granite, i.e., aplite (Fig. 3.14).

Metamorphism can affect younger Phanerozoic rocks due to burial depth, high temperature-pressure, contact with new magma, and tectonic plate convergence. The lowest degree of metamorphism, shown by zeolite, may be simply due to increasing overburden pressure and temperature with depth. It may also occur in areas of tectonic collision. The emergent lava at the Mid Atlantic Ocean Ridge may become a low grade 'greenschist,' upon encountering hot hydrothermal fluids.

The intrusive magma can also cause "contact metamorphism" in surrounding host rocks. Some high temperature minerals—garnet, pyroxene, andalusite, may form in the host rock nearest to the magma body. Also, intense shearing action between moving fault

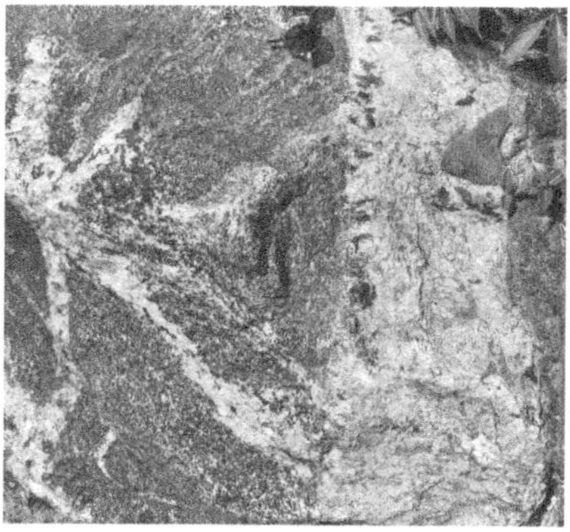

3-14 A migmatite rock showing a pink pegmatite dyke on right, and several off-white aplite offshoots into the adjacent dark diabase with abundant biotite mica and hornblende. Lake Muskoka lake front,, Ontario, at the southern edge of the stable Canadian Shield.

blocks may pulverize rocks, creating a 'mylonite'—a recrystallized metamorphic rock.

Subduction zone metamorphism occurs where a plate is diving beneath another. The friction between plates continuously scrapes off materials from the subducting plate. The intensely deformed, jumbled mass of all sizes, create geological 'mélange.'[14] The bluish amphibole mineral—'glaucophane' in 'Blueschist,' is the result of high pressure-low temperature metamorphism. It is diagnostic of mélange rocks that are present in the USA, Colombia, Venezuela, Jamaica, Dominican Republic, among others.

The metamorphic minerals of low to high grade reflect changes due to increasing subsurface pressure-temperature conditions: chlorite, biotite, garnet, staurolite, sillimanite. Pressure-temperature in the deep subsurface also imparts a new look to schists with distinctive 'foliation,' i.e., alignment of elongated, platy minerals.

In summary, minerals always try to reach equilibrium under the existing pressure-temperature condition, be it deep inside Earth or at the surface. Rocks undergo metamorphism to attain stability at a new fluid-pressure-temperature condition, as nature eliminates or reforms some old minerals and creates the new stable ones.

Fig. 3.15 shows some interesting metamorphic rocks of different origins. We recall metamorphic rocks are pre-existing rocks (igneous or sedimentary) cooked under variable subsurface temperature, pressure, and fluid conditions.

Deformation and Uplift—Faults, Folds & Tilted Rocks

A cursory look at any mountain or rock outcrop, from the mighty Himalayas to the commonplace one anywhere, reveals that rock layers are not always horizontal. Many are tilted because of deformation and violent upward push. Rocks that create mountains of great heights suffered tectonic deformation, undergoing uplift from deep underground. Deformation affects all rocks—sedimentary, igneous, and metamorphic.

3-15 Some common metamorphic rocks. **1** A granite gneiss with folded texture, Ontario, Canada. **2** A granite slab showing foliation of parallel dark & white minerals, Tennessee, USA. **3** A pink granite gneiss showing distinct foliation, i.e., a linear arrangement of black mica biotite and amphibole (hornblende), Ontario, Canada. **4** A weakly metamorphosed schistose and deformed shale, Newfoundland, Canada. **5** Columns of the famous Parthenon in Greece built with marble of the Cretaceous Acropolis Formation. **6** Banded marble floor tiles of a church in Santorini Island, Greece. **7** A marble statue of the Ephesus ruins, near Izmir, Turkey. **8** A folded quartzite and sandy schist, Cretaceous Barquisimeto Formation, Venezuela.

Rock Deformation and Structures—Buried Layers Buckle Under Heat and Pressure

Rocks show structures because they had reacted to forces of nature in the deep crust. Geological structures may vary in size and shape, depending on type of forces acting on them. Large regional forces may involve tectonic blocks pushing against each other, i.e., geological plate movements, a term we will use often. Rock deformation depends on two simple concepts—stress and strain.

Stress—When a rock is deformed, the force applied is known as stress. The Stress is of three types—compression, tension, and shear. Compression may happen when two tectonic plates collide. Example: The Indian plate clashed headlong with Eurasia—the Himalayas rose.

The tension stress is the opposite of compression, it results from two diverging forces pulling apart rocks. Example: Plate movements along the Mid-Atlantic Rift are moving the Americas apart from Europe/Africa. The shear stress-related deformation happens when two rock bodies are grazing, scraping, and screeching past each other along a large fracture or fault.

'*Strain*' is the opposite of '*Stress*', that is, it defines what happens to a rock when under stress. The simplest case is when one sees a set of inclined beds. The beds were originally horizontal on land or on the seafloor but got strained and deformed deep underground. They show up as inclined beds after suffering vertical stress.

The deformative stress gives rise to several geologic structures—folds, faults, joints, simple fractures, and foliation. Two other useful concepts are brittle and ductile deformation. The 'brittle deformation' is easy to understand. If anyone accidentally drops a china plate on the floor, it breaks into many pieces because the china clay is brittle. An example of ductile deformation is how Mexican tortillas are prepared. The doughball of corn flour is flattened into a disc shape by placing it in a tortilla press and simply applying pressure from above. It will come out as a circular disk of flattened dough ready for a griddle. Ductility is plastic deformation. Sandstones and limestones are naturally brittle, whereas pliable shale, salt and schist are prone to ductile deformation. However, temperature and pressure are also crucial elements in how a rock behaves deep underground. Both high temperature and pressure will render a sandstone or limestone more plastic, more ductile.

Faults—Don't Chide Geologists for Their Faults

Geologists routinely find faults in field outcrops and subsurface seismic images. A fault is a vertical or inclined planar surface showing displaced strata on two sides (Fig. 3.16a). One block will drop down or go up relative to the other across the fault plane in normal and reverse faults respectively. The hanging wall separates the block above the fault plane, and footwall is the one below. When the block above the fault plane (hanging wall) goes down, it is a normal fault. When it goes up along the fault slope, it is a reverse fault (Fig. 3.16c).

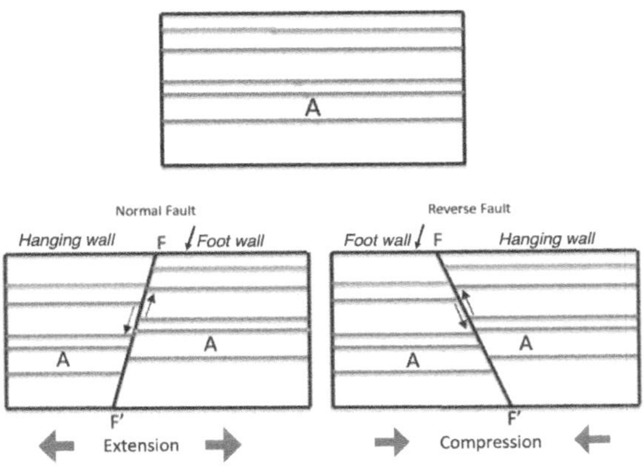

3-16a top Strata without any faulting. **bottom** Examples of a *normal* (left) and a reverse fault (right). In a normal fault, the block above the hanging wall goes down, whereas in a reverse fault, it goes up as in figure 3.16c

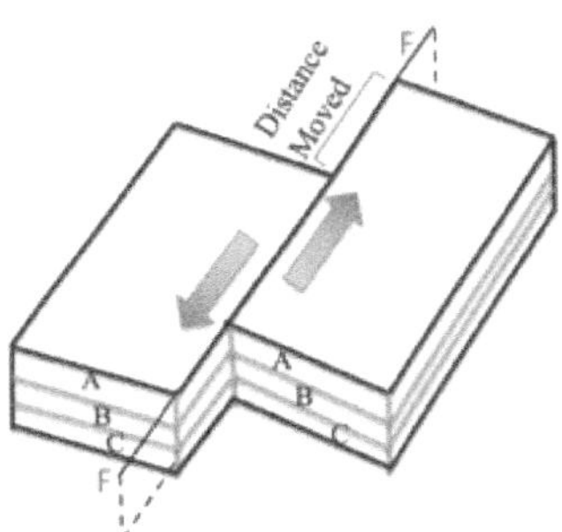

Figure 3-16b Strike-slip (Transform fault). Blocks or plates move past each other along the fault. The Pacific and North American plates, slide past along the San Andreas fault..

3-16c In a reverse fault, the hanging wall strata overrides the foot wall shown by white arrows. The solid and dash lines, mark two reverse faults in deformed Andean strata, Venezuela. Bed A shows offset.

Often faults and folds may occur intimately juxtaposed like one can see in the Andean outcrops (Fig. 3.18, Image 4).

In a strike-slip fault, the blocks may slide past each other along a fault plane. It is a special case (Fig. 3.16b) where the blocks are simply like two oppositely moving trains crossing each other. Frequent fault movements, often associated with plates, usually create earthquakes. In 2023, the devastating earthquake along the North Anatolia strike-slip fault in Turkey and Syria, killed over 60,000 people. Two major earthquakes in Morocco and Afghanistan in 2023, also occurred along fault lines at the junction of the Eurasia-Africa and Eurasia-India plates respectively.

Folds—Even Solid Rocks Buckle When Under Pressure

Folds are arched up or buckled down strata that were originally horizontal (Fig. 3.17). A fold appears when two mutually directed forces act on strata. The two limbs dip away from the crest in anticlines.

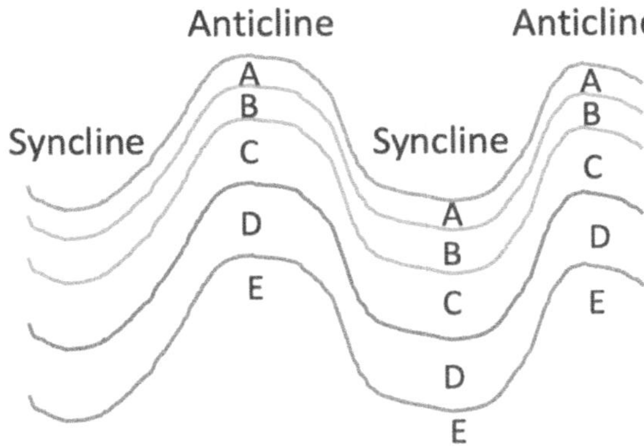

3-17 Beds A-E: Anticline and saddle-like syncline, with imaginary fold axis along the crest and trough respectively. *See image 3 of 3-18*

Trough-like synclines, in contrast, show two limbs dipping toward each other. An imaginary fold axis separates a fold into two equal halves or limbs. It is important to note that folding occurred at great depth many thousands of meters underground, due to tectonic deformation in the geologic past. Rocks, otherwise brittle, will become ductile when subjected to high underground heat and pressure. The original planar beds become pliable, adjust to new reality by simply 'folding' or buckling, crumpling, and flowing plastically (Fig. 3.19). Folded strata show lateral shortening. If one takes a thin plastic sheet, and puts inward pressure from two ends, it will form a 'fold,' and the ends will be closer than before. Folds, like faults, can occur at any scale. Orogenic fold belts may show up on a continental scale. The Valley-and-Ridge Province, west of the Blue Ridge Mountains, eastern USA, is a textbook example of folds shaping modern topography. Here the Appalachian folded belt of elongated, subparallel, broadly N-S ridges separating deep valleys, extends for nearly 2,000 km from New York to Alabama.

The compressive forces of the Paleozoic Appalachian orogeny shaped the straight ridges and valleys now as eroded anticlines/synclines (Fig. 6.1).

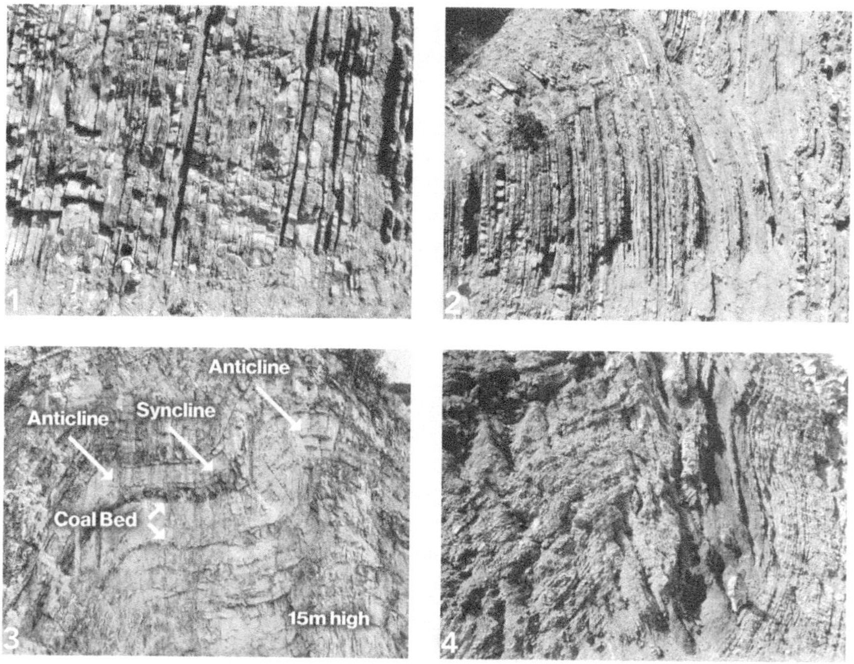

3-18 1 Near vertical beds due to deformation is common in tectonically active areas like the Andes, Venezuela **2** Vertical beds appear folded, crunched, and faulted **3** A syncline flanked by anticlines in a sandy 15 m-high outcrop with coal, Bocono, Trujillo **4** Highly deformed vertical beds and sheared off folds, showing fault disruption.

A Monocline is a simple step-like fold, showing an inclined segment between two horizontal ones. It is like an inclined escalator connecting two horizontal floors. Monoclines are 'drape folds' as they may drape over fault-displaced beds below.

Earth's Antiquity—Way Older Than Many Believe

How do geologists estimate relative ages of rock strata? Multiple tools exist to help geologists to this end. Some important tools like seismic data, paleontology, vertical order of strata, and radiometric age dating are commonly used now. The newer advanced techniques—novel electron spin resonance, thermoluminescence, and infra-red stimulated luminescence, help to date recent Holocene landforms including terraces.

Geological Unconformity: May Mean Millions of Years of 'Geologic Amnesia'

The concept of unconformity implies a discontinuity in a geological sequence, i.e., some strata missing from a vertical rock sequence (Fig. 3.20). It is akin to the missing page(s) in an interesting suspense story. Seismic images can bring out details of unconformities over large areas in the subsurface. Absence of some rocks due to erosion or non-deposition may represent millions of years of time record. The Grand Canyon, USA, exposes the 'Great Unconformity' surface, representing erased rock record of ~500-Ma time interval.[34]

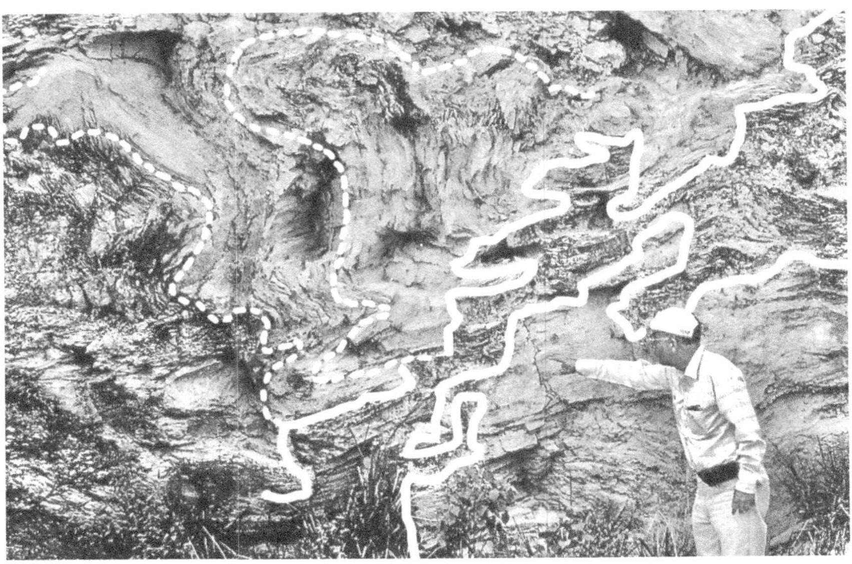

3-19 Plastically deformed, reddish shale (solid line), zigzags through the folded (dashed lines) sandy Jurassic strata; La Quinta Fm., Venezuela.

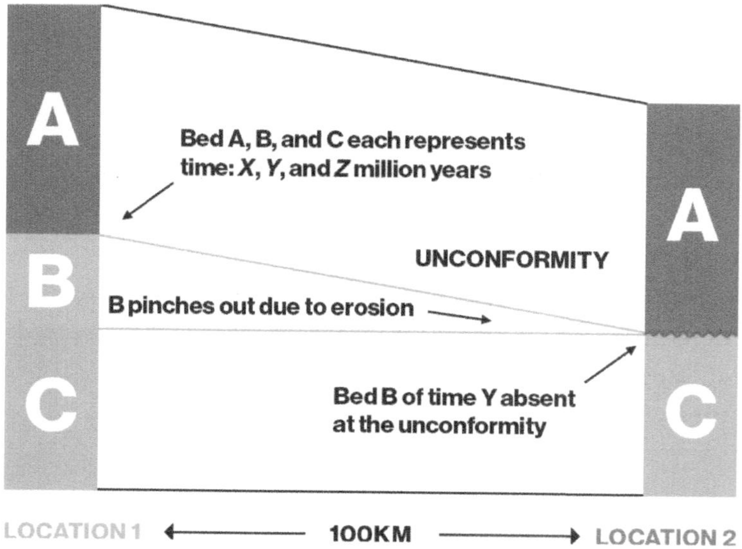

3-20 A geological unconformity means both physical and temporal break in a sequence. Bed B of Location 1 absent in Location 2 due to erosion.

Global Time Sequence: Eon, Era, Period—A Matryoshka Doll Analogy

Geoscientists have established and fine-tuned a hierarchy of divisions of ancient time in astonishing detail. Our timescale is like the wooden Russian Matryoshka dolls of varying sizes that snugly fit inside one another (Fig. 3.21). It helps us to visualize the nested chronicle of Earth history from its violent birth to the Present.

Eons, Eras, Periods and Epochs denote that each subsequent division spans a shorter time range. A rule of thumb—the farther back we delve into the past, the less details we have at our disposal. We remember more vividly incidents of the last year than the first ten years of our lives. We will have fewer subdivisions for the remotest past, whereas the stratigraphic column of the recent past is split into finer and finer levels—periods, epochs, age/stage etc. The word 'Eon' is used in astronomy and geology to denote a unit of time spanning many hundred million years. The four geological Eons from the old-

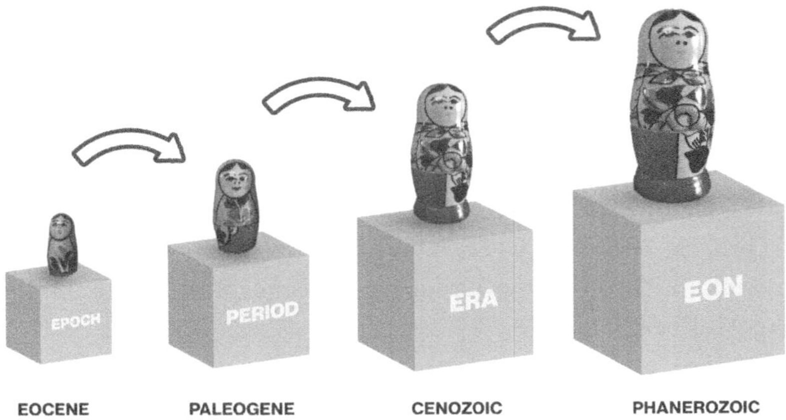

3-21 Matryoshka dolls—nested time scale of Epoch, Period, Era, and Eon. The Epochs fit into a Period, and so on.

est to the youngest are Hadean, Archean, Proterozoic, and Phanerozoic; the first three comprise the Precambrian. The International Commission on Stratigraphy has formalized the Geologic Time Scale from Eons to Stage/Age.[21]

All Eons, except the Hadean, have a connotation to life (Greek suffix 'zoic' means life). The Archean or Archeozoic denotes the time when early simple unicellular *archaea* and bacteria thrived. The Proterozoic shows somewhat better developed but still primitive unicellular and multicellular life forms, and later some complex fauna without hard shells. The Cambrian Period ushers in the youngest Phanerozoic Eon (the shortest). 'Phanerozoic' means new, visible life; abundant and diverse hard-shelled fauna debuted, optimum for preservation as fossils.

Pioneer Sir James Hutton—The Present Mimics the Past

British geologist James Hutton (1726–97) established the early scientific concept of relative geologic time. Hutton, known as the 'Father of Modern Geology,' made a lasting contribution to our understanding of earth processes. He was versatile in several professions as geologist, chemist, farmer, physician, etc. He clearly laid out his seminal ideas in his writing. It marked a turning point in geological thinking of the day.[15] He was the first to state that Earth's antiquity was unbelievably old. He commented in 1785 that the world had no beginning nor end.[16] His painstaking field studies produced the first scientific evidence that our planet was ancient, unlike the prevalent belief of a few thousand-years-old Earth. The creation date in the King James Bible—October 23, 4004BC—is based on Archbishop James Ussher's biblical chronology. Hutton's philosophical concept of endless time found its formal expression in the concept of "Deep Time" of John McPhee, 1981.

Hutton's theory of 'Uniformitarianism' greatly influenced the renowned English geologist Charles Lyell (1797–1875). Hutton's universally accepted 'Uniformitarianism Principle,' meant that Earth has been going through repeated cycles of change—destruction and renewal, just as we observe the earthly processes operating today. Charles Lyell echoed the phrase "The Present is the Key to the Past" to support Hutton's seminal message of uniformitarianism, later becoming an everyday geological dictum. Modern processes make rivers flow, form deltas, and cyclically flood extensive land areas or create deep-sea fans; all these processes operated equally in the past. The idea of uniformitarianism worked very well for geologists to interpret the planet's complicated geologic history. The short-term events—annual monsoons in Asia, hurricanes in the Americas, etc., regularly strike Earth. Similarly, long-term events—earthquake, volcanic eruption, asteroid impact, etc., periodically cause major impact to parts of the globe. The same events we witness today, such as tides, affected Earth with equal or greater vigor during the geologic past (Fig. 3.22).

Hutton's other contribution is the theory of 'Natural variation' of species. He clearly foresaw the vital importance of natural selection in the survival of different species. He succinctly stated the principles of natural selection, though never used the same phrase. He argued that species evolve constantly to adapt better to the changing environment around them.[16] His idea sounds eerily like what Charles Darwin and George Edward Wallace propounded in the famous theory of evolution decades later. They were both deeply influenced by Hutton's ideas. To note, Darwin and Wallace independently advanced the theory of evolution.

3-22 Tidal effects on sands at contrasting times. **left** The image shows how two sets of perpendicular ripple trains developed due to tide and wind (the area near hammer handle base) in the 1.5-billion-year-old Vindhyan sandstone, India. **right** View of two sets of modern ripple marks—one highlighted by dashed lines, while the other discontinuous short ones perpendicular to the first. Lucaya Beach, Grand Bahama Island.

Vertical Order of Strata—The A-B-C of 'Stratigraphy'

Earlier, geologists used the principle of the 'Law of Superposition' to estimate the relative age of given strata. James Hutton studied rocks in many places in Scotland and concluded that each strata represents a given interval of time. He interpreted the bottom layer to be the oldest in a sequence of strata, and each succeeding layer progressively younger toward the top. It is like a bunch of multicolored bricks stacked one above another starting with a basal red one. This sounds rather elementary today, but it was a novel smart idea 250 years back. Many layers show distinctive physical characteristics such as color, thickness, mineral and fossil content. The method helped Hutton to correlate physically distant outcrops in the UK.

Tools of the Trade—We Read Earth History Written in a Novel Language

Earth's origin and subsequent history is complex, yet colorful, multi-layered, and vibrant with life almost from the beginning. In geology, the older the rocks are, the less details we capture. For example, we know a lot more of the last few million years than the first billion years of the nascent Earth. Having said that, our understanding of the early Earth has increased exponentially during the last several decades, thanks to the new high-resolution tools and instruments. Earlier, geology used to be an observational science unlike other basic sciences like physics and chemistry. But it is no longer true today; the routine use of laboratories in multi-disciplinary procedures enables geoscientists to precisely document the subtle age differences of important events in Earth history. Scientists in the fields of biology, geochemistry, geophysics, and astrogeology have joined forces with geologists to unravel nature's mysteries.

Contribution of Seismic Studies—
A Paradigm Shift, Solving Countless Mysteries

Seismic data can add valuable input toward building a regional geologic picture across land and ocean. In comparison, outcrop study and borehole geologic data analysis at a specific location is limited in scope. Let us imagine a sequence of inclined beds exposed somewhere, amenable to detailed analysis. But the inclined beds may disappear in the subsurface a short distance away because the outcrop is limited in areal extent. The outcrop is like the tip of an iceberg, it continues underground for a long distance. A geologist will have no way of knowing the details of the sequence below, because it is unreachable. Similarly, a borehole may penetrate thousands of meters of strata underground. Data from the borehole may yield useful information about the subsurface geology of only the borehole site. In comparison, subsurface seismic images can provide data of a vast three-dimensional area of hundreds of cubic kilometers. It gleans information of deep subsurface, otherwise, totally out-of-reach of boreholes.

How is the subsurface seismic data acquired? An array of acoustic sources on the ground sends seismic waves by detonating explosives in shot holes. The generated shock waves travel through the underground. When the sound wave hit rock boundaries, it is reflected to the surface where receptor geophones capture it. Thus emerges an image of deeper strata, many thousands of meters thick. Seismic data from both land and oceans can throw light on past sea level variation and location of major erosional unconformities. It can obtain visual images of ancient river channel, delta, marine shelf, deep-sea fans, faults, folds, and salt structure. Seismic data is an indispensable tool for oil, gas, and mineral exploration now-a-days, often helping to discover large oil fields in the Gulf of Mexico, offshore Atlantic, North Sea, and all continents.

Paleontology—The Study of Fossils: Evolving Life Immortalized in Stone

A very useful item in the geological toolbox is paleontology; it is the systematic study of fossils—the vital pillars of stratigraphic time sequence. Sadly, fossils were virtually nonexistent during the earliest 'Hadean' and Archean eons. However, the remains of organic life soon became an integral part of the geo-history as Earth slowly matured.

Early on, some geologists noted that rock layers are arranged in a predictable fashion, and specific strata can be identified accurately over large distances by the type of fossils they contain. British geologist William 'Strata Smith' (1769–1839) had a lasting impact in the field of geology and especially used paleontology in geologic correlation.[17] He collected systematic rock samples, especially fossils from all localities he visited in England and neighboring regions. Using the fossils as markers, he could observe the same vertical succession of rock strata in many places, and successfully correlated them all over Great Britain. His work came to fruition when he published the first geological map of Great Britain in 1801, a remarkable achievement for the time. Geology students and geologists all over the world follow his methodology even today in preparing geological field maps.

Since William Smith's time, the discipline of paleontology has developed spectacularly. Smith could only work with fossils visible to the naked eye, i.e., macrofossils. Now other branches, such as micro-paleontology and palynology, have developed into powerful tools to aid in rock correlation. The study of microscopic fossils such as foraminifer (single-celled zooplankton), nannofossil (including coccolith—microscopic alga from chalks), etc., is essential for the oil industry (Fig. 3.23). The same holds true for palynology—the branch of paleontology dealing with ancient spores and pollens; dinoflagellates, a type of protist (like algae, diatoms), is useful in analyzing marine sequences.

Fossils are now being employed with great success in all kinds of geological activities such as mining, oil exploration, correlation of

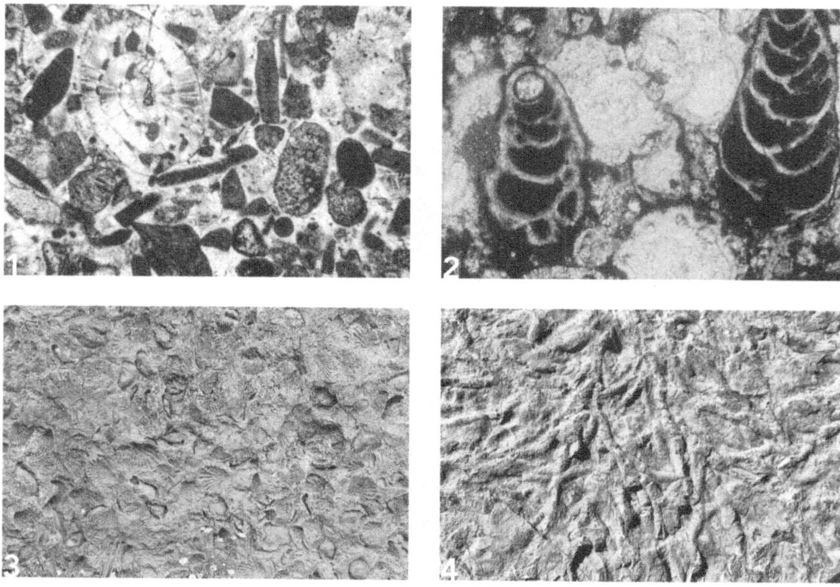

3-23 Microscopic view of foraminifer, macrofossil, and animal tracks and trails **1** A large, disc-like benthic shallow marine foraminifer (<1 mm) in a granular limestone of fossils, pellets and dark-coated intraclasts **2** Black dead oil highlights the uniserial chambers of 2-mm-size benthic (bottom-dwelling) foraminifers. The white grains are sub-millimetric planktic (floating) foraminifers (globigerina) with calcite-filled chambers. Late Cretaceous hardground, Venezuelan Andes **3** The undersurface of a Cretaceous bed showing abundant Pecten-like (scallop) bivalve impressions, Villa de Leyva, Colombian Andes **4** Tracks and trails of marine animals on Paleozoic carbonate shelf, Ontario, Canada.

coalfields and bedded mineral deposits. The marker fossils can put a date stamp on a stratum that can be correlated locally or even across continents. Ammonite (earlier known as snake stone) macrofossils, and various microfossils have been used successfully in global correlation of strata. We know the dinosaur is one of the exclusive indicators of the Mesozoic. Similarly, Permian-Carboniferous flora and coal forests are distinctive everywhere. The once-joined and now-very-distant landmasses of the ancient Gondwana mega-continent bear testimony of the extensive Permian-Carboniferous coal forests. Many organisms sometimes left their foraging and dwelling traces and trails on bedding surfaces. Geologists learn from these traces about the past physical, chemical, environmental conditions, and even animal behavior of the time (Fig. 3.23).

Indeed, the ancient Gondwanaland is a very good example of global correlation of strata both past and present. Gondwana or Gondwanaland was a huge landmass, named by British geologists after the 'Gond' indigenous tribe in India in the 1900's. Gondwanaland existed as a mega-continent for 350 Ma (550–180 Ma) from the Neoproterozoic to Early Jurassic. Gondwana landmasses occupied about two-thirds of the present continental areas in Africa, Madagascar, Antarctica, Australia, South America, India, and Arabia. Gondwanaland breakup would have coincided with the Jurassic (Toarcian) Karoo-Ferrer volcanism ~185–175 Ma ago. The extensive, younger Deccan lava eruption happened in the Late Cretaceous.

Presently all the ex-Gondwana landmasses are thousands of kilometers apart. How do we know they were once welded together? Geologists have found the same fossils in coeval rock strata in all the continents of the ancient Gondwanaland. Two plant fossils *Glossopteris* and *Gangamopteris* occur in all ex-Gondwana continents. Similarly, ancient glacial tillites are widespread in the Gondwana continents.

Radiometric Dating—Unlocking the Age of Ancient Rocks

How does it work? We may recall that an atom consists of a nucleus of proton(s) and neutron(s). The number of protons is always fixed which makes its atomic number constant. Many elements, but not all, also have their number of neutrons fixed. Neutrons lend weight to atoms. Some atoms of a given element may have a varying number of neutrons, so will have different atomic weights, i.e., some heavier than other(s). Isotopes are atoms of the same element with differing atomic weights. The isotopes are of two types—stable and unstable. Let us consider a common element like carbon. Carbon has three naturally occurring isotopes ^{12}C, ^{13}C and ^{14}C, containing 6 protons in all, and 6, 7 and 8 neutrons in the three respective nuclei. Carbon in nature occurs as ^{12}C (99%), ^{13}C (1%), and traces of ^{14}C. Both the ^{12}C and ^{13}C are stable isotopes, i.e., they do not change through time. Bombardment of nitrogen atoms by cosmic rays carrying neutrons, constantly generates radioactive ^{14}C in the atmosphere. The newly formed ^{14}C freely mixes with ambient ^{12}C and ^{13}C in the atmospheric carbon dioxide. Eventually, all living animals and plants assimilate ^{14}C, and maintain a constant ratio of stable to unstable radioisotopes. However, the parent radioactive ^{14}C constantly decays into daughter ^{14}N by emitting beta particles, thus reducing its original ^{14}C content. All living trees have the same amount of ^{14}C. Any tree, upon death, will gradually lose its stored ^{14}C content through spontaneous radiometric decay. Let us take the example of a dead tree that lived hundreds of years. By knowing the rate of decay and the remnant ^{14}C in the dead tree, we can estimate the age of a sample of wood. How do we know the rate of decay? ^{14}Carbon is a radioisotope with a half-life of 5,730 years; half-life simply means the time it takes to decay one-half of the radioactive isotope in a sample, in this case a sample of wood. By measuring the sample's remnant ^{14}C amount and comparing that with a similar living tree, one can estimate the antiquity of the tree.

One practical limitation of ^{14}C is that it can only date up to 30–50-ka-old samples. A sample 50-ka-old may not have any ^{14}C because of complete radioactive decay into ^{14}N. This may be ideal for archeological research, but of no use for old rocks. So how do we date rocks several billion years old? Unlike carbon isotopes, some other isotopes found in natural rocks show a rate of decay—slow to very slow. These are ideal for use as geological time clocks.[18]

The parent isotope is the one that decays, daughter isotope is the new decay product. For example, parent uranium-238, with a half-life

Table 3-1 Some Radioisotopes for Age Determination[18]

Parent Isotope	Daughter Isotope	Half-Life (Ga)
Uranium-238	Lead-206	4.5
Uranium-235	Lead-207	0.704
Potassium-40	Argon-40	1.25

of ~4.5 Ga, undergoes radioactive decay to finally produce daughter Lead-206—a non-radioactive and stable isotope. It means the original U-238 will take ~4.5 Ga to lose half of its radioisotopes in a rock. By counting the remaining U-238 isotopes and the amount of new decay product Pb-206, physicists can determine the age of the rock sample with reasonable accuracy. The parent-daughter isotope ratio yields the date of mineral formation within magma. The same process happens with the potassium-argon method, with the decay of parent potassium and generation of daughter argon.

How do we estimate age of sedimentary rocks—sandstone, shale, and limestone? The age dating normally does not give us a direct estimate of antiquity of these rocks—all product of rivers, oceans, and deserts. Why? Because a sedimentary rock consists of loose grains of different minerals coming from erosion of older source rocks. However, the dating of a heavy mineral like a common zircon in a sandstone sample, will tell the relative age of the sandstone. The zircon's radiometric age will show that the host sandstone bed is younger than the dated zircon, because it crystallized from igneous magma in an older rock. The detrital zircon is now one of the grains of the sandstone.

A volcanic ash bed interstratified with sedimentary layers, may be useful to evaluate the relative antiquity of a sedimentary strata. Volcanic ash appears from the explosive venting of lavas from volcanoes. The fine powdery ash particles may travel hundreds of kilometers by wind, and finally settle on land, a riverbed, and elsewhere on top of older strata. A radiometric age of the ash bed will mean that the strata above the ash are younger, and the ones below are older. The volcanic ash layers of the archeological Olduvai Gorge deposits in Tanzania, Africa, dated by K-Argon method, yielded crucial ages for hominin evolution. The famous Louis Leakey team discovered there (1964) the 1.88-Ma-old fossil of our ancestor *Homo habilis*.[19] Like ash beds, authigenic or in situ minerals—illite and allied green glauconite, are also useful tools in dating sedimentary layers.

One of the oldest ages on Earth comes from the Canyon Diablo meteorite in the Barringer Crater, Arizona. The ~4.55 Ga age of the meteorite validates the 4.57–4.58 Ga antiquity of our solar system, including Earth.[20]

Life on Earth—Origins and the Prokaryote vs. Eukaryote Debate

It is a million-dollar question fundamental to our understanding of the amazing evolutionary chain of life from the nascent Earth to where we are today.[22] Some competing theories exist about the primordial origin of life. How and where did it all start? The known microbes on Earth are archaea, bacteria, fungi, protists, and viruses. Archaea and bacteria are the oldest, both are prokaryote, i.e., their cells lack a nucleus. The name 'prokaryote' comes from *pro* (before) and *karyon* (nucleus), whereas eukaryote means 'true' nucleus in Greek. A eukaryote microbe has an organelle or DNA nucleus within a membrane. Evidently, it is more complex than a prokaryote like a cyanobacteria. Both have DNA, but the prokaryote contains it in a nucleoid without a membrane. Microbial mats and stromatolites are the handiwork of lowly cyanobacterial prokaryotes during Earth's Archean childhood. While the prokaryote dates from ~3.8 Ga, the eukaryote is a relative newcomer, 2–1.5 Ga or younger. Scientists opine that the latter evolved from the prokaryote. Protists like alga, amoeba and dinoflagellate are eukaryotes. These appeared later and may be both multicellular and unicellular living in large colonies. The eukaryote emerged only once, likely formed by endosymbiosis of prokaryotes. The word 'endosymbiosis' simply means one organism living within another, having a symbiotic relationship. The eukaryote shares genes with prokaryote bacteria and archaea, i.e., symbiosis of two prokaryote cells creates a eukaryote. The mitochondrion of the first eukaryote was once a prokaryote, englobed within another prokaryote.[23] Each cell has one nucleus but hundreds of mitochondria.

The mitochondria (~0.75-3μ long, where μ = 0.001 mm) within membrane are the essential energy-producer in eukaryotes, making the cell's chemical energy in the form of ATP or adenosine triphosphate. Scientists think all cells, existing or that ever existed, descended from a single ancestor cell ~4 Ga ago. The million-dollar question—how did the first cell trigger the explosion of life?

Proton Gradient: Sparking Life in Deep Oceans

The nascent semi-molten Earth was totally devoid of life. If this is so, then how did it start? Nobel laureate Peter Mitchell proposed an idea about how the living world would have emerged out of the primordial chemical soup of the Archean Earth. His idea explains how life started from a 'proton gradient' phenomenon across a membrane. It also describes the evolution of eukaryotic cell, that just happened only once, by chance, in Earth's age-old biotic history.[24] A unique physical-chemical environment catalyzed proton gradient. A NASA geochemist postulated that deep-sea hydrothermal vents would have been the ideal setting for the natural proton gradient to happen. The much-needed proton gradient materialized because of the acidity-alkalinity imbalance between the primordial ocean and hydrothermal vents spewing fluids into the ocean. The emerging super-alkaline (pH 10–11) hydrothermal vent fluids created a proton gradient with the mildly acidic seawater (pH 6), aided by heat. In chemistry, pH means potential of Hydrogen. Recent experimental results also support this idea. The complex organic compounds formed due to the chemical reaction of hydrogen and carbon dioxide.[25] A suitable analog is the Lost City Hydrothermal Field, 15 km away from the Mid-Atlantic Ridge, showing how and where the first spark of life could have happened. The Lost City sits on the 4,000-m-tall Atlantis Massif, where the vents have erected limestone towers rising 60 m above the seafloor at an ambient temperature of 88°C. Limestone samples from the chimneys showed thriving microbial communities. Here olivine-bearing mantle peridotite reacts with water and CO_2 from the vent fluid to produce methane, hydrogen, and hydrocarbon, in addition to the alteration mineral serpentine. The pH difference sets up a natural proton gradient between the highly alkaline, proton-bearing vent fluid and proton-poor ambient seawater. The proton gradient develops across mineral membranes in pores of precipitated carbonate. Archaea type microbe population is thriving in biofilms on the inner walls of the carbonate chimneys, built with calcium and CO_2 out of the vent. The microbes draw nourishment from methane and hydrogen-rich anoxic fluids.[26] Terrestrial hydrothermal crater lakes may also be the incubator of first life. Experts documented thriving microorganisms in the Poas crater, Costa Rica.

The discovery of deep-sea primitive life on Earth, may open the possibility for life beyond Earth. Similar alkaline vents and potential proton gradients might have triggered microbial life in other planets and moons. The U.S. space missions have documented icy moons of our distant solar gas planets—Jupiter and Saturn. They might hold such vents in the oceans.[25] Particularly, Jupiter's moon Europa, and Saturn's moon Enceladus are promising as evidence points to the presence of oceans of water under an icy canopy there.

Eternity to *Homo sapiens*: A Play of Countless Acts and Fleeting Protagonists

The previous two sections gave us an idea about how and where humble life would have started in the hazy early Archean Eon. But that was just the beginning. Life evolved continuously in the long Proterozoic Eon, and then through the Paleozoic, Mesozoic, and Cenozoic eras of the Phanerozoic Eon. It was not all smooth sailing, though. Five notable mass extinctions had happened, but these proved to be temporary (geologically speaking) dampers. The amazing parade of diverse life marched on till the debut of *Homo sapiens* on the scene.

Table 3.2 is an attempt to distill the complex data in a simple chart highlighting the important milestones along the way. Details of life up to the Neogene will appear in each time slice (Chapter 5–10) under the heading 'Miracle of Life,' while life in the Quaternary discussed in Chapter 12.

Mars May Help Us Understand Ourselves Better

The red planet orbiting the Sun always intrigued our ancestors. Mars played a preeminent role in all ancient cultures including the Egyptian, Babylonian, Roman, and Greek. Roman God of War 'Mars' becomes 'Ares' in Greek. Mars happens to be a son of Earth ('Bhumi') in Indian mythology. Astronomer Galileo Galilei first viewed Mars with a telescope in 1609, but our forefathers probably knew about it several thousand years back. They would have noticed that Mars moves in the sky from night to night, unlike the scintillating silver dots of countless stars.

Table 3-2: The Tapestry of Life Through the Ages

Time	Description of Life
Late Cenozoic 2.58–0.00 Ma	**Quaternary Period:** • Modern species: Most living species, including polar species and today's penguins, evolved. • *Homo genus*: Continued evolution of the *Homo genus*. Emergence of *Homo sapiens*, the "crown jewel" of evolutionary development (see Fig. 12.1 for details).
Late Cenozoic 23–2.58 Ma	**Neogene Period:** • Mammals: Evolution of the Panthera family (big cats like lions, tigers, and leopards). Woolly mammoths and widespread grazing animals such as deer thrived. • Reptiles: Crocodiles and snakes diversified. • Marine life: Sharks appeared in locations like Nicaragua Lake. Dugongs, the giant Megalodon shark, and pinnipeds (seals, sea lions) were abundant. Kelp grass supported diverse marine ecosystems. • Flora: Deciduous trees and grasslands expanded. • Primates and early humans: Lesser apes like gibbons and non-human great apes (orangutans, gorillas) evolved. *Pan genus* (chimpanzees and bonobos) appeared. Transitional hominins emerged, leading to the first hominin, *Australopithecus*.
Early Cenozoic 66–23 Ma	**Paleogene Period:** • "Age of Mammals": Lemurs and tarsiers emerged among primates. Whales evolved from the land mammal Pakicetus. Birds dominated the skies. • Land mammals: Early horses and Cambaytherium. Paleomastodon and the giant Asian rhinoceros (Indricotherium). Elephants, rhinos, and Old World monkeys appeared during the Oligocene epoch. • Flora: Cashew, rose, and leguminous plants evolved. Angiosperms (flowering plants and fruits) thrived. • Climate impact: Fauna and flora turnover occurred due to cooling temperatures.

Late Mesozoic 145–66 Ma	**Cretaceous Period:** • Faunal diversity and dominance: Dinosaurs, ammonites, rudists, pterosaurs, plesiosaurs, ichthyosaurs, pliosaurs, and Archelon turtles thrived. Numerous unhatched dinosaur eggs found in fossil records. • Marine microorganisms: Diatoms, coccolithophores (algae), and foraminifers flourished. • Plant evolution: Angiosperms (flowering plants), grasses, and shrubs emerged. Flowers introduced vibrant color and scent to ecosystems. • End-Cretaceous: Fifth mass extinction event, marking the end of the dinosaurs and many other species.
Early to Middle Mesozoic 252–145 Ma	**Jurassic Period:** • Dinosaurs: Thrived and diversified, including sauropods (*Diplodocus, Brontosaurus, Apatosaurus, Titanosaurus*). • Bird-reptile transition: Archaeopteryx. • Marine reptiles: Marine crocodiles and mosasaurs. • Amphibians: Frogs, toads, and salamanders. • Plants and invertebrates: Thriving gymnosperms, First Kalligrammatid insect. • Marine ecosystems: Glass sponge reefs flourished. • Marine reptiles: Ichthyosaurs and Plesiosaurs dominated the seas. **Triassic Period:** • Reptiles: Archosaurs, ancestors of birds, dinosaurs, and pterodactyls. • Firsts: First dinosaurs, First mammals, small and rodent-sized. • Marine reptiles: Ichthyosaurs and Plesiosaurs dominated the seas. • End-Triassic: Fourth mass extinction event.
Late Paleozoic 359–252 Ma	**Permian Period:** • Marine life: Abundant fauna with diverse ammonoids. • Tetrapod evolution: Amphibian *Eryops*, Carnivorous reptile *Cynodont*, Synapsid (*Dimetrodon*), a mammal-like reptile. • Plants: Flourishing gymnosperms, including cycads, gingkoes, yews, and conifers, *Glossopteris* plants spread widely. • End-Permian: Third mass extinction event. **Carboniferous Period:** • Land vertebrates: First reptiles and the development of amniotic eggs. • Gigantic arthropods: *Meganeura*, a dragonfly with a 70 cm wingspan. *Arthropleura*, a millipede reaching 1.5 meters in length. • Flora: Huge ferns, early scale-trees *(Lepidodendron)*, and other lycopods (ancestors of club mosses). • Widespread coal forests.

Middle Paleozoic 444–359 Ma	**Devonian Period:** • Marine life: Stromatoporoids (extinct sponges) and colonial reef corals flourish. • Vertebrates: First tetrapods and bony fish appear. • Acanthodian fish with jaws and well-defined sharks evolve. • End-Devonian: Second mass extinction event. **Silurian Period:** • Acme of life: Rapid diversification and expansion of life forms. • First jawless fish (Agnatha), similar to lampreys. • Marine arthropods: "Scorpions of the sea." • First vertebrate fossils appear. • Terrestrial arthropods emerge: insects, arachnids, centipedes, and millipedes.
Early Paleozoic 539–444 Ma	**Ordovician Period:** • First terrestrial organisms: Arthropods, plant microfossils, spores, and cuticles. • Marine life: Crinoids and echinoderms continued from the Cambrian. Early sharks appeared. • End-Ordovician: First mass extinction event. **Cambrian Period:** • Emergence of hard-shelled animals: Arthropods, brachiopods, pelecypods, gastropods, trilobites, corals, and graptolites. • Notable species: "Penis-worm" and anomalocarids ("abnormal shrimp").
Proterozoic 2500–539 Ma	• 630–539 Ma: Soft-bodied animals, including mollusks, sponges, and arthropods, thrived. • 900 Ma: Evidence of burrows • 1.2 Ga: Emergence of algae and fungi. • 1.8 Ga: Appearance of acritarchs (unicellular microfossils). • 2.5 Ga–539 Ma: Cyanobacteria dominated.
Archean 4000–2500 Ma	• ~3.5 billion years ago: Appearance of stromatolites, formed by microbial life (e.g., in the Pilbara Hills, Australia). • ~3.8 billion years ago: Earliest known fossils, including cyanobacteria and archaea.

Mars and Earth have a lot in common, but there are differences too.[27, 28] Mars has two very small satellite moons (diameter 12, 22 km). Mars is 227 million kms away from the Sun (Earth 149 million kms away), and its orbit speed is 22 km/sec (Earth's ~30 km/sec). Mars has a 25-degree tilt axis, a day length of 24h37 min. It has an onion-ring-like internal structure: core radius~1,800 km, mantle 1,240–1,800 km, and crust ~50 km. The density of Mars is 3.93 gm/cm^3. Mars has varied rock types—sandstone, conglomerate, clay beds, dust stone (wind-origin) and sand dunes, other than basalt, andesite, and peridotite.

The red planet is nearly half in size, and it has about one-third the gravity of Earth. Martian atmosphere is 96% CO_2, with minor gases—argon, nitrogen, traces of oxygen, and water vapor. Atmospheric pressure is 1/100th of Earth. The Martian climate is very cold and wind-swept, mean temperature hovers around minus 60°C. Erosion has been the dominant natural process for the last 3.5 Ga. Experts argue that 3.5–4 Ga back, the Martian climate used to be like ours today, though no water flows there now. Water-driven erosion, transport and sedimentation was common in the lake, river, delta, and ocean on Mars. The absence of a normal atmosphere is due to the disappearance of its protective magnetic field about 4 Ga ago. NASA's MAVEN probe established that fierce solar wind and extreme ultraviolet radiation had stripped away most of the atmosphere and water resources.[28] The present Martian magnetic field is very weak. Here on Earth, the playful Aurora Borealis is a visual reminder of the strong magnetic field, battling and successfully warding off the harmful solar wind and charged particles.

Mars Missions from Earth—Ancient Martian Climate and Geography

Scientists have been curious about Mars for a long time, probing it by telescopes in the 19th century, then came NASA's exploration missions of Mariner in the 1960s, and Viking series in the 1970s. Experts have long suspected the presence of water on Martian surface (Fig. 3.24). Space vehicles and surface landers took many good quality pictures of stream-like channels and deep canyons. The test results of

3-24 A NASA image of the Martian surface littered with boulders of all sizes (sedimentary or impact ejecta origin?). The backdrop is a symmetrical hill of unknown origin.

the Viking mission experiment in 2012 suggested the possible presence of earlier microbial life on Mars.[29]

NASA's Mars Perseverance Rover landed on the Jezero crater on Feb 18, 2021. It might be a game changer in terms of its potential to finally identify bio-signatures in Martian rocks, if any. The rover carried the most sophisticated scientific tools for any astrogeology mission ever. It landed on the 3.9-Ga-old, ancient lake-stream-delta deposits within the Jezero impact crater. High resolution images reveal impressive details of sedimentary structures of lake and river-deltaic origin. The far end of the lake-delta ('pro-delta') is an optimum site for yielding ancient clues to bio-signature preservation.[30] The rover had collected 30 samples in 2023, to retrieve them later for more detailed analysis in labs on Earth. NASA expects that the results would give valuable insight into the paleoclimatic conditions during lake sedimentation, and later fluvial delta formation in the crater. Additionally, the rover would be able to probe up to 10 m below Martian surface to detect ice, water, etc., with the help of Rover's onboard special GPR system.[31] NASA Mars Reconnaissance Orbiter (MRO) had sent images documenting the presence of a >100-m-thick ice layer.[32] Mars digital model confirms that an ancient ocean existed about 4–3.7 Ga ago. The Martian polar high latitudes show evidence of thick water and ice-bearing layers, covering an area of nearly 5 million km^3. The presence of permafrost layers and methane in the atmosphere may also be a clue to the past organic history of Mars.

CHAPTER HIGHLIGHTS

- Geology 101: Key concepts for understanding and enjoying the book.
- Earth's Structure: Core, Mantle, and Crust.
- Rock Types: Igneous, Sedimentary, and Metamorphic—origins and examples.
- Rock Deformation: Folds form under stress, heat, and pressure; faults result from strata rupture and displacement.
- Earth's Antiquity: Stratigraphy, the geologic time scale, and unconformities.
- Pioneers of Geology: Trailblazers and their foundational concepts.
- Geoscientist's Tools: Outcrop analysis, drill cores, seismic imaging, paleontology, and radiometric dating.
- Origin of Life: Key concepts include proton gradient, prokaryotes, and eukaryotes.
- Evolution of Life: A chronicle of development of life.
- Mars Geology: Insights into Mars' geology and potential signs of ancient life.

WHAT'S NEXT?

Chapter 4 unravels the miracle of Plate Tectonics—hard-to-believe, geologic tales of wandering continents. Plate Tectonics affects climate, sea level variation, mountain building, biogeography, continental drift, lava eruption, earthquakes, etc.

An artistically sculpted chariot made of local Hospet granite (seen on the hill behind) - Hampi archeological site, Karnataka, India.

4

Restless Earth: The Jigsaw Puzzle of Plate Tectonics

PLATE TECTONICS IN A NUTSHELL

History of Plate Tectonics

No other geological event has impacted Earth so deeply as the workings of plate tectonics. We are all living on active plates. The U.S. cities on the Pacific plate like Los Angeles and San Diego, California, are blissfully unaware that they are sliding past the rest of North America including San Francisco and Sacramento, albeit very slowly (4 mm/year). The wide acceptance of the plate tectonics concept by earth scientists has been a paradigm shift in geological thinking since the 1960s. This scientific theory helps us to form an integrated view of the ever-changing Earth with its many moving parts. It convincingly explains many natural events, not understood earlier. It is the driver of large-scale movement of a dozen or so crustal plates. These have been constantly moving for a long time—over 3 Ga, revealing the subtle changes taking place in the deepest part of the oceans and the deep sub-surface. As well they provide a clue to the mystery of the tallest peaks. Earlier, German geophysicist and meteorologist Alfred Wegener planted the seed of the seminal concept in 1912. He daringly proposed the idea of the continental drift, just looking at the relative positions of the Americas and Europe/Africa on the map. The amazing match of the Atlantic boundaries of South America and Africa was nothing fortuitous for him.

For Wegener, the close match of the continental borders meant only one thing—the continents were once glued together in one supercontinent 'Pangea,' and since then, they parted ways, creating the Atlantic and other oceans in the space between. A little later in 1930, British geologist Sir Arthur Holmes espoused Wegener's model, and proposed mantle convection as the driver of continental drift. But it took a long time for the geoscience community to finally embrace Wegener's drifting continent model. Researchers refined it with new emerging data from geology and geophysics and expanded it into the elegant plate tectonics model we take for granted today. In the 1950–1960s, several prominent geoscientists—Profs. Harry Hess of Princeton University, USA, Tujo Wilson of Univ. of Toronto, Canada, their students, and other academics in America, Europe, and elsewhere, established plate tectonics theory on a firm footing. The plates are huge chunks of landmasses resting on plastic, solid lithosphere, i.e., uppermost mantle. The huge, combined plate-lithosphere slabs are mobile, somewhat like rafts, or icebergs floating on the mantle 'ocean.'

How Does Plate Tectonics Work?

The mantle appears to be the driver of plate movement. It is really the topmost mantle sliver—the lithosphere slabs that are afloat, the plates just passively ride on them. Below the rigid lithosphere is a viscous, plastic upper layer of the mantle ('the Asthenosphere') with the consistency of perhaps silly putty. The tectonic plates simply hitch a ride on the lithospheric slabs, that sail on the vast ocean-like asthenosphere below.

What is the motor of plate movement? It is the heat from mantle and core, due to continuous decay of radioactive minerals—uranium (U), thorium (Th), and potassium (K). The heat propels the mantle to form a convection cell of magma. The hot magma from deep mantle may well up vertically on Earth's surface splitting apart plates. Conversely, two plates may collide, or one dives under the other, when they approach each other.

This process continues in Earth's gut *ad infinitum*. To note, plates are mortal, and all activities take place only near the plate borders. The Paleozoic Iapetus Ocean, a precursor of today's Atlantic, disappeared around 450 Ma ago, due to plate collision between adjacent Laurentia (N. America) and Baltica (Europe).

The United Plates of the World—Seven Large, Many Small

The major tectonic plates are North and South America, Pacific, Africa, Eurasia, Antarctica, and India-Australia (Fig. 4.1). The smaller ones are the Caribbean, Arabia, Philippines, Nazca, Cocos, Scotia, Iran, Anatolia, etc. Some continental plates sit atop larger lithospheric slabs—The Americas, India-Australia, Africa, and Antarctica. Other slabs carry the oceanic Pacific, Nazca, Cocos, and Caribbean plates. A curious fact—Turkey falls on two adjacent tectonic plates.

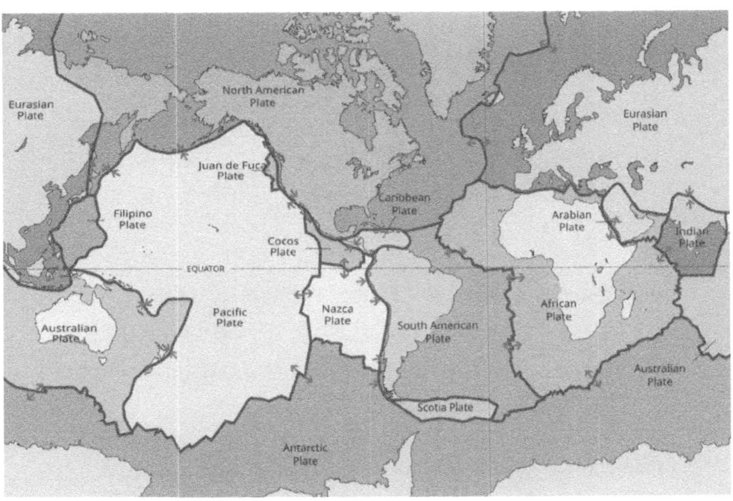

4-1 Tectonic plates of the world (USGS). Black lines mark lithospheric plates. Africa, Eurasia, North and South Americas, India-Australia, Antarctica, are continental plates. Arrows—divergent (< >) and convergent plate (> <) boundaries. India, Australia and surrounding areas are part of the larger Indo-Australia tectonic plate.

Most of Turkey including capital Ankara, is on the southerly Anatolia plate. A thin sliver of Turkey including the Istanbul City falls on the Eurasian plate.

The Wilson Cycle: Birth and Demise of Plates

Prof. Tuzo Wilson proposed a novel scheme in the 1960s.[1] He posited how ocean basins form, and eventually perish through plate tectonic activities. The tectonic plates come together to form supercontinents, only to break up later into discrete continents over geologic time. The tectonic cycle goes through varied plate activities—rifting, creation of ocean basins, plate subduction, continental collision, etc.

The six stages of plate movements are:

1. *Initial rifting:* Mobile magma from a rising mantle plume pushes apart plates. Modern example—East African Rift System.
2. *Seafloor spreading event:* Birth of a nascent ocean basin. Example—Modern Red Sea.
3. *Ocean basin creation:* Rifting and continental drift create it. Examples—Paleozoic Iapetus, and the modern Atlantic Ocean.
4. *Subduction initiation:* Denser oceanic crust buckles, drags down oceanic plate under continents. Example—Pacific plate subducting beneath the Americas and Asia.
5. *Closure of an ocean basin:* Two plates converge, wipe out the ocean in between. Example—demise of the Paleozoic Iapetus Ocean.
6. *Continental collision:* Example—India-Eurasia collision, creating the Himalayas, and obliterating the Cretaceous Tethys Ocean.

Once the cycle runs a full circle, a new rifting starts. Example: The Atlantic Ocean rifted, i.e., opened in the Jurassic, where the Iapetus Ocean existed earlier.[1] The Atlantic will also disappear some day in the geologic future. The East African rift is the harbinger of a future ocean basin.

PLATE MOVEMENTS

Three Types—Divergent, Convergent and Strike-slip

Earlier, we briefly mentioned key concepts to understand divergent and convergent plate movements. A *divergent* drifting scenario means two plates move away due to rifting like seafloor spreading. In a *convergent*, i.e., butting-head scenario, two plates collide, or one dives (subducts) under the other. In a third—*strike-slip*, scenario, two plates just slide past each other horizontally. The Pacific and the North American plates move in opposite directions along the San Andreas Fault, occasionally creating major earthquakes.

DIVERGENT MARGIN 'OPEN SESAME'—PLATES DRIFT APART

Scenario 1: Seafloor Spreading—Where Oceans Part in the Middle

New magma spills on the ocean floor at divergent margins along mid-ocean rifts. The Americas and Africa are two parting plates at the Mid-Atlantic Rift. Prof. Harry Hess and R. Dietz, USA, called this process 'seafloor spreading' in the 1960s. It happens when magma erupts along a long central rift—a deep, long fissure on the ocean floor. The magma spills on both sides.

The Mid-ocean-ridge is notable for its length of 65,000 km, the longest mountain chain on Earth. It is a pair of parallel ridges with a narrow median rift valley, where fresh magma is continuously welling up. The ridge crest is 2,600 m below ocean surface and rises about 2,000–3,000 m above the ocean floor. The Mid-Atlantic Ridge goes all the way from the Greenland shores, passing through Iceland, to the southern tip of South America.

Evidence of Seafloor Spreading—North and South Magnetic Pole Flip

At Mid-Ocean-Ridge (MOR), hot mantle rock rises and melts due to 'decompression.' The erupting magma cools quickly, becomes rock, and moves away from the ridge vicinity to make room for new upwelling magma. As the process continues, ocean floors show progressively older rocks furthest from the Mid-Ocean-Ridge crest. The oldest seafloor is of Jurassic age, occurring along the two continental margins.

The magnetic field at any location on the seafloor is the sum of two measurements—the expected average magnetic field strength due to the present dipole of Earth (created by the liquid outer core), and the local magnetic field strength because of the iron-rich basalt on the sea floor. 'Dipole' means pair of equal and opposite magnetic or electric poles separated by a distance. Geologists compared the expected magnetic field strength at each seafloor location with the measured strength there. The difference is magnetic anomaly–positive or negative. In positive anomaly, the observed strength surpasses the average strength, whereas negative anomaly implies that the observed strength is less than the average strength. In the next step, when scientists systematically plotted the data, starting from the Mid Ocean Ridge sideways, they found a miraculously matching zebra-pattern of anomalies (Fig. 4.2). The alternate positive (as the present normal anomaly), and negative anomaly stripes occur on both sides of the ridge. Later, a similar anomaly pattern emerged in other oceans. The presence of paired, matching magnetic stripes on either side of the rift axis is characteristic of seafloor spreading. The iron-rich magnetite minerals in mafic rocks like basalt act like mini-compass needles that faithfully preserve a record of Earth's polarity when they cool down from magma. The stripes that match today's positive anomaly are defined as normal polarity, whereas the negative stripes had reverse polarity, i.e., Fe-rich minerals were aligned, pointing toward the south magnetic pole. Geoscientists realized that polarity rever-

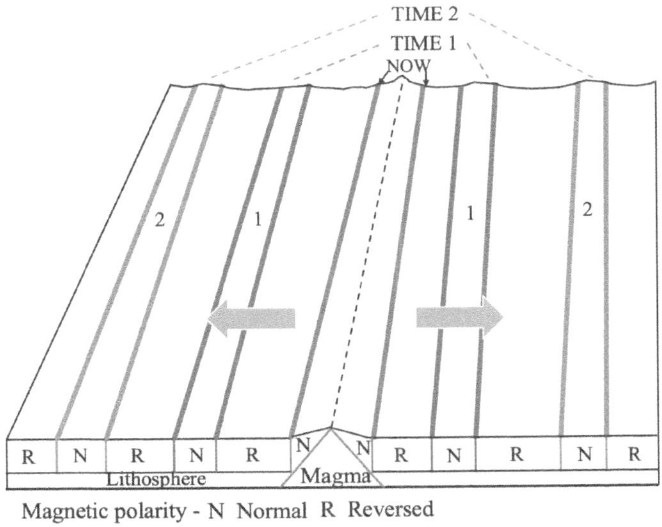

4-2 Cartoon of the ocean floor, showing alternate zebra-pattern, normal-reverse magnetic polarity and seafloor spreading. Arrows point to older events. The polarity now is normal.

sal was the norm throughout the geologic time when poles swapped position many times (Fig. 4.2). Earth experienced two pairs of alternate normal—reverse polarity intervals in the last 5 Ma[2]: Brunhes normal chron: 0–0.78 Ma; Matuyama reverse chron: 0.78 Ma–2.58 Ma; Gauss normal chron: 2.58 Ma–3.6 Ma; Gilbert reverse chron: 3.6 Ma–5.2 Ma. In Greek, 'chron' means 'time.' The present northerly normal polarity has lasted 0.78 Ma. Earlier, the Matuyama reverse polarity persisted far longer—1.8 Ma.

Plates Also Age and Fall Apart Like All Things Natural

Geologically, rocks at the mid-ocean-ridge are rather transitory. Today the age of rocks in world ocean floors confirms that. Why is it that no pre-Jurassic rocks exist on the ocean floor, even though plate tectonics have been occurring since the Precambrian era? The Cretaceous to Pleistocene rocks are more abundant today in the oceans, because all older rocks had subducted into the mantle earlier. Fig. 4.3 shows the different plate tectonic activities like rifting, subduction, etc.

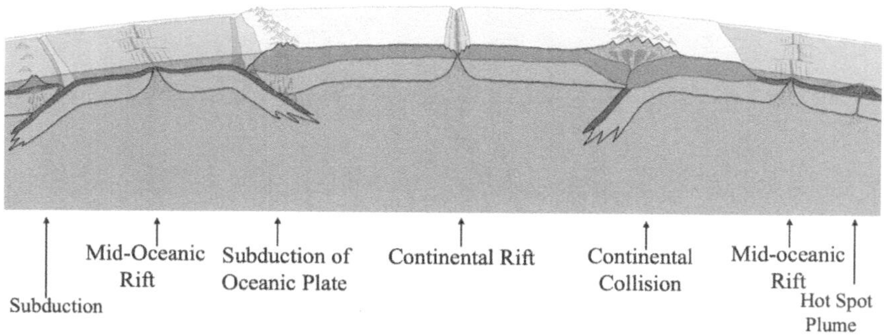

4-3 A simple scheme of tectonic plate interactions: subduction, rifting, collision, hot spot plume, on land and ocean. The very thin and dark oceanic plate on both sides is unlike the grey, thicker continental plate in the center below the rift.

What Moves Plates Apart?

What hidden force pulls apart the Americas and Africa across the Atlantic Ocean floor other than the mantle convection? Some geoscientists invoke a special mechanism for the drifts. The 'Ridge Push' or slope gravity theory suggests that gravitational push drives the extruding lava over the edge of the spreading ridge[3]; it is because the seafloor is higher on the ridge than on the abyssal plains. As extruded lava cools on the ridge it becomes denser, gravity pulls it away from the spreading center. Cooler magma moves away, as fresh hot magma from deep below rises at the ridge axis and the cycle goes on (Fig. 4.4).

The 16,000-km-long, 1,000–1,500-km-wide, Mid-Atlantic-Ridge rises 2–3 km above the sea floor, while its crest is about 2.6 km below sea level. The abyssal plains are 4–5 km below sea level, extending up to 800 km away from the ridge axis.[4] So, the erupted ridge magma travels downslope nearly 800 km to the abyssal plain, where it cools down, becomes denser, finally subducts into the mantle along the deep-sea trench.

Scenario 2: Continental Plates Split—A Rift Basin Forms

When drifting involves two or more continental plates, it will form a trough-like depression between them (Fig. 4.3). A continental scale example of rifting comes from the Gondwanaland. In the Mesozoic, the southerly Gondwanaland developed major rifts by rising mantle plume eruptions ('Karoo-Ferrar' Igneous Province). The Deccan traps came later. The rifting finally broke apart Gondwanaland into discrete continents—Africa, South America, Australia, Antarctica, and India. Only Antarctica stayed put, while the rest wandered away. Madagascar, now an Indian Ocean Island, broke off from Africa. Many other rift basins existed in the geologic past like the Rio Grande Rift in New Mexico and Colorado, and the rifted large Newark Basin, New Jersey, USA.

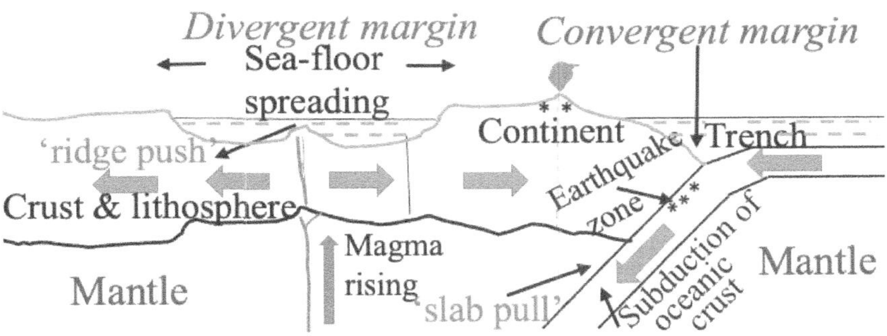

4-4 Seafloor spreading, and oceanic plate subduction, where 'ridge push' and 'slab pull' are the plate drivers. Arrows show relative movements of plates at both divergent and convergent margins.

A modern textbook example is the East African Rift (EAR) Valley system where a series of elongated narrow lakes formed in the low between the rifted blocks. The EAR is the southward extension of the Red Sea and Gulf of Aden rift systems. The EAR running NE-SW near the Horn of Africa separates continental Africa from the Somali Peninsula, Kenya, and Tanzania to the east. Lake Victoria sits between the two N-S trending rift valleys of the EAR. Lake Tanganyika falls in the western rift valley. The Death Valley, California, and Lake Baikal in Asia are two other modern rift basins.

CONVERGENT MARGINS—WHEN TWO PLATES BUTT HEADS

Scenario 1: Continent-Continent Collision

The continents with nearly the same density and buoyancy present this situation. As they collide frontally, rocks are crunched, deformed, and folded into high mountains (Fig. 4.3). In continent-ocean convergence, the mafic-rich oceanic plate is denser and less buoyant, so naturally subducts under the lighter continental crust.

Products of Continental Clash—The Majestic Everest

The outcome of a continental duel is an orogeny which spawns towering mountain belts. The clash of India and Eurasia is a textbook example, where the intervening Tethys Ocean was a hapless victim and disappeared. Earlier, the East African Orogen had developed during the east and west Gondwana collision at the End-Precambrian. The rise of the Himalayas and the Appalachians of North America, involved intense deformation like folding, faulting, crumpling, and underplating. The horizontal tectonic push during collision produces crustal shortening. Scientists estimated a ~700 km shortening of the Himalayan fold-thrust belt.[5] An enormous, tapering wedge of the Indian plate (700-km-long and 20-km-thick) underplated the Tibetan crust since the lower Eocene Epoch (Fig.4.5).

As collateral damage, large sections of colliding plates may end up fragmented, displaced and transported opportunistically to some adjacent free space like an ocean. Some argue that the southeast Asian Archipelago formed in this way. Likewise, the face-off of the Eurasia-Arabian plate resulted in the squeezing out of the Anatolian plate with Turkey riding on it, towards the Mediterranean Sea.[6]

Tall mountains like the Himalayas control weather pattern in a significant way; they act as an insurmountable wall between the Indian subcontinent and the drier, colder northern areas. It stops the monsoon rain and wind from reaching the other side of the range. Deserts like the Gobi and Taklamakan developed in the northern rain shadow zone.

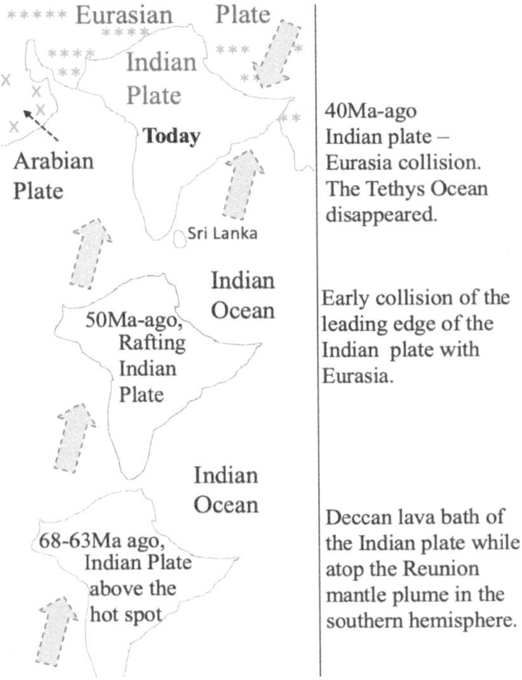

4-5 A simplified cartoon of the collision of the Indian and the Eurasian plates during the Eocene Epoch.

Moisture in high mountains condenses to become snow and water producing rainfall. Rivers like the Indus, Ganges, Brahmaputra, and their tributaries, carry the water downstream to keep the foothills humid and green; they support bountiful agriculture in the plains.

What is Orogeny?—It's Simply Tectonic Plate Collision

Orogeny is a concept we will refer to often. Orogenesis is a mountain building event due to plate collision. The Greek words—'Oros' + 'gen' literally mean 'mountain + genesis.' Plate collisions have recurred since deep-time, leaving behind many deformed, lofty mountain ranges.

The Precambrian Grenville orogeny brought the Rodinia Supercontinent together. The Paleozoic Taconic, Caledonian and Hercynian orogenies had built the Appalachians. The Laramide gave birth to the Rockies in North America. The Andean, and Alpine-Himalayan orogenic belts are majestic testimony of nature's forces acting on rock formations.

The 1.1-Ga-old Grenville Mountain belt in its prime and the Paleozoic Appalachian belt of North America, were originally taller than the Himalayas, but not anymore. One sees more evidence of the youngest and latest orogeny, compared to the earliest one, documented in the hazy Precambrian past. The reason—two opposing natural forces act on an orogenic edifice. As a mountain actively rises due to plate interaction, the forces of weathering (rain, wind, erosion, and landslides etc.) tend to progressively diminish the majestic stature of mountains. The destructive activity of erosion and weathering have not yet overtaken the effect of tectonic rise of the youngest orogenies—Himalayas, Alps, Rockies, and the Andes, etc. In contrast, the older belts have lost their towering heights, because weathering has outstripped the mountain building momentum. The pared-down Precambrian orogenic mountains are now part of stable continents.

Many Supercontinents Appeared and Vanished. Why?

Supercontinents came together through plate collision and resultant mountain building.[7] Some notable supercontinents of Precambrian age are the Archean Vaalbara ~2.9 Ga, Kenorland ~ 2.5–2.7 Ga, Mesoproterozoic Columbia/Nuna ~1.5–1.8 Ga, Neoproterozoic Rodinia 1.1–0.75 Ga, and the fleeting Pannotia ~ 0.6 Ga. Finally, the Phanerozoic collision of Laurasia and Gondwana landmasses sculpted the 335–175-Ma-old, last supercontinent—Pangea. The southerly Gondwana mega-continent included Africa, South America, Antarctica, Australia, and India. Laurasia englobed North America, Greenland, and the rest of Eurasia.

Of note old supercontinents came together and then went asunder. Now we have only seven continents, thanks to incessant actions of plate tectonics. The break-up of the Gondwana and Laurasia mega-continents spawned the modern continents. The mostly submerged Zealandia may be the 8th continent, including New Zealand and New Caledonia.

Collisions Bring Out Gemstones from Earth's Entrails

The collisional mountains are often sites of precious and semiprecious stones. Many faults and fissures develop during the mountain building. Deep, hot, hydrothermal fluids, rich in mineral ions, rise along cracks and may precipitate attractive minerals. The Kashmir sapphires are world famous. Alluvial gold is mined in many places including the Indus River plains. The flawless Burmese ruby is a coveted item for many. Other minerals—beryl (aquamarine, emeralds), green tourmaline (verdelite), selenite, pink quartz, garnet, cats-eye (chrysoberyl), and many others, occur in the Himalayas. America's second gold rush of 1829 happened in the Appalachians of northern Georgia. The Rockies and the Andes also host many attractive gemstones and semiprecious stones.

Scenario 2: Continental—Oceanic Plate Face-off, Ocean Subducts, i.e., Buckles

The scenario is commonly known as subduction process where oceanic plates and overlying ocean basins disappear in the subduction zone (Fig. 4.3 and 4.4). The site where a plate goes under is a mega-scale 'trench' on the seafloor—a long, narrow, deep, fissure-like depression. The oceanic Juan-de-Fuca plate is going under the continental American plate at the Cascade subduction trench on the American west coast. The Deep Challenger at the southernmost point of the U-shaped Marianas Trench is the deepest trench in the world. Here the Pacific plate is diving under the Philippines. The horizontal 'U' is facing westerly toward the southeast Asian archipelago.[8] A submersible of the U.S. Coastal and Ocean Mapping Center measured the depth of the trench—11,034 m[9] which is much greater than the height of Mount Everest (8,848.86 m).

Why Subduction? Denser, Less Buoyant Oceanic Plate Bows

Plate subduction occurs at convergent junctions for two reasons–density and temperature. The dense, metal-rich colder oceanic plate sinks underneath the buoyant, lighter, warmer continental plate at journey's end (Fig. 4.3). It disappears below the continental plate, along the sloping surface of the trench. Imagine the N-S Peru-Chile Trench at the junction of the Pacific and South American Plates.

The 'Slab-Pull' mechanism suggests how subduction may happen (Fig. 4.4). The gravitational pull of the heavier oceanic plate is the real driver of it. At the plate junction, the continental plate stays afloat, while the leading edge of the oceanic plate will sink into mantle pulling the rest of the plate downward.[3] Once the plate starts to descend, gravity takes over. The diving slab moves 4-times faster than a non-subsiding plate, due to the magic of gravity.[10] Experts debate about the definitive driving force behind plate movements—Ridge Push or Slab Pull, though lately the latter may be more accepted.

Ophiolite—Deep Secrets of Oceanic Crust Uncannily Perched Atop Continents

Ophiolite represents a special case where large chunks of oceanic crust and upper mantle of a down-going slab may break loose and find its way onto the overlying continental crustal plate. The presence of serpentinite, pillow basalt, and deep marine sedimentary rocks, testify to the oceanic origin of ophiolite on continental plates.

A complete ophiolite complex occurs in the Bay of Islands Complex of Newfoundland, Canada.[11] The outcrops confirmed the 1966 theory of Prof. Tuzo Wilson about an ancient ocean that has since vanished. The ophiolite came from the "the Iapetus Ocean," the ancestral Atlantic Ocean of Cambrian-Ordovician age. The old ocean formed by rifting, only to disappear later during the Paleozoic collision. Later, the Jurassic opening of the Atlantic Ocean emplaced the ophiolites on Newfoundland, Canada, and Portugal/Spain in western Europe.

Similar ophiolite complexes are common in the Himalayan belt, occurring extensively from Oman to northern India.[12] The Tethys Sea, its sediments and lavas, largely subducted under the Eurasian continent. Some Tethyan ophiolite and sediments were scraped off during the India-Eurasia collision. It got stitched to the continental crust of the rising Himalayas, now outcropping at high altitude.

Oman holds the world's largest ophiolite patch of Tethys origin.

The Alps and Appalachians also contain deep-sea ophiolites. In South America, notable ophiolite rocks occur in the Coast Range of Venezuela.[13] The out-of-place Jurassic ammonite, sedimentary rocks of deep-sea origin, pillow basalt, and serpentinite, occur in a 'mélange' assemblage. A chaotic juxtaposition of allochthonous rocks characterizes a mélange.

Products of a Subduction—Lavas, Blueschist, and More

What is the relation between subduction activity and lava eruption? As a plate's edge disappears in the mantle, new lava wells up at the distal mid-ocean ridge. In the end it is a zero-sum game. The subducting plate melts in the red-hot mantle at great depth, and magmas resurface as lavas in volcanoes. The Pacific plate, subducting under the Americas, eastern Asia and beyond, created the 'Ring of Fire' chain of volcanoes all around the Pacific Ocean (Fig. 9.15).

The long Andes Range owes its origin to the subducting Pacific plate, along the oceanic Peru-Chile trench west of South America. Modern islands, plateaus, and ocean-floor sediments—all ride along on the subducting oceanic plate. The process may go hand in hand with low-temperature high-pressure metamorphism, producing diagnostic glaucophane (amphibole) mineral in blue schists in the subduction zone. Eclogite develops in the deepest part of the upper mantle. The subducting plate can cause large-scale natural disasters such as volcanic eruption, earthquakes, and tsunamis. The friction between two opposing slabs has created an earthquake-prone zone up to 660-km-deep in the mantle.

The Scenic Santorini Volcanic Caldera, Greece—Fraught with Lava and Uncertainty

Visitors to Santorini Island (or Thira) in the Aegean Sea probably pay little attention to the geologic backdrop and how the world-famous tourist hotspot came into being. Plate tectonics is implicated in the upheavals in the area. The subduction of the African plate (5 cm/

4-6 1 The >300 m high Santorini caldera walls show 650-ka–0-ka-old layers of lava, tephra including tuff, and ferruginous paleosols. Houses of Santorini Island line the rim of the vertical caldera walls **2** NASA satellite image of Santorini caldera, the Nea Kameni volcano, and other Aegean islands **3** A view from above of the new, 2-km-across, Nea Kameni volcano at the center of the caldera, active since 1707 **4** Domed churches and houses adorn the steep cliffs of Santorini Island.

year) under the Aegean microplate (a fragment of Eurasia) along the Hellenic arc trench is the driver of frequent earthquakes and volcanic events in the area. The latest earthquake of July 1956, near the Amorgos Island, caused severe destruction in Santorini. Twelve powerful eruptions marked the area during the last 350 ka years. The islands of Thira, Thirasia, Aspronisi are remnants of the original Santorini Island after the volcanic event of 1600 BCE (Fig. 4.6).

An exceptionally large Plinian (i.e., explosive) volcanic event, >4 times stronger than the 1883 Krakatoa eruption, destroyed the earlier island, sculpting the imposing caldera, ejecting nearly 100 km^3 of debris, and forming the Santorini Island we know. The volcano taps a 4–8 km deep magma chamber. The 3600-BP eruption destroyed the Minoan civilization here, and more importantly in Crete, 140 km away. Attractive paper-white houses, and blue-domed churches of the Santorini town Fira, and villages Oia, Akrotiri, Pyrgos, etc., neatly frame the >300-m-high rim of the humongous caldera (8X11 km).

In 1707, the volcano's central peak (Nea Kameni) resurfaced within the flooded caldera. Since then, it has spewed lava many times, including thrice in the last century alone. One can clearly see the dark recent lava next to pale older and outer parts (image 3, Fig. 4.6).

Recent tomographic images of the nearby 1500-m-subsea Kolumbo volcano (~6 km from Santorini), reveals a hitherto undetected shallow active magma chamber about 2–4 km below the sea floor. It is growing perilously @ 4 x 106 m^3/year, since its last eruption in 1650 CE, leading experts to think that it may affect the area sometime in the next 150 years.

The near vertical, laterally continuous caldera walls with spectacularly exposed layered volcanic strata, offer a rare geological gift to decipher the complex repetitive development of the volcanic edifice. The caldera walls attest to several Plinian eruptions, showing multiple lava flows, pumice horizons, ash, tephra beds, and oxidized ferruginous paleosols. Common lava tubes, dykes, veins, and normal faults occur at all levels. Experts posit such destructive Plinian effusive occurs every 20–30 ka, alternating with periods of lesser lava flows.[27, 28, 29]

PLATE TECTONICS AND ITS MYRIAD GLOBAL EFFECTS

Plate Tectonics May Affect Climate Pattern, and Deeply

Plates go places. Think of Antarctica now—a cold, windy, snow-covered, desolate, and rather inhospitable place. However, one would have witnessed a more congenial Antarctica if one could travel back to the Permian (299–252 Ma ago) in a time machine. It was enjoying a congenial tropical to subtropical climate, supporting exuberant foliage and thriving forests. The petrified vegetation now occurs as coal seams in the Trans-Antarctic mountains in the icy continent. The same Antarctica experienced glacial climate during the earlier Devonian era (419–359 Ma).

Volcanoes and mountain chains control climate in an intimate way. Volcanoes erupt lava, and spew large amounts of gases such as CO_2, SO_2, nitrous oxide, water vapor, etc. The excess of CO_2 creates a greenhouse effect, while sulfur aerosols prevent sunlight from entering Earth's atmosphere, making it cooler. Tall mountains can modify air circulation. They create rain shadow zones on the back side, while causing copious rainfall and large rivers at the front end. Plates drifting toward polar latitudes may create permanent ice caps, changing the weather pattern. The northward movement of the South American plate in the Paleogene created the Antarctic Southern Ocean. The Antarctic Circum-Polar Current along the Southern Ocean prevents warm northern currents from reaching Antarctica.

Sea Level Ups and Downs—Plate Tectonics Partly to Blame

Sea level oscillation has been very common in Earth history, and plate tectonics is a direct accomplice in this. Sometimes the continents got fused together, later they splintered due to plate interactions. In the process, they shut and open seaways, impacting the rise and fall of sea levels.

The fluctuation of sea level had a far-reaching influence on all living creatures through geologic time. The highest ever level was >220 m above the present sea level (PSL) during the End-Cambrian/Ordovician time, 488 Ma ago.[14] Just imagine how far inland the sea will encroach today if sea level rises another 220 m. Soon after, sea level dropped during glaciation at the End-Ordovician. The lowest sea level (slightly lower than today) was 250 Ma ago at End-Permian hyper-thermal time, a juncture coinciding with the largest mass extinction ever. The hot, dreary Triassic Period was a weary testimony of this arid landscape. Then came the second highest sea level ever, ~220 m above PSL, not too long thereafter, when large swaths of coastal areas went under water in the upper Cretaceous. This was the time with no permanent ice anywhere and when forests and dinosaurs thrived from pole to pole. Fast forward, and *Homo sapiens* were eyewitnesses to the impact of both sea level rise and fall during the late Pleistocene. About 125 ka back, sea level was 4–6m higher than today.[15] During the Last Glacial Maximum (LGM) deep freeze (21–16 ka ago), the Gulf of Mexico (GOM) sea level was nearly 125 m below PSL.[16] The Mississippi delta shoreline receded 200 km southward from the present GOM shoreline, well into today's deep Gulf of Mexico waters.

At the end of the Last Glacial Maximum, however, sea level was rising steeply till about 5 ka ago, when it suddenly flattened out. Since then, the rise has been minimal as glacial melting had almost ceased. However, data shows that the 20th century global sea level has been markedly higher than the previous few thousand years. It was rising ~1.8 mm/year due to climate warming.[15] In the 21st century, the rate of rise has gone up even more.

What Else May Cause Sea Level Oscillations?

Temperature variation is a proven culprit. Two contrasting factors are: 1) thermal expansion of water in prolonged warm climate causes sea level rise, and 2) draw-down of seawater through ice cap formation triggers sea level fall. Plate tectonics has a hand in both events.

A continent-continent collision will reduce sea level, as the intervening sea between the continents will be the first casualty (Fig. 4.3). The Tethys Ocean disappeared when India docked into the Eurasian plate. The rate of seafloor spreading at mid-ocean-ridge (MOR) may be another cause. The larger the MOR spreading, the greater will be the volume of erupted magma on the seafloor and more marked will be sea level rise.

Does Plate Tectonics Influence Biogeography?

The birth of the Panama Isthmus in Central America is a major milestone, causing migration of fauna and flora across continents. The Caribbean Ocean kept the two Americas separate, prior to the creation of the ~3-Ma-old Panama land bridge. Many endemic species were distinct in each continent at that time. But once the land bridge was open for business, there was free interchange of flora and fauna between continents. Another unintended consequence of the land bridge is that it stopped the equatorial current flow between the Pacific and Atlantic Oceans. The warm equatorial current morphed into the Gulf Stream, headed northward along the east coast of North America toward Europe. Thanks to the bracing Gulf Stream, the U.K. enjoys a rather milder winter than Canadian regions, though both are at about the same northern latitude. Panama Isthmus proved to be a natural bridge between continents, yet a barrier between two mighty oceans.

Plate Tectonics Explains Almost Everything—Hot Spots and Seamounts Are Exceptions

Plate tectonics theory can explain almost everything, except for a few natural events that still defy a plate tectonic rationale. For example, most volcanic activity is concentrated at two plate-controlled areas—mid-ocean-ridge, and volcanic arcs at subducting plate margin (Fig. 4.3). There is also an altogether new suite of volcanic centers that occur within plates, not at the margins (Fig. 4.3, at far right). The 'Large Igneous Provinces' (LIPS) occur far from the continental or ocean plate margins and include many features—the hot spot, ocean plateau, submarine seamount, and massive continental lava flows.[17]

Geologic Hot Spots—What Are They and How Do They Form?

It was Prof. Tuzo Wilson who had the first clue of hot spot origin, as he was studying the Hawaiian chain of islands (Fig. 4.7). He noted a string of dead volcanoes ending up with an active volcano at one extreme—in this case, the Big Island of Hawaii. Prof. Wilson reasoned that the phenomena is due to the motion of the Pacific Plate over a deep, fixed source of heat in Earth's interior, below the Island of Hawaii. The fact that only the last volcano, i.e., the Island of Hawaii is active, suggests it is now sitting right atop the heat source, whereas the other islands with inactive volcanoes had successively traveled over it in the past. The Big Island will be no exception, becoming extinct sometime in the future, and a new volcano will make its debut farther southeast. The unknown heat source turned out to be a mantle plume, a deep long vertical magma conduit reaching up to the

4-7 Testimonies of Hawaiian volcanism **1** A lava flow in 1790 in the island of Maui **2** Solidified lava flows on the Maui coastline **3** The >300-m-high Diamond Head Caldera formed after the collapse of a volcanic cone, 300,000 years ago, Honolulu, Oahu **4** Black sand beach from the disintegration of basalt lava on the Big Island of Hawaii coast.

ocean floor. However, as reasonable as it might seem, not everybody accepts the theory.

The Yellowstone National Park is a continental example where numerous geysers and hot springs tap the heat source from a rather shallow magma chamber lying below.

The Island of Iceland is another unique setting (Fig. 4.8), it is sitting on top of a hot spot located on the Mid-Atlantic-Ridge. The rift is splitting apart Iceland at the rate of 2–5 cm/year. Iceland may become two distinct islands 2 to 5 km apart in a million years. Awakening of the Geldingadalir volcano, Iceland in 2021, was a new eruption happening along the ridge after many centuries, sourced by the deep mantle hot spot.

Oceanic Plateaus—Submarine Volcanoes Taller Than Everest

The oceanic plateaus are much larger than the point-source man-

4-8 The Mid-Atlantic rift is splitting apart Iceland; here the water-filled rift is a few meters wide. Photo credit: Omar Colmenares.

tle hot spots. The Pacific Java-Ontong plateau is larger than Alaska, USA. It is a 120–90-Ma-old, mafic igneous province. It was originally a large, 1.5-million-km^2 plateau (25-km-thick), now broken up into several smaller ones due to tectonism.[18,19] The Kerguelen Plateau on the Antarctic plate covers an area of 2.2 million km^2. A mantle plume beneath the Kerguelen apparently gave rise to the long Ninety-East Ridge on the Indian Ocean floor near India.

Hawaiian Seamounts—Mauna Kea Rises 10,205 m from Seafloor

Seamounts above mid-plate hotspots are more common than plateaus. The tallest mountain in the world, the Mauna Kea ('White Mountain') is a seamount from a dormant volcano. The majestic peak is about 10,205 m tall, measured from the seafloor (~6,000 m below sea level and 4,205 m above).[20] There are 100,000 seamounts higher than 1,000 m. They are consequential as they divert ocean currents and are important for deep-sea life. They also may pose a danger to submarine navigation. Mining companies are targeting seamounts as a potential source of critical minerals like cobalt and nickel of deep-sea hydrothermal vent origin.

Large Igneous Provinces (LIPs)—Basalt Floods on Land and Sea

LIPs of massive continental lava flows defy explanation with standard plate tectonics models. LIPs had developed in a very short time, less than a million years, in settings like the Deccan Trap, India, Siberia in Asian Russia, Columbia Plateau, USA., Karoo in South Africa / Antarctica, and Parana in Brazil. Scientists argue that a deep mantle plume had triggered these massive outpourings of flood basalt.[21]

Why and how does the mantle undertake the mammoth job of flooding continents and oceans with multi-km-thick lava flows? Why does it spew enormous quantities of toxic gases into the atmosphere, disrupting organic life? The MOMO theory, i.e., 'Mantle Overturn, Major Orogeny' may explain the gigantic flows. When a cool oceanic subducting slab descends to a depth of 660 km at the base of mantle asthenosphere, it may break through the density barrier and enter the hot mantle layer below, creating huge overturns (*splash*). It possibly upsets mantle's predictable convection cell activity. The lower mantle may react, creating multiple huge plumes of magma, inundating continents with thick blankets of flood basalt.

To note, three huge eruptions had coincided with the End-Permian, End-Triassic, and End-Cretaceous mass extinctions in Earth history.

Epeirogeny—Huge Craton-Centered, Rolling Arches and Sag-Depressions

Another geological feature that baffles plate tectonics is 'Epeirogeny.' The process involves regional vertical uplift and warping, forming broad circular or elliptical arches and equally large shallow depressions (basins), spread over hundreds of kilometers in cratons, i.e., stable interior of continents. Distinctive sedimentary packages develop in the basins, separated by unconformities.

One practical effect of epeirogeny is that it may trigger marine flooding in areas of the sag depressions, and retreat of seas over large areas around the elevated arches. A thick pile of shallow marine limestone, sandstone and shale is characteristic of the basins. Prof. L. Sloss in the 1960s recognized six Phanerozoic mega-sequences, the product of repeated marine incursions in North America.[22] The southern Rocky Mountain suffered epeirogeny and a consequent 1,300–2,000-m uplift of strata, during the Cenozoic Era.[23]

In summary, epeirogeny is vertical movement of stable craton areas in continental interiors. What causes it? The unequal mantle heat distribution may explain the gentle crustal undulations. Models of mantle current data suggest that the broad arches arise where the crust is directly above the warm mantle currents. Likewise, sag-basins rest above cool mantle currents.

POLAR WANDERINGS—EVIDENCE OF PLATE TECTONICS AND MANTLE ACTIVITY

Apparent Polar Wandering—A Case of Transferred Epithet

The Apparent polar wandering (APW) and True polar wandering (TPW)) are the two types of polar wandering in geological discussions. The former is the supposed movement of Earth's magnetic poles with respect to some 'fixed' earthly features like a tectonic plate or a continent. Earlier, scientists had the notion that continents were fixed, and the magnetic poles were wandering through time. Why did they assume this? They knew that the magnetic minerals always align in the direction of magnetic poles when they crystallize from cooling magma. They also noted that successive strata on a continent show different orientation of magnetic minerals. The scientists carefully plotted all the pole positions of different strata on a map. They deduced (incorrectly) that the different mineral orientations were due to the wandering of poles over time. If the poles were wandering, data from all the continents with respect to the magnetic poles should have been the same. However, the scientists were in for a big surprise when they compiled data from the different continents. The Apparent Polar Wandering of each continent was different. This could only happen if the continents were wandering, instead of the magnetic poles. It is a case of wandering continents, not poles. That is why they coined the moniker—Apparent Polar Wandering. The premise was wrong, but something truly important came out of this. It confirmed that continents and plates are constantly wandering around like rafts on floating lithosphere slabs. Alfred Wegener was prophetic when he proposed the theory of the 'continental drift' in 1912.

An example of how continents drifted around in time comes from the west coast of North America.[7] The west coast's telltale time-lapse latitudinal positions are illustrative: Near the Equator—180 Ma ago (Jurassic), 24⁰N–150 Ma ago (Late Jurassic), 33⁰N–120 Ma ago (Cretaceous), 45⁰N–60 Ma ago (Paleocene), 37⁰N–Present day.

The True Polar Wandering—Blame the Mantle

The True Polar Wandering, on the other hand, represents the true shift of Earth's vertical rotational axis. Let us imagine a spinning top to understand this. The top's spin axis will deviate from the vertical line when it just starts spinning or when finally slowing down. Similarly, the North Pole wanders in a circle, does not go far from the vertical spin axis. It tends to move around in a small area like a spinning top. This motion appears to be less than 6 m/year.[24]

Researchers first tied True Polar Wandering to the onset of the ice age.[25] Others found that Earth's spin axis remained at a stillstand between 48–12 Ma ago, based on the Hawaiian hot spot data.[26] Earth moved relative to its geographic north pole within last 12 Ma, causing Greenland to be closer to the North Pole, triggering the Pleistocene ice age about 3 Ma ago. What happens inside Earth, especially in the mantle, may manifest itself on the surface. Mantle is in the driver's seat, it is 2,900 km thick while the crust is very thin, <100 km thick. Normally, Earth is not a perfect sphere—it is about 42 km wider at the equator, because extra mantle material is pulled toward the equator reflecting Earth's centrifugal force. Scientists think that the true polar wandering takes place when mantle material builds up abnormally at higher latitudes.[26]

CHAPTER HIGHLIGHTS

- Plate Tectonics: History, key concepts, major plates, mobility, and role in Earth's changes.
- Plate Tectonics Mechanism: Seven major plates and the Wilson Cycle of plate formation and destruction.
- Plate Movements: *Divergent*—plates drift apart (e.g., Mid-Atlantic Rift), *Convergent*—plates collide (e.g., India-Eurasia collision), *Strike-slip*—plates slide past each other (e.g., North America & Pacific Plate).
- Plate Tectonics Impact: Shapes climate, land-ocean configurations, supercontinents, mountains, sea levels, volcanism, and biogeography.
- Volcanic Features Beyond Plate Tectonics: Hot spots (e.g., Hawaii), seamounts, large igneous provinces (LIPs), and epeirogeny.
- Polar Wanderings: Apparent and true polar shifts linked to plate tectonics and mantle dynamics.

WHAT'S NEXT?

Now that we know the secret workings of our restless planet, the layer-by-layer development of the Precambrian Earth will be our focus in Chapter 5.

Intricate sculpture on Precambrian soapstone of the 800-900-year-old Belur temple, Karnataka, India. Soapstone or steatite is a soft, talc-chlorite schist of metamorphic origin containing talc, mica, quartz, tremolite amphibole, etc.

PART TWO

EARTH'S EVOLUTION THROUGH TIME

5

Precambrian Eons: Early Earth Convulsions

PRECAMBRIAN EONS— HADEAN, ARCHEAN, PROTEROZOIC (4.57 Ga–539 Ma)

Throughout geologic time, many earth-shattering events happened. Soaring mountain range, high plateau, vanished ocean, exceptional marine incursion, meteorite bombardments, are just some of them. Similarly, the deep burial of strata many kilometers inside the boiling womb of Earth, has made the task daunting, yet rewarding, to unravel the secrets of 'Deep Time.'

We will highlight the unfolding earthly landscape layer by layer. Fig. 5.1 details the geologic time-ladder, i.e., stratigraphy. To note, the exact age of different eons, eras, etc., are somewhat tentative, as the numbers keep shifting with new data.

The changing geological panorama will be a recurring theme in the book. Chapters 5 to 12 deal with the evolving tale of Earth's long journey. It is amazing how our planet matured from the Hadean hazy, nascent stage to the full-grown Earth at the final visible milepost of the Holocene Epoch today. *Homo sapiens* has been a witness, and more importantly, a protagonist of the very last period where he has been playing a preeminent role.

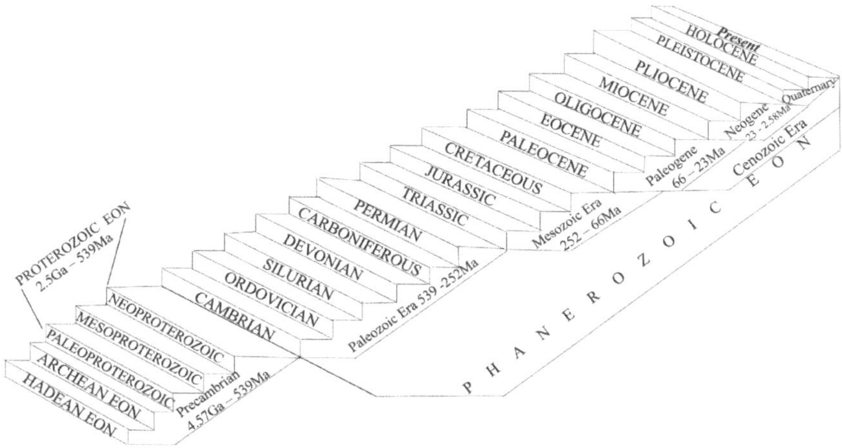

5-1 Earth's 4.57-Ga-old geologic time-ladder (data—ICS-2023). The Proterozoic Eon consists of 3 eras. The Quaternary Period (Pleistocene-Holocene): 2.58 Ma-Present.

Precambrian to Phanerozoic Eon— All-around Evolutionary Change

The chapter unravels the closely guarded secrets of the Hadean-Archean-Proterozoic Eons. The significance of progressive changes in these and younger Phanerozoic intervals will become clear as we climb up the deep-time ladder. We will discuss the journey under three headings in every chapter. The *Paleogeographic Highlights* deals with geographical features, including landscape, contours of land and ocean at different geological periods. We will note climate variation, and the changing distribution of Earth's surface features—rivers, flood plains, coasts, deep sea realm, snow-capped poles, or in extreme cases, deserts.

The *Stony Testimony* of rock assemblage of each time slice is a silent, yet eloquent record of the ever-changing scenery. The diverse rock types hold tell-tale clues to their specific geo-history. The *Miracle of Life* is the evolutionary story of the marvelous continuity of biota on Earth. It starts from the earliest lowly, unicellular microbes in the Archean to progressively complex species, culminating in the *Homo sapiens* in the Quaternary. Each geological period boasts of myriad organisms, some even unique; those that preceded and succeeded them, complete the strand of evolution. Often there is lacuna, or a geological *unconformity*, erasing millions of years of rock and fossil records. But science found a way to skirt the problem. Let us start with the oldest.

Our discussion of the Precambrian (~4 Ga) is at a more macro scale, spanning billions to hundreds of millions of years (Eons and Eras). In contrast, the story of the Phanerozoic Earth (the last 539 Ma) will be down to Periods and Epochs, because of relative abundance of data.

Hadean Eon (4.57–4.0 Ga): Planet Earth, A Glowing Fireball

PALEOGEOGRAPHIC HIGHLIGHTS

The three Precambrian Eons—Hadean, Archean and Proterozoic, precede the lowest rung of the Phanerozoic Eon—the Cambrian Period. Eons again are subdivided into several geological Eras. For instance, the Archean Eon is subdivided into Paleo-, Meso-, and Late Archean Eras, though we will discuss the above eras all together for brevity. The Hadean Eon (4.57–4 Ga) is treated as one because of the lack of data. It's the oldest and least understood Eon because of its antiquity; the name comes from 'Hades'–the Greek god of the underworld.

Protoplanet Thea (Theia) Smashed onto the Hadean Earth

The giant planet impact hypothesis posits that the Mars-like planet 'Thea' impacted early Earth. The momentous collision likely created a new permanent cosmic body—our satellite Moon, from the splintered planetesimals of both the alien planet and Earth. We recall the details of the Thea impact hypothesis from Chapter 2.

Nascent Earth—Scorching Amid Nonstop Cosmic Bombardments

The early Earth felt severe birth pangs amid literally hellish conditions of the burning solar environment. A 'period of heavy bombardments' marked the first few hundred million years of Earth's childhood, like all its sister planets. A continuous rain of cosmic planetesimals mercilessly pounded and pulverized the primordial Earth surface. Ages of weathering and erosion have erased most of the scars. The hot atmosphere cooled down enough to have water within ~10 Ma after birth.[5a]

Toxic Gases and Water in the Primordial Earth

The early Hadean atmosphere was toxic, heavy with carbon dioxide, methane, hydrogen, other gases, and water; it essentially lacked free oxygen. Water was present despite the prevailing ambient temperature of 230^{0}C.[1a] A very high atmospheric pressure, related to heavy CO_2 concentration made the presence of water possible. Volcanic outgassing, ice from impacting comets, atmospheric water vapor concentration, and rain—all likely contributed to the water budget in the oceans and land. The evidence of liquid water in the oldest dated zircon crystal nucleus is significant.[5] Over time, the atmospheric CO_2 level would have gone down. Two natural processes likely reduced the ambient greenhouse gases—dissolution of CO_2 in water, and its removal by the plate subduction process.

The Earth Engine at Work from the Get-go: Mantle Convection and Plate Tectonics

Earth's internal structure started stabilizing with gradual cooling. Fractionation happened, i.e., heavy elements like iron and nickel gravitated to the core, and the lighter silicon and aluminum floated to the crust. The mantle convection cell probably started due to heat from the decay of radioactive elements in Earth's interior. It likely triggered plate tectonic movements, and later subduction and assimilation of most of the original Hadean crust into the mantle. Some scattered remnants of the Hadean crust lie exposed sporadically as rare witnesses.

STONY TESTIMONY: WITNESSES TO FUZZY HADEAN WRINKLES

Scarce Relicts of the Hadean Rocks, Here and There

Researchers used extinct radioisotopes to date basalts from an outcrop in the northwest Quebec coast of Hudson Bay, Canada.[2] The radioisotope *samarium* and its alpha-decay product *neodymium* are unique elements; Samarium-146 is an extinct isotope that only existed during the first 500 Ma of the early Earth. Its presence serves as an invaluable timekeeper of the primordial crust as it cooled from a molten state. The 4.28-Ga-old samples imply that the rocks formed only about 300 Ma after the birth of the solar system.[3] It is significant because the original ~4.57-Ga-old crust has since been mashed and remobilized beyond recognition. The dated rocks are part of the metamorphosed 'Nuvvuagittuq' greenstone belt.[4] But the surprise is not yet over. Some xenocrysts (older crystals within younger rocks) of zircon in the Ewokrama Formation of Guyana Shield, are dated 4.22 Ga.[1]

Granitic Rocks Formed Early on the Hadean Earth

A geochemist confirmed an astoundingly old radiometric age of 4.375 Ga +/- 6 Ma of a tiny zircon crystal from the Jack Hills Range, Australia. The oldest mineral on Earth occurred in a metamorphosed, 3-Ga-old conglomerate. The zircon came from a pebble of a much older water-rich granitic rock. The presence of trapped water as fluid inclusion in the zircon nucleus attests to the very early presence of water on Earth's surface, and the formation of silicic continental crust. All this happened just 100 Ma after the impact of 'Thea' creating our Moon.[5] Indeed, the Hadean marks the moment when our Moon debuted as Earth's lovely companion.

Hadean Life? The Essay Will be Brief

The ambient condition during the Hadean was inhospitable, to say the least. The early atmosphere was totally unlike today. The Eon yielded little direct evidence of palpable fossils. However, the presence of light carbon ^{12}C isotope and water in the Hadean rocks suggests that ultra-primitive life might have existed even in the nascent Earth. The earliest (extremophile) life likely varied between 3.8–4.1 Ga. Scientists argued that the organisms only used ^{12}C for their skeleton, none of ^{13}C. Absence of ^{13}C in the zircon nucleus may be an indicator of primitive life in the Hadean.[6] Other researchers disagreed—graphite in zircon nucleus is of abiotic origin, according to them.[7] However, if modern-day microbes survive without oxygen, the early microbes could have lived in the infernal Hadean time too.[8]

Archean Eon (4.0–2.5 Ga): Respite from the Hadean Inferno

PALEOGEOGRAPHIC HIGHLIGHTS

The Archean (Greek—'beginning') Eon lasting 1.5 Ga, spans about one-third the age of the 4.57-Ga-old Earth. From the hot Hadean as we crawl up into the archaic Archean, the haze of uncertainty is somewhat lifted. The clarity of more data helps geologists to paint a reasonable picture of this Eon, though many clues are still missing. Archean rocks are much more common than the preceding Hadean, occurring in the Americas, Russia, Greenland, India, China, Australia, and many other places. There are 35 large Archean cratons present all over Earth, but they form only 7–10% of the exposed continental crust.[9] The tectonically and seismically stable 'cratons' (Greek god *kratos*—'strength') occur in the interior parts of continents. The word 'shield' refers to an outcropping craton, while a platform is a craton with a sedimentary cover. The craton or shield mainly forms the silica-rich crust, i.e., granite family of rocks. The Canadian Shield of mostly metamorphosed granites constitutes the core of the North American Continent.

The Cratons—How Did They Form?

Earth's internal workings control what transpires on its crust, including craton formation. Processes like plate tectonics, volcanism, and erosion by natural agents shaped the early Earth. The intense radioactive decay of elements, and higher heat flow from the mantle, was the motor of tectonic movements. The heat would have triggered plate tectonic movement, intrusive magmatism, and volcanism including deep hot spot plumes. The atmosphere already existed during the Archean, though it virtually lacked free oxygen. Rainwater and wind would have produced sediments from weathering, erosion, and transport. The long and complex history of craton/shield formation involved repeated plate tectonic orogenic assembly. The creation and destruction of basins, rafting and collision of plates, mountain building, and renewed drifting, were all part of the process. Every continent has important shields and cratons.[10a, 10b]

A supercontinent is one solid mass of continents englobing all disparate existing landmasses. Two Archean supercontinents—the Vaalbara and Kenorland, formed through rafted plate collisions, only to be shattered later (Fig. 5.2). The Hadean-Archean Terrane of the 4–3.58-Ga-old, Acasta gneiss, Canada, predated the Vaalbara craton. Both the South African Kaapvaal and Australian Pilbara cratons were part of the Vaalbara. The name *Vaalbara* comes from the last four letters of Kaapvaal in Africa and Pilbara in Australia.

The Pongola glaciation, the first one in Earth history, happened ~ 2.8–2.9 Ga ago, when the Vaalbara supercontinent was disintegrating.

Kenorland was the next supercontinent to appear on the scene. It included the new 2.65-Ga-old 'Arctica' continent (englobing the North American Shield, Greenland, and Baltica). But the restless Kenorland soon splintered into Laurentia (America), Baltica (ancestral Europe), and W. Australia.

The Atmosphere—Tough Times, A Far Cry from Later

The Archean had a toxic atmosphere with abundant carbon dioxide (CO_2), ammonia (NH_3) and methane (CH_4). The O_2 level was very low, only 0.4–7% of the present value (21%), free oxygen being virtually non-existent. Nitrogen content was somewhat similar to today. The CO_2 level was 10–2500 times higher, while CH_4 was 100–10,000

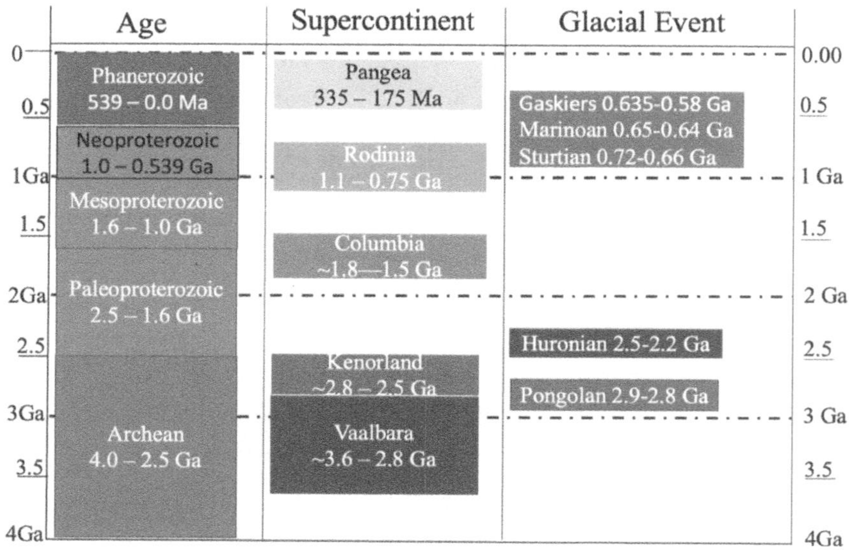

5-2 Precambrian supercontinents and timeline of five glacial events (Huronian, Marinoan, Sturtian, etc.). Pangea is Phanerozoic in age.

times the modern values.[11] Greenhouse gases moderated the climate, and temperature probably varied between 0–40°C, up to 80°C according to others. The early Archean Sun had luminosity of about 77% of today because of the pervasive methane droplets in air.

STONY TESTIMONY

Archean Shield—Forged from Repeatedly Metamorphosed Rocks

It is time to recall the story of rocks, their names and origin. Early in Earth's life, the dark heavier minerals and metals gravitated to the center of Earth through fractionation, forming the metallic core, enveloped by the heavy peridotite-rich mantle. The iron-magnesium-rich olivine in the mantle crystallizes at very high temperature. Earth's crust, in contrast, consists of a less dense, silica-rich fraction of low temperature origin. It is the Archean crustal rocks that suffered metamorphism, being subjected to variable heat and pressure in the subsurface.

The Archean Shield or craton consists of buoyant, relatively lighter, igneous rocks—granite and diorite. Extensive lava flows like pillow basalt, metamorphic greenstone and schist also occur. Other common rock types are deep oceanic rocks—greywacke ('dirty sandstones'), mudstone, banded iron formation, and chert beds. These rocks were variably metamorphosed, i.e., cooked, and recooked. The Canadian north is a virtual museum of Archean rocks.

Another distinctive rock is metamorphosed komatiite, discovered first in South Africa, later in the Abitibi greenstone belt, Ontario, Canada.[13] The nickel and copper-rich komatiite is an olivine-rich, high-temperature volcanic rock. It extruded profusely during the Archean and Paleoproterozoic Era. It rose from the deep mantle to the upper crust at a time when heat flow was much higher than today. Now it is part of the metamorphic greenstone belt in Ontario, Canada.

Banded Iron Formation—
Lowly Primitive Microbes Had a Hand in It

The Banded Iron Formation (BIF) is unique to the Archean and Paleoproterozoic. It occurs extensively in Australia, Africa, India, Brazil, Venezuela, Russia, Ukraine, Canada, and the United States. It is a rhythmic, color-banded unit of reddish hematite (Fe_2O_3), minor magnetite (Fe_3O_4), and white bands of silica (SiO_2). The silica may occur as one of its polymorph varieties—quartz, chert, chalcedony, or jasper, as thin bands varying from 1 mm to a few centimeters. The exact origin of the BIF is not fully clear. One likely explanation is that the iron oxide precipitated from seawater, when photosynthesis by cyanobacteria released free oxygen. The Archean atmosphere was very O_2-poor, as most available oxygen was locked up in water molecules. Metallic iron picked up the newly available free oxygen to produce the Fe_2O_3 and minor Fe_3O_4. Once the biomass of cyanobacteria rose to a critical level, the oxygen level also went up. The available iron could not use up all of it. As a result, free oxygen build-up would have reached a high, toxic level; at that point the biomass suffered and died. The seasonal absence of biomass triggered precipitation of silica. The sedimentary/chemical cycle of the alternating hematite and silica bands went on for a long time, nearly 800 Ma.[14] A rare analog is a Quaternary Banded Iron Formation, showing how organisms would have mediated the Archean BIF genesis.[15] The Banded Iron Formation serves as a widely used iron ore worldwide, including the Lake Superior region of the USA.

The scarcity of limestone in the Archean is noteworthy. Why is that? The early Earth, including the primitive ocean, was saturated with carbon dioxide. The Archean CO_2-rich marine water was probably too acidic for calcium carbonate, the main ingredient of limestone, to precipitate.

MIRACLE OF LIFE

Archean Life—Single-Celled Primitive Life

The evidence of Archean life is more palpable, unlike the Hadean. The spontaneous complex chemical reactions produced simple cells from carbon molecules that did not need oxygen to survive during 3.5–2.1 Ga period. The cells would have contributed a minor amount of oxygen to the environment through photosynthesis.[12]

Paleobiologists documented primitive biota from several localities in the world. Like the Hadean, the early Archean life forms preferentially would have incorporated the stable isotope ^{12}C in their cells rather than the ^{13}C. Natural processes seem to fractionate the stable carbon isotopes in certain organisms.[16.]

All Archean bacteria and archaea were single-cell prokaryotes. Prokaryotes are primitive microbes without a membrane enclosing the nucleus. South Africa documented terrestrial, fossilized 3.22-Ga-old microbes.[17] A 3.47-Ga-old baryte mineral yielded primitive sulfate-reducing bacteria.[18] A 3.48-Ga-old stromatolite in Pilbara sandstone, Australia, contains the earliest identifiable fossils.[19] Stromatolites are microbial mats in shallow oceans secreted by the cyanobacteria, earlier known as 'blue-green algae.' Experts found 40 morphotype of putative microfossils in stromatolites. While Archean life is ~3.5 Ga,[20] carbon isotope ratio suggests life is 3.8 Ga or older, as discussed before.[6]

Proterozoic Eon—(2.5–0.539 Ga): Paleoproterozoic, Mesoproterozoic, and Neoproterozoic Eras

A Slow but Steady Change in the Air

The Proterozoic tectonic collision and coalescence of continental plates produced larger, stable cratons/shields, while smaller plates/microplates characterized the Archean. Oceans produced more $CaCO_3$ on broad shallow marine shelves, forming limestone and dolomite.

The three eras of the longest lasting (~2 Ga) Proterozoic ('earlier life') Eon are again divided into periods. The manifold divisions mean geologists had more information to work with. Several standout features of the Proterozoic include atmospheric oxygen, glacial events, making and unmaking of the Rodinia Supercontinent, common occurrence of limestone, transition from unicellular to complex multicellular life, etc.

Single-celled organisms continued to live in the Proterozoic. However, the Eon witnessed atmospheric oxygen build-up, and the game-changing debut of eukaryotic cells, i.e., cell nucleus encased in a membrane.

Milankovitch Cycle—
Clues to Variation of Solar Radiation on Earth

The Milankovitch Cycle is an important natural event affecting Earth's climate (see Fig.11.15). According to the theory, the orbit of Earth around the Sun controls the amount of incident light on Earth surface. Three factors affect Earth's orbit and sunlight intensity on it: *precession* or wobble around its vertical rotational axis, *eccentricity* or varying orbit path, i.e., distance from Sun, and Earth's *obliquity* or axial tilt angle.

The gravitational pull of the Moon and Sun causes Earth to wobble influencing earthly climate. The wobbling Earth is just like a spinning top's late-stage circular movement. Earth completes one wobble or *precession* cycle around it's vertical axis every ~26 ka. The two hemispheres will experience alternate extreme solar radiation every 13 ka. The wobble also determines which star will be the future North Star.[33]

Due to its wobble, Earth's celestial pole now points toward Polaris, the North Star or Pole Star, but it pointed towards Thuban in Draco constellation in 3000 BCE; more surprise—the North Star will be Vega in the Lyra constellation ~ 13.7 ka from now.[34] In Hindu astronomy, the Pole star is '*Dhruva nakshatra*,' the constant or steadfast star. Nature is not so constant, after all.

Eccentricity refers to the shape of Earth's orbit around the Sun. The orbit path varies from less elliptical (nearly circular) to a slightly elliptical orbit path. It is elliptical due to gravitational pull of two larger planets—Jupiter and Saturn. More solar energy reaches Earth, when its orbit path is nearly circular because it is nearer to the Sun. Currently, Earth's eccentricity is at its least elliptical (i.e., shortest) path around the Sun. The orbit path becomes most elliptical every 100 ka.

Earth's axial tilt angle or *obliquity* varies from about 22^0–24.5^0; the greater the tilt angle, the more sunlight the two poles receive. Earth's current axial tilt is 23.4^0 and slowly decreasing. It takes about ~40 ka to complete a tilt cycle from 22^0 to 24.5^0. According to NASA, Earth's last maximum tilt angle happened 10.7 ka back, and its future minimum will occur about 10 ka from now. The variable sunlight intensity influences the mundane hot and cold cycles.

Modern estimates show that one complete Milankovitch cycle now lasts about 100 ka, though it was different in the past. For instance, a study of 1.4-Ga-old deep-water rhythmites reveals interesting data for the Precambrian: The Earth-Moon distance was 340,900 km +/-2,600 km; now 380,000 km. A day was about 18–19 hours, now 24 hours. The wobble cycle used to be ~14 Ka, now 26 ka; the Precambrian eccentricity cycle was ~131 ka, now 100 Ka.[34a]

Proterozoic Atmosphere—Oxygen Palpable, Ozone Debuted

The Proterozoic atmosphere showed several prolonged mercury oscillations during its nearly 2 Ga time span. Climate scientists have debated about the origin of the long-lasting cold snaps, especially in the Neoproterozoic. According to one theory, Rodinia disassembly caused increased rainfall, rock weathering and erosion. Enhanced weathering would have consumed more atmospheric carbon dioxide and its accumulation in seawater, making the atmosphere leaner in CO_2 and colder.[26]

Another novel explanation suggested that volcanic activity could have caused atmospheric temperature change—both warming and cooling.[29] The eruptions increase atmospheric temperature that melts or thins existing glacial ice sheets. The thinned ice sheets will take away pressure on top of the buried volcanoes, thus activating them. The result—more CO_2 and heat, and the consequent atmospheric warming will put an end to the ice age. The SO_2 buildup can also cause a volcanic winter.

The variation of the atmospheric oxygen level was significant in the Proterozoic. From the 3.5 Ga to early Proterozoic (~2 Ga), only sulfate reducing bacteria existed. Microbes like archaea and primitive bacteria received energy by reducing sulfur to hydrogen sulfide (H_2S). The O_2 level even later during the 1.6–0.8-Ga period was very low, about 0.1% of modern value, a fact that might have delayed the emergence of animals.[30] Oxygen production was strictly mediated by organisms like cyanobacteria (obsolete term 'blue green algae'). The cyanobacteria itself went through an evolutionary cycle from the earlier non-photosynthetic to fully photosynthesis-capable groups about 1.8 Ga ago. Cyanobacterial photosynthesis[31] likely controlled the sudden spike of oxygen in the Neoproterozoic, probably up to 10% of the present atmospheric level.

Ozone genesis was an important breakthrough in the early Earth as O_3 creation depends upon the presence of free oxygen in the atmosphere. How does it form? The O_2 absorbs energy from Sun's ultraviolet rays in the 10–16-km-high zone within the 50-km-wide stratosphere; the ultraviolet rays break down many oxygen molecules into individual atoms. An ozone molecule (O_3) forms when a free oxygen atom (O) combines with a molecule of oxygen (O_2). A moderate amount of ozone was present (about 10% of modern value) only during the End-Neoproterozoic, about 0.60 Ga back.[32] All life in the Proterozoic could have existed only in the ocean. The presence of ozone was a boon for evolving organisms, they could finally migrate to land during the earliest Paleozoic.

PALEOGEOGRAPHIC HIGHLIGHTS

Paleoproterozoic Era (2.5–1.6 Ga)

First Taste of Proterozoic Icehouse—Huronian Glacial Event

The Paleoproterozoic experienced the global 2.5–2.22-Ga-old, Huronian or Makganyene deep freeze event. The 2.4-Ga-old 'Great Oxygenation Event' was probably linked to glaciation. Icehouse condition removed atmospheric greenhouse-causing methane gas, while adding free oxygen to the Paleoproterozoic atmosphere. The Gowganda Formation tillites in Ontario-Quebec, Canada, is an example. South Africa and Australia, among others, also endured the oldest glacial event.

Transient Supercontinents NENA & Columbia

The Paleoproterozoic saw colliding landmasses of the Laurentia and Baltica forming the continent Nena (acronym for North Europe-North America).[10] Then North America was next-door neighbor of both Europe and westerly Antarctica. After the Nena event, the crustal blocks were impatient again, danced around through collisions and drifting for a long time. Finally, the cratons of Nena, including Baltica, East Antarctica, and Arctica formed the transitory supercontinent Columbia (1.8–1.5 Ga). The eye-catching rapakivi granite is linked to the supercontinent Columbia (Fig 5.7). Crustal shield rocks, the most stable part of all continents, reveal a multigenerational tectonic history.

The crust partly formed in the Paleo- to Mesoproterozoic in India, China and South America including Amazonia. The main four Indian cratons probably matured and stabilized at this time. The same is true for 'Baltica,' the precursor of the European continent.[21]

Consolidation of Earth's Crust—Some Examples

South America provides an example of Earth's layered crust formation. The Guyana Shield is spread over Venezuela, Guyana, and Brazil, reaching its peak development and stability in the Proterozoic. The Venezuelan Imataca Complex is Archean, but the Pastora supergroup, and later volcanic rocks are of Paleo- to Mesoproterozoic age. The towering Proterozoic table mountains of the Roraima Group display many waterfalls (Fig. 5.3). The Angel Falls is the highest waterfall in the world (>900 m), it descends from the top of one of the isolated Table Mountains—the Auyantepuy, Venezuela. American aviator Jimmy Angel daringly landed on top of the mountain in 1937 in search of gold, and hence the name of the waterfalls. Famous English explorer Sir Walter Raleigh explored the neighboring 31 km^2 Roraima Peak, during his 1595 expedition. His book detailing the search for the mythical 'El Dorado' gold deposits centered around the peak.[22]

5-3 The world's highest waterfall (>900-m)—Angel Falls, coming down from the rim of Auyantepui Table Mountain, Bolivar State, Venezuela. Note the cave-like openings from where the water is gushing out.

Mesoproterozoic Era (1.6–1 Ga)

All Landmasses are Transitory—The Supercontinent Columbia Broke Up into Seven Cratons

The Columbia supercontinent came together during the transition from the Paleoproterozoic to Mesoproterozoic (Fig. 5.2), it likely had disintegrated by 1.2 Ga. The Congo craton, Laurentia, Antarctica, India, Australia, Siberia and Baltica—were all part of Columbia. In India the era displays metasedimentary rocks interstratified with some volcanic rocks. On a lighter note, the first white tiger (a cub) was a gift of the Mesoproterozoic Vindhyan Mountain in central India where the author had worked. The present white tiger population in the world zoos owes a debt to this orphan cub found in 1951.

Grenville Orogeny and the Debut of the Rodinia Supercontinent

The next major highlight of the Mesoproterozoic Era was the 'birth' of Rodinia (Russian 'Rodina' = 'giving birth') around 1.1 Ga, due to repeated collision of large crustal blocks. The splintered fragments of Columbia had supplied landmasses for the latest supercontinent. It englobed all major cratons and lasted 350 Ma before breaking up 750 Ma ago (Fig. 5.4). The Grenville Orogeny of North America marks the last phase of this supercontinent buildup. The older crustal blocks within the Rodinia supercontinent are reminiscent of the shape of modern continents. However, their geographic locations 1 Ga back, had been topsy-turvy relative to today. North America was part of Laurasia along with Greenland in one model.[16] Other cratons were Australia, Antarctica, Amazon (precursor South America), India, Siberia, Baltica (precursor Europe), Congo, Kalahari (both ancestral Africa), and West Africa. Paleomagnetic data throws light on the latitude of the blocks but cannot pin-point the longitude of the same. A breakthrough came in 2011 (Geol. Soc. America press release). The radioactive lead (Pb) isotope composition in samples of the Red Bluff Granite, Franklin Mountains, West Texas, matched lava rock sam-

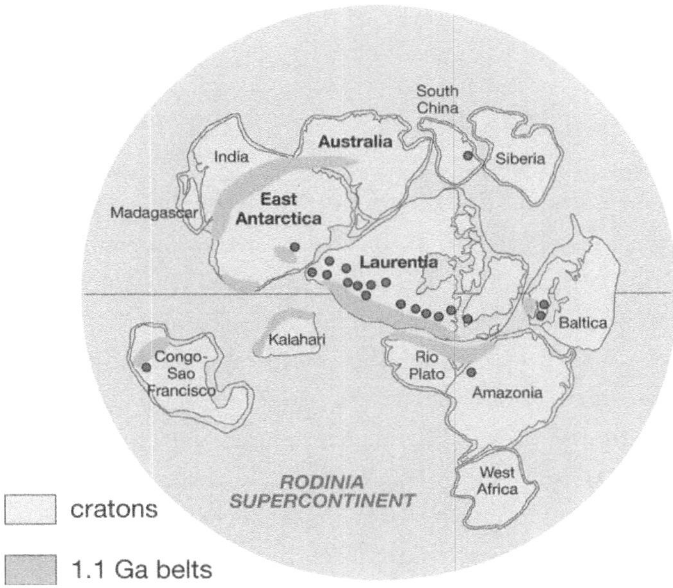

5-4 The break-up of the Rodinia Supercontinent 750 Ma ago. Grenville orogenic belts in darker grey (1.1 Ga belts). Dots: 1.3–1.5-Ga-old granite in old continents.

ples from the present-day Coats Land region of Antarctica. Samples from an area near the Great Lakes of North America also showed a similar match. North America and Antarctica were physically linked 1.1 Ga ago. [23a, 23] Antarctica had the unique distinction of being adjacent to ancient northerly Laurentia, and part of the southerly Gondwana mega-continent during its life.

Neoproterozoic Era (1.0–0.539 Ga)

Rodinia Break-up

The Supercontinent Rodinia broke up into three crustal plates about 0.75 Ga ago. Later, a short-lived supercontinent Pannotia appeared during the Ediacaran Pan-African Orogeny. Soon Pannotia also splintered into two: the northern Laurasia (including North America), and the southerly Gondwana mega-continent. The Gondwana englobed South America, Africa, Antarctica, Australia, and India. The two landmasses later coalesced in the latest Permian into Pangea, the last supercontinent to date.

Snowball Earth Vista—Whitewash from Pole to Pole

The Neoproterozoic Era experienced several marked glacial episodes. The Cryogenian Period (0.85–0.63 Ga) separates a younger Ediacaran (0.63–0.539 Ga), from an older Tonian Period (1.0–0.85 Ga).

The Cryogenian icehouse of the Sturtian and Marinoan glacial events are well documented the world-over. Joe Kirschvink (1989) coined the term 'Snowball Earth' referring to the Neoproterozoic glacial event and rocks. The glacial deposits outcrop in various places—South Africa, Namibia, Australia, Canada, and Norway, etc. The Cryogenian glaciation was the most severe of the geologic record, when ice reached down to equatorial regions of Africa, North America, and other continents. The wide extent of the Snowball Earth event is significant.

Australian geologists were the first to document the glacial events. The 'Sturtian' was older (W. Howchin, 1908), while the 'Marinoan' event came later (D. Mawson, 1949).[24] The U-Pb isotope ratio in tiny zircon crystals precisely dated the 717–659-Ma-old Sturtian, and the 650–635-Ma-old Marinoan events.[26]

Most geologists now accept the Snowball Earth scenario as new evidence surfaces every year, aided by radiometric dates. Prof. P. Hoffman documented in Namibia the coexistence of Neoproterozoic stratified glacial tillite and warm, shallow marine carbonate deposits.[25] The juxtaposition of glacial deposits and overlying warm, shallow water dolomitic limestone, proves that the Snowball Earth event extended over the entire globe from the poles to the equator. Prof. Hoffman cited similar sites in Svalbard, Norway, Arab-Nubian Peninsula, China, etc.

Newfoundland, Canada provides the evidence of Gaskiers glacial event of Ediacaran age (579.63–579.88 Ma). The 300-m-thick section shows diagnostic glacial striated clasts. Coeval glaciation occurred in eight other places worldwide, some close to the equator between latitude 10^0 to 30^0.[27] Radiometric data reveal that only 9 Ma separated glacial Gaskiers from the overlying Ediacaran warm water fauna.

A major discontinuity surface in the Grand Canyon, USA, indicates that the Snowball Earth erosive event removed nearly 5 km of strata.[28] A 500-Ma-old sequence sits directly above a billion-year-old rock sequence. Imagine, the geological temporal record of 500 Ma is missing.

STONY TESTIMONY

The Proterozoic Eon

Rocks are Largely Metamorphosed

The Cuddapah and Vindhyan rocks of India are Paleoproterozoic to Mesoproterozoic in age, the latter may extend up to the Neoproterozoic. The Cuddapah rocks of probable rift origin show evidence of metamorphism in marbles. It is dominantly sandy and argillaceous. S. Patranobis, 2018, reported extensive development of microbial stromatolite in the Cuddapahs. Earlier, the author had examined abundant stromatolite in the Vindhyan limestone.

Anatomy of Table Mountains

If one wants to summarize distinctive Proterozoic rocks, what will those be? More than 80% of the crust had formed by the Proterozoic time. Earth had cooled down from the hot Hadean-Archean time as there was less magmatic activity. The cratons became larger, and progressively more stable through accretion of new tectonic plates/ microplate landmasses.

We recall an important towering landform of Proterozoic age (1.8–1.5 Ga and younger) is the cliff forming table mountains in the old Gondwana continents—South America, Africa, India, and Australia. These are massive cliffs of sandstone, product of ancient river, delta, and shallow marine origin. The author had the opportunity to work extensively on these formations in India and Venezuela (Fig. 5.3). The Venezuelan table mountains were once a continuous blan-

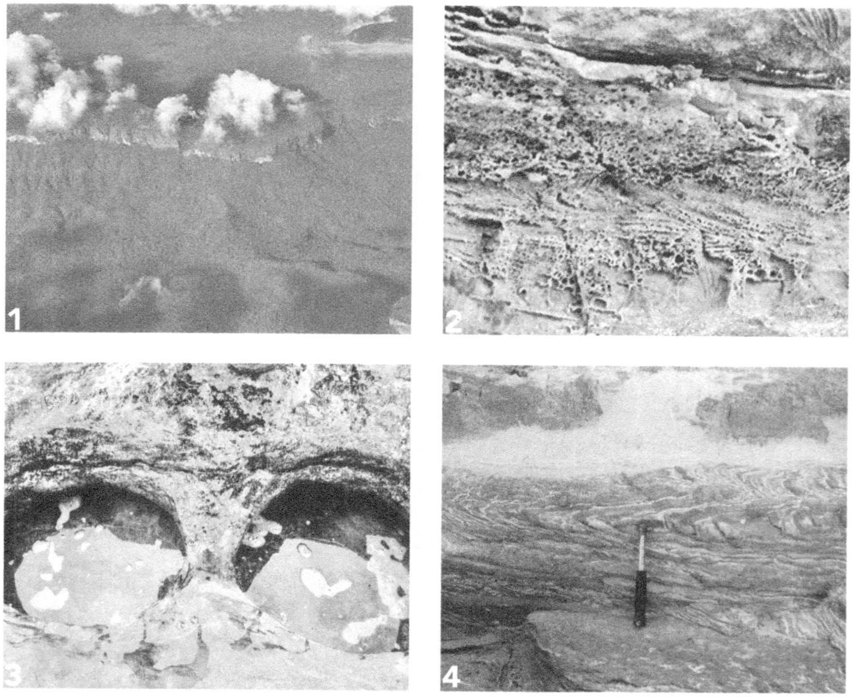

5-5 Features of the Proterozoic Roraima Group, Bolivar State, Venezuela **1** Aerial view of the isolated flat-topped Kukenan tepuis **2** Wind and rain erosion created karstified, pock-marked hard sandstone textures **3** Stream-formed potholes resembling eyeglasses, Roraima tepui **4** Oxidized dark laminae highlight soft-sediment folding, Gran Sábana **5** Prolonged erosion along vertical joints shaped the tombstone-like panorama on Auyantepui.

ket of sandstone, now occurring as isolated 'tepuis,' due to tectonism and erosion (Fig. 5.5). Similar 'tepuis' occur in neighboring Brazil, Guyana, and Suriname, and even in far off west Africa. Sandstone 'karst' development is one of the striking features of the mountains. Copious dissolution of rocks gave rise to impressive cave systems and sinkholes. Weathering and erosion have sculpted features that often mimic human and animal forms, to the delight of the lucky few who can visit the remote sites (Fig. 3.9). Quartz veins of hydrothermal origin show beautiful prismatic quartz crystals in the 'Valle de Cristales' on the Roraima tepuis.

Transported Deep-Sea Rocks Violently Collided and Merged with the North American Craton—The Grenville Orogeny

The 1.1-Ga–750-Ma-old Grenville Orogeny was the architect of Rodinia, adding new landmass and exotic rocks to North America and elsewhere. Since then, North America and Greenland became the nucleus of the future Laurentia continent, with Baltica (Europe) in the vicinity. The Gondwana landmasses were disperse across the globe at that time.

The Grenville Range, 350-km-wide in North America, is the longest Mesoproterozoic belt (~10,000 km) in the world.[35a] The vestiges of the Grenville orogenic belt (Fig. 5.4) continue both north and southward from present-day Canada. Hugging the eastern seaboard, it extends beyond Texas, ending in the Oaxacan terrane of Mexico. It reaches up to Scotland in Europe. Erosion and prolonged weathering had pared down its majestic original height, which was taller than the Himalayas.

The Grenville Belt consists of three prominent zones, each with a unique tectonic and rock signature: North American Superior Craton (NASC), Para-Autochthonous Zone (PAZ) in the middle, and Allochthonous Zone (AZ) of deep-ocean origin to the southeast (Fig. 5.6).[35]

The Grenville Front, a major tectonic boundary, bore the brunt of the orogeny, shielding the North American Superior Craton (NASC) from major deformation.[36] It separates the stable NASC of Archean-Proterozoic age, from the deformed Para-autochthonous Zone (PAZ).

The in-place Para-Autochthonous Zone of piled-up rocks was originally part of the undeformed North American Superior Craton. The tectonic jolt deformed the in-place PAZ rocks.

The Allochthonous Zone (AZ) at the southeast end arrived from far off. How did that happen? A subducting oceanic plate normally dives under a lighter continental plate. However, occasionally an oceanic plate may have other thoughts. Some fragments of it may break

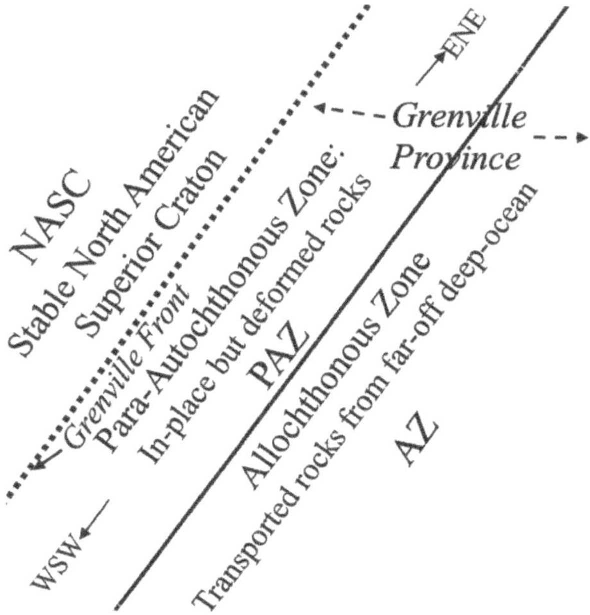

5-6 The North American Superior Craton (NASC), and two Grenville orogenic belts—Para-Autochthonous Zone (PAZ), and Allochthonous Zone (AZ). Black line marks the *thrust contact*, i.e., where the out-of-place AZ rocks rammed against the fixed continental landmass PAZ.

loose and jump upward, override the lighter continental plate due to tectonic push. Thus, the out-of-place 'ophiolite' and deep oceanic sediments came to rest on the continental crust in the Allochthonous Zone (AZ). The AZ is a potpourri—jumbled mass of metamorphosed and tectonized rocks piled upon the ancestral North American Shield more than 1 Ga ago. It is a 'mélange' of granite, gneiss, syenite, gabbro, peridotite, and anorthosite (75% of the world occurrence).[36] Also, marble, iron formation, and graphite are present.

The Grenville Province is a great natural laboratory where geologists can delve into an ancient orogenic belt. The 'cooked' rocks from a high pressure-temperature zone came from dozens of kilometers deep inside Earth. These are now outcropping on the surface due to the tectonic uplift and erosion. The Grenville is rich in economic and metallic mineral deposits.

Other Proterozoic Rocks—Banded Iron, Silicic, and Carbonates

The unique Paleoproterozoic Banded Iron Formation of our earlier discussion is related to the 2.4-Ga-old Great Oxygenation Event.

New evidence exists of a much younger, 750-Ma-old, Neoproterozoic banded iron formation.[37] The rhythmic hematite-silica layers likely precipitated during a transient warm interval of the Cryogenian Snowball Earth.

The Ediacaran strata occur widely. Well-preserved marine, fossiliferous sandstone and siltstone occur in the White Sea region of Russia. The host rocks at the famous 0.565-Ga-old Mistaken Point fossil locality of Newfoundland, Canada, are tilted and cleaved dark mudstone of deep-water origin. The Ediacaran fossils also occur in siltstone of the Upper Miette Group, British Columbia, Canada. The Clemente Formation in Sonora, Mexico, reveals Ediacaran fauna in thin-bedded marine sediments.

The eye-catching rapakivi granite of anorogenic origin (unrelated to orogeny) is typical of the middle Proterozoic rocks of the Columbia supercontinent, now scattered in places like USA, Finland, Labrador, Canada, and Amazonia and Guyana shields in South America (Fig.5.7).

5-7 Polished slab of Precambrian rapakivi granite—the ovoid large crystals are of orthoclase (K-feldspar) with a thin dark rim of oligoclase (Na-feldspar). Many small black crystals of biotite (mica) and amphibole in the groundmass surround the large feldspar crystals.

MIRACLE OF LIFE

The Proterozoic Eon

Field Day for Cyanobacteria—Supplanting Anaerobic Life

The Paleoproterozoic primitive anaerobic bacteria and archaea existed because they could survive without oxygen. They used sulfur/sulfates to draw energy and thrive. Some authors opine that the Great Oxygenation Event was toxic to primitive anaerobic bacteria, as a result, most perished at the time. Not long after, the cyanobacteria, already present in the Archean, held hegemony during the Paleoproterozoic and Mesoproterozoic Eras. The latter is better documented the world over when cyanobacterial stromatolite in carbonate rocks became more common. Cyanobacteria was a game changer by the Archean standard. The intricate, ubiquitous, organo-sedimentary structures in carbonate rocks in Canada, Australia, India, Russia, etc., are testaments of cyanobacterial activity in the Proterozoic (Fig. 5.8). However, cyanobacterial dominance waned toward the end of the Neoproterozoic. Today, only a few places like the Shark Bay, Australia, and the Great Bahamas in the Caribbean, show stromatolite mounds. Both red and green algae also appeared in the Proterozoic.

Thriving Fauna in Ediacara Hills—Enigmatic Marine Life

The cyanobacteria ('Blue-green algae') ruled till the latest Ediacaran Period when a major biotic change happened on Earth. The earliest oxygen-dependent fauna appeared at this time. It was a great time when a sudden burst of life was evident worldwide. The complex, soft bodied, multicellular Ediacaran macrofossils characterized the post-stromatolite period.

What sets apart the Ediacaran fauna from the younger Cambrian biota? The main difference is that the Ediacaran fauna (630–539 Ma) lacked hard skeleton and shell. It preceded the more evolved Cambrian fauna of trilobite, sponge, snail, etc., and plants like seaweeds.[38]

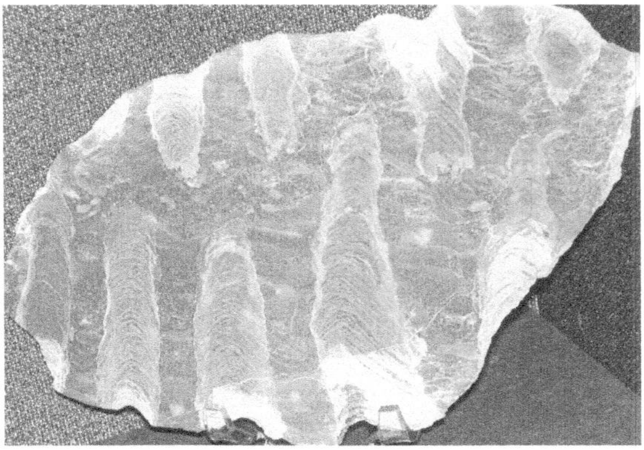

5-8 Precambrian stromatolite from Bashkiria, Russia, showing slender conical microbial buildup.

The mineralized calcium carbonate or silica skeletons of fauna were a Cambrian evolutionary hallmark, arming them from predators and the vagaries of climate.

The Ediacara Hills in South Australia unearthed the first Ediacaran fauna, perfectly visible to the naked eye, including the famous 539–570-Ma-old '*Dickinsonia*.' It is common in the Neoproterozoic rocks everywhere. The Ediacaran discovery confirmed that Darwin was right as he predicted the existence of a long history of metazoan evolution. The abundant, somewhat strange fossil remains, tracks and trails left on the ancient seafloor by worms and animals, occur widely in Canada, Russia, Mexico, India and elsewhere. Some faunas may belong to an extinct separate phylum, some are like the family of mollusks, sponges or sea urchins, or soft arthropods like crabs, ticks, etc. The fossils also include algae, lichen, protozoan (like single-celled amoeba, flagellate, etc.). The 555-Ma-old Russian White Sea fossils are taxonomically and timewise very similar to the Ediacaran fossils. New data on Australian macrofossils offer valuable insight about the Ediacaran multicellular organisms.[39]

Mistaken Point, Newfoundland, Canada—Diagnostic Ediacaran Fauna

In 1967, Shiva Misra, then a student, discovered the amazing Ediacaran fossil assemblage in the Mistaken Point Ecological Reserve, Newfoundland.[40, 41] The dark mudstone, about ~565-Ma-old, hosts

5-9 Several imprints of complex Ediacaran fauna without hard mineralized skeleton that got preserved in dark mudstone in the Mistaken Point Ecological Reserve, Canada. Coin ~2cm.

beautifully preserved, complex, multicellular fossil assemblage showing myriad shapes—spindle, frond-like, radiating, dendrite-like, etc. (Fig. 5.9). Fauna lived in deep-water of the ancient oceans. The contemporaneity of fauna from distant places is significant—Australia, Russia, Canada, India, and elsewhere. The Ediacaran faunal exuberance was truly a global event.

Here comes the riddle. How on Earth did the soft-bodied biota get fossilized and preserved? Normally only hard shells and bony parts survive as fossils. The exquisite preservation of Ediacaran biota was probably because scavengers were absent that early in the geological time. Both the sessile (fixed to seafloor) and motile animals settled on the seabed after death, but their remains were left intact and ultimately got perfectly preserved. Additionally, an oxygen-poor, deep-water, sheltered paleoenvironment also helped in the preservation.

- Introduction to Chapters 5–12, detailing each time slice through three themes: Paleogeographic Highlights (tectonics, landmasses, oceans, climate), Stony Testimony (distinctive rocks), and the Miracle of Life (extant life).
- The ~4-billion-year-old Precambrian: Three Eons—Hadean, Archean, and Proterozoic.
- The superheated Hadean Earth faced intense meteorite bombardment, including the Moon-forming Theia impact, leaving few surviving rock remnants.
- Discussion of the Archean-Proterozoic Eons highlights plate tectonics, climate, craton formation, transient supercontinents, diverse rocks, water's origin, oxygen scarcity, and notable Archean banded iron formations (BIFs).
- The Neoproterozoic Era features the unique Snowball Earth, the Grenville orogeny, the formation of the Rodinia supercontinent, and its eventual breakup.
- Life starting in the Archean—debut of primitive single-celled sulfur-reducing bacteria and archaea, and later, common cyanobacterial stromatolite structures.
- The End-Proterozoic proliferation of abundant and diverse multicellular, O_2-dependant soft-bodied animals. Thriving Neoproterozoic Ediacaran fauna going extinct just before the paradigm-shifting Cambrian Period.

WHAT'S NEXT?

Chapter 6 introduces the Paleozoic Era, marked by record-high sea levels, icehouse-greenhouse cycles, vast coal forests, and two major mass extinctions. The Cambrian Period, the lowest rung of the Paleozoic, burst onto the scene with astounding varieties of complex life blooming everywhere. The eruption of life with hard shells was a paradigm shift, aided by the presence of oxygen in the air and plentiful calcium ions in the oceans.

top Emperor Shah Jahan had the iconic Taj Mahal (1632–1648 CE) built in Agra, India, in memory of his beloved queen Mumtaz, using the 1.5-Ga-old Proterozoic Makrana white marble from Rajasthan, India. **bottom** Modern craftsmen in Agra create souvenirs of Taj Mahal-type inlay work with semiprecious stones (carnelian, jade, agate, turquoise, lapis lazuli, bloodstone, i.e., chalcedony, garnet, onyx, etc.) set on the Makrana marble. Note the two intricate marble lattice works in the small plate.

'Uluru' or Ayer's rock, Central Australia, has been a 10,000-year-old, sacred site for indigenous culture. The ~550-Ma-old, vertical Cambrian arkose sandstone is of fluvial fan origin, later submerged in an ancient inland sea. The Uluru 'Island mountain' is 345 m high with a perimeter of ~10 km. *photo: Shanti Ghosh*

6

Paleozoic Era: Cambrian to Permian Period

Life's Explosion and the Scars of Two Mass Extinctions

PHANEROZOIC EON —539 Ma TO PRESENT
PALEOZOIC ERA—539 TO 252 Ma

Earth had suddenly changed so deeply at the beginning of the Paleozoic Era, which is the lowest of the tripartite Phanerozoic Eon. The Precambrian phase of intense tectonic activity, which was the making and unmaking of supercontinents, was finally over. The 'Cambrian Explosion' was a game-changer in terms of faunal diversity and abundance. The most important change was, of course, the appearance of novel fauna and flora that we can mostly relate to at the present time. The essential ingredients for life were in place—a benign habitable atmosphere with water and enough oxygen to breathe. The superhot temperature of earlier times was now a distant memory. Geoscientists succeeded in establishing a more detailed time-ladder ('stratigraphic column') for the Paleozoic, Mesozoic, and Cenozoic Eras of the Phanerozoic Eon, compared to the Precambrian. The Eras are farther detailed into Period, Epoch, and Stage. The Cambrian Period is the bottom interval of the Paleozoic Era.

Phanerozoic Time-Ladder in a Nutshell—Rock Sequence, i.e., 'Stratigraphy'

Fig. 5.1 shows the major divisions and subdivisions of the Phanerozoic history, as per the International Commission on Stratigraphy (ICS) Chart, 2023. The concept of 'Stratigraphy,' used in geology and archeology, simply means rock successions of different geologic ages.

Several geologic events constrain the assigned ages of Eon, Era, etc. The ICS criteria is the first appearance of a species, extinction of one or multiple species, major global climate change/shifts, catastrophic events, the formation and/or break-up of supercontinents, and global magnetic polarity change, among others.

Phanerozoic (539 Ma) Is Better Known Than Precambrian (~4 Ga)

The Eon forms only a miniscule part (12%) of the entire Earth history (4.57 Ga). The Paleozoic Era spans about one-half of the total Phanerozoic Eon (Fig. 5.1). If we imagine a calendar year representing Earth's age (4.57 Ga), then the Precambrian lasted 10.8 months, the Phanerozoic 1.2 months, and the Paleozoic 0.6 months. Surprisingly, we know much more of the last 1.2 months compared to the preceding 10.8 months of the special Earth year.

A Rock Sequence May Hide Record of Missing Strata and Time

To recall, the sequence of rock strata is not always continuous, often marked by an unconformity of missing strata, representing long time intervals. An unconformity is a physical discontinuity in time, fossil, and rock record, in an otherwise continuous vertical succession (Fig. 3.20). A simple analogy will illustrate the concept. Suppose we have 10 books in a series, where each one continues into the other. Suppose again some pages of the book #3 is missing. When someone starts to read the series, he/she will not know the content of the missing pages, though the reader can get an idea of the chronology of events by reading the rest of the book #3. An unconformity is like the missing pages that narrate the story for a time interval. It may make it difficult to decipher the complete physical and faunal/floral geological history in a particular region. Geoscientists can piece together the sequence of events, despite some missing pages in the unfolding Earth book. They can reasonably extrapolate the content of the missing section from other areas of the country or the world where the strata is intact and verifiable. It is somewhat like borrowing the same book from another library if the local one does not have it.

At the Siccar Point, Scotland[1], an angular unconformity separates the vertical, 435-Ma-old Silurian sequence below, from the horizontal, 370-Ma-old Devonian sequence above. The unconformity hides the 65-Ma-long missing rock/time record of the vanished Iapetus Ocean.

Big Picture: Paleozoic Era

The Paleozoic means 'Ancient Life' in Greek, in contrast to the Mesozoic (Middle Life), and the Cenozoic (Recent Life). The ~1.1-Ga-old Grenville orogeny gave rise to the Rodinia supercontinent lasting several hundred million years, but not forever. About 750 Ma ago, Rodinia broke up into fragments, thanks to plate tectonics and extensive volcanism. No single supercontinent existed during the lower two-third of the Paleozoic; only separate ex-Rodinia continents were present, clustered around two distinct northerly and southerly blocks.

The southerly Gondwana block of Antarctica, India, Australia, South America, and Africa, were bunched together near the South Pole. It witnessed two glacial icehouse events—one at the end-Ordovician, and the other in the Permo-Carboniferous. North of the equator, the ancestral North America and Baltica (Eurasia) coexisted, accompanied by a distant ancestral China. Plate tectonic movements slowly but steadily brought all the discrete landmasses closer together throughout the Paleozoic Era. Finally, the Gondwana, northern Laurentia and Baltica collided about 250 Ma ago at the End-Paleozoic. The collision produced the new supercontinent Pangea, surrounded by the Panthalassa Ocean. Pangea also did not last forever, as renewed plate tectonic activity broke it into pieces in the Mesozoic Era (Jurassic Period).

The Paleozoic marked a paradigm shift in terms of the paleogeography of Earth, other than the Pangea supercontinent. Mountain-building and orogeny continued, oceans appeared and disappeared through rejuvenated plate tectonic activities. The Era saw wandering continents, a more benign climate, and the appearance of new organisms with shell and skeleton. A Paleozoic hallmark had been the explosion of plants and lush dense vegetation around swamps. The extensive shallow marine carbonate shelves hosted teeming new life.

We discuss the geological highlights of the Paleozoic under 3 headings: Early (Cambrian-Ordovician), Middle (Silurian-Devonian) and Late (Carboniferous-Permian). Table 6.1 shows the Paleozoic periods and relevant details.

PALEOGEOGRAPHIC HIGHLIGHTS

Early Paleozoic: Cambrian-Ordovician Periods

An Eventful Start of the Phanerozoic Eon

The Early Paleozoic (539–444 Ma) craton margins showed development of shallow marine carbonate platforms. Sea level rise and marine incursion caused a gradual, steady flooding of lands. The Late Cambrian-Early Ordovician set the global record of maximum marine encroachment in Earth history, >220 m above the present sea level.

By the earliest Cambrian, the short-lived Neoproterozoic Pannotia supercontinent disintegrated into two large continents. Laurentia (ancestral N. America), Siberia, and Baltica existed in the north, and the Gondwana mega-continent remained as a separate southerly landmass.

A regional paleo-karst unconformity between the middle and upper Ordovician was a significant feature in North America and even in China. The erosional hiatus developed due to a drastic sea level fall, later marked by a new sea level rise. A karst develops when acidic freshwater dissolves calcium carbonate, producing many solution features during sea level fall.[2,3]

The last noteworthy Early Paleozoic event to follow was the Late Ordovician glaciation that caused sea level fall.

An important tectonic event—the Taconic orogeny, happened during the upper Ordovician when a volcanic island arc collided with the eastern North American craton. Examples of modern volcanic island arcs are the string of Japanese Islands, and the Aleutian Islands of Alaska, USA, both related to the subducting Pacific plate.

Table 6-1: Paleozoic Era—Six Periods

Periods	~Duration (Ma)	Etymology
Permian	299–252 = 47	'Perm' region, Russia
Pennsylvanian	323–299 = 24	Upper 40% of Carboniferous Period from Pennsylvania, USA
Missisipian	359–323 = 36	Lower 60% of Carboniferous Period from the Mississippi River, USA
Carboniferous*	359–299 = 60	Marked by vegetation and coal of U.K.
Devonian	419–359 = 60	'Devonshire' England
Silurian	444–419 = 25	Latin 'silures,' ancient southeast Wales people
Ordovician	485–444 = 41	Celtic tribe 'ordovices'
Cambrian	539–485 = 54	Cambria, Latin for Wales

**Carboniferous Period = Mississippian + Pennsylvanian, U.S. nomenclature*

The Taconic event had created high mountain ranges, since largely eroded by weathering and erosion. The reduced Taconic range now forms part of the New England area and the Appalachian Mountains of North America. Mount Greylock (1,064 m), Massachusetts, and Mt. Equinox (1,175 m), Vermont, represent the Taconic range.

Middle Paleozoic: Silurian—Devonian Periods

Glacier Melts, Sea Level Rises

The Silurian (444–419 Ma) was a coveted goldilocks interval from an organic evolution point of view. It straddled between the cold earlier glacial climate of the late Ordovician, and the somewhat drier climate of the succeeding Devonian Period (419–359 Ma). The high sea level, partly due to the melting of Ordovician glaciers, was conducive to abundant carbonate precipitation in the photic zone of shallow seas. Marine incursion extensively flooded adjacent continents.

Mountain-Building in Full Gear

Intense tectonic activity marked the Middle Paleozoic. The Caledonian, and later the Acadian Orogeny (Devonian), impacted Europe and North America. While Laurentia and Baltica stayed near the mid-Paleozoic equator, the Gondwana mega-continent drifted closer toward the South Pole. The Caledonian orogeny involved collision of Laurentia, Baltica, and Avalonia Island Arc, creating Euramerica. The Avalonia microplate broke away earlier from the Gondwana mega-continent. As ancestral North America and Baltica (Europe) collided, the intervening Iapetus Ocean disappeared. The remains of the old Iapetus Ocean strata now outcrop on both sides of the Atlantic. Rocks of the Caledonian orogeny occur in UK, Norway, Sweden, and the northern Appalachians including Newfoundland, Canada.[10] The collision of Laurentia— Avalonia Island Arc gave rise to the Devonian Acadian mountains, forerunner of the Appalachians. The old Avalonia fragments now lie scattered over New England, USA, and eastern Canada.

The Antler orogeny rocked the western edge of North America during the late Devonian—Mississippian.[11] A major mountain-building event running from the Nevada-California border northward, emplaced the deep marine 'Roberts Mountain Allochthon' on the carbonate shelf.

Late Paleozoic: Carboniferous—Permian Periods

The Carboniferous Period: Sea Level Fall, Swamp Lands, and Coal Forests

The Carboniferous Period (359–299 Ma) witnessed several new events affecting Earth, its fauna and flora. Both the northern and southern landmasses moved around, slowly coming closer, and eventually merged in the Permian (299–252 Ma). Climate-wise, the Carboniferous experienced greenhouse and icehouse conditions. An extensive warm shallow ocean covered the Mississippian (36 Ma out of 60 Ma, Carboniferous)—an ideal greenhouse environment for abundant carbonate precipitation and faunal growth. The upper Carboniferous (Pennsylvanian) revealed a contrasting face of Earth. A marked sea level fall exposed extensive marine areas where swampy conditions produced abundant coal forests. The mutual approach of Laurasia and Gondwana reduced the intervening oceans causing sea level drop. Secondly, growth of ice sheets near the South Pole extracted much water from the atmosphere and oceans.

The Variscan (Hercynian) Orogeny— The Birth Certificate of Pangea

The Variscan orogeny marked an important milestone in the upper Paleozoic. The Variscan or Hercynian orogenic belt rose when Laurasia and southerly Gondwana finally collided. The Variscan belt is extensive in Europe, running from southern Iberia to easterly Bohemia. It continued into North America later as Alleghenian orogeny and gave rise to the Appalachian and Ouachita Mountain belts. The tight folds in the strata of resistant sandstone and softer shale/siltstone

6-1 The aerial (plane window) view of the Appalachian valley and ridge region of Pennsylvania, USA; it mimics the underlying geology of the area. Weathering and erosion of folded strata resulted in the resistant sandstone ridges (darker areas, highlighted by dashed lines), and softer shale-siltstone valleys (off-white patches—cultivated valleys and houses). Two rivers marked 'R,' meander in the valleys near the upper-middle part.

gave rise to the characteristic valley and ridge physiography (Fig. 6.1). The combined Variscan, Alleghenian and earlier orogenies shaped the Permian Supercontinent—Pangea. The Anti-Atlas Mountain of northwest Africa used to be part of the Appalachian Mountain belt. Now these mountain belts lie on two separate continents because of the Jurassic opening of the Atlantic Ocean.

The Permian Period: A Climatic Pot-Pourri

Permian rocks document the cataclysmic climax of the Paleozoic. They are widely distributed—the 'Perm' region in the Ural Mountains of Russia, Permian basin in Texas and New Mexico, USA, and in South America, China, India, Australia, etc. The Permian Earth saw major climatic swings—glacial icehouse, balmy greenhouse, and an arid terminal phase. The early Permian was still glacial in some parts, but the climate gradually ameliorated. A drier climate marked the Late Permian, unlike the generally moist Carboniferous. Howev-

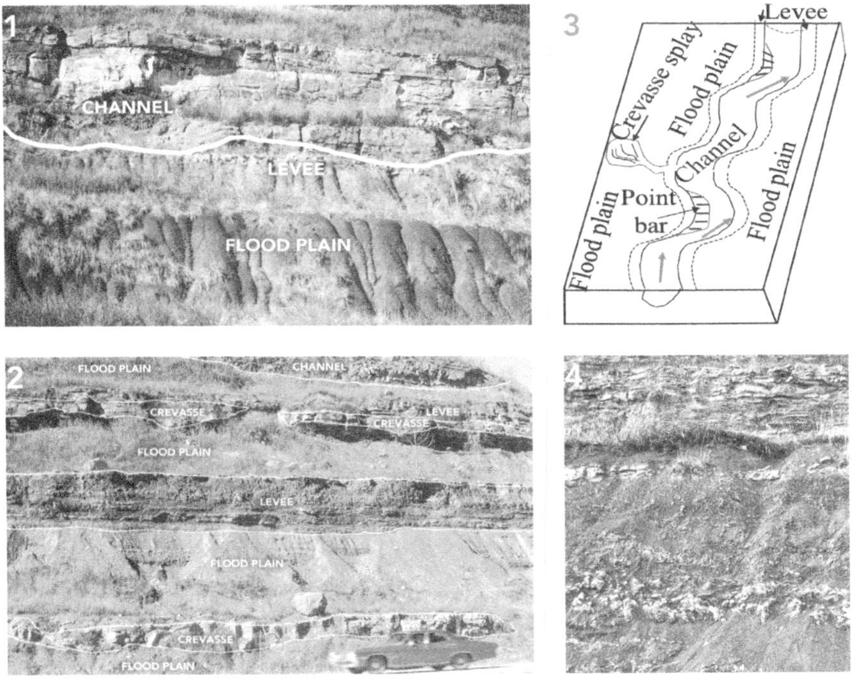

6-2 1,2 Alluvial plain sediments of channel, levee, crevasse splay, and extensive red flood plain mudrock. **3** The illustration shows how a river channel can erode the cut-bank (grey arrow) and create a sandy point bar deposit on the opposite bank. **4** A black and white image of a reddish paleosol sequence about 1.5 m thick. It shows the paleo-soil horizon with irregular, white, caliche or calcrete, a natural hardened calcareous rock—Permian Dunkard Group, W. Virginia, USA.

er, extensive growth of fern, moss, conifer, and coal forests during the middle-upper Permian, marked some Gondwana continents—India, Africa, and Antarctica. The arid End-Permian registered the sharpest sea-level fall, coinciding with the worst mass extinction ever.

In Laurasia to the north, the Permian was unlike the southerly Gondwana mega-continent. The Permian was arid and rather dry in North America, giving rise to extensive alluvial plains with well-developed oxidized red flood plains, crevasse, levee, and fluvial channel and common paleosols including calcrete and caliche horizons (Fig 6. 2). A crevasse is a small channel that may breach the levee during the flood stage of rivers and drains into the flood plain.

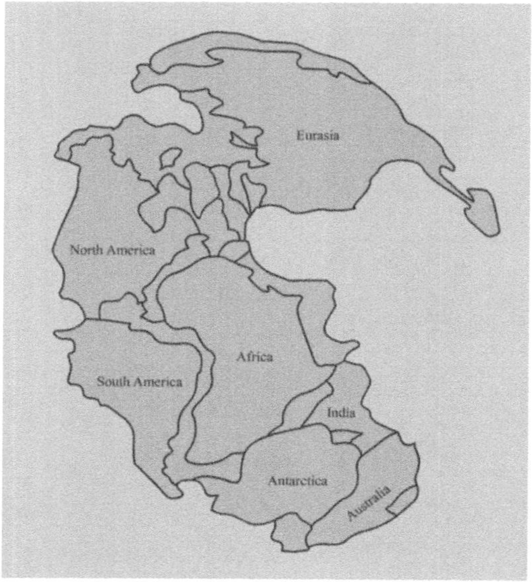

6-3 The Pangea Supercontinent became a reality at the End-Permian when the Gondwana (Antarctica, Australia, India, Africa, South America) and Laurasia (N. America, and Eurasia) landmasses came together. Pangea was the only Phanerozoic Supercontinent.

The Supercontinent Pangea: a Testament to the Magic of Plate Tectonics

The Variscan orogeny and the eventual continental collision between Laurasia and Gondwana put the final touches to the giant Pangea supercontinent (Fig. 6.3). It encompassed all existing landmasses at the End-Permian time when land connection extended all the way from pole to pole. Two oceans, the westerly Panthalassa (precursor of the Pacific Ocean), and the easterly Paleo-Tethys, flanked the supercontinent. Cathaysia microcontinents (China plates) broke apart from Gondwana mega-continent at the End-Paleozoic. The microcontinents remained as islands on the eastern margin of the Paleo-Tethys Ocean during the Pangea assembly. The southern part of North America was basking in a subtropical climate in the Upper Paleozoic. In contrast, all Gondwana continents remained bunched together in the southerly Antarctica region.

Carboniferous-Permian Hallmark—The Icehouse-Greenhouse Cycle

The Upper Paleozoic shows excellent greenhouse-icehouse cycle.

While the Early–Middle Carboniferous remained in a warm and humid greenhouse, the Middle and Late Carboniferous–Early Permian displayed icehouse conditions in the Gondwana continents. Thereafter, the Permian Period again showed a distinctly warm greenhouse phase, completing the greenhouse-icehouse-greenhouse cycle. The myriad rock layers and fossils show evidence of alternating Paleozoic climate cycles.

The Gondwana Icehouse-Greenhouse Cycles—The Cause? Plate Wandering and High Latitude Proximity

What triggered the icehouse condition? Firstly, only landmasses within the Gondwana continents experienced icehouse conditions because they happened to be at the high southern latitudes near the South Pole. The Gondwana glaciation was more widespread than Earth's latest Pleistocene event. The onset of the Gondwana icehouse condition probably was not simultaneous in all the landmasses. South America (what is now Argentina, Brazil), and eastern Australia were the first to experience icehouse conditions in the Carboniferous.[26] Soon after, the icehouse deep-freeze extended to Antarctica, Africa, and India during the earliest Permian (Sakmarian Stage), when Antarctica happened to be over the South Pole. Earlier, the author had worked on the Permian glacial deposits and succeeding greenhouse sequence in India. The icehouse of Antarctica, Africa, India was rather short-lived, because warming climate produced vast coal forests later.

According to the USGS data, Antarctica stores coal seams of the Permian, Triassic, Cretaceous and Paleocene-Eocene, highlighting the profound effect of climate and continental wandering.

Likely Other Causes for Extreme Climate Swings

Lava eruption also could have been an important trigger for climate change. Volcanic ash or dust, a mixture of sulfur dioxide and water, may shade solar radiation and raise Earth's reflective power (albedo). The net effect will be cooling of the Earth. When volcanoes mainly spew water vapor and carbon dioxide, it will have the opposite effect, that is, warming of Earth. Other causes of climate change are mountain-building phases, like the Himalayas, or entrapment and burial of organic material as peat, coal, bituminous shale, and oil. The carbon-sink of coal forests would have produced an abundance of atmospheric oxygen, leading to cooler conditions.[27] The waxing and waning of faunal life on Earth may also control the variation of carbon dioxide in the atmosphere. Some even wonder about extra-terrestrial factors such as interstellar clouds in space between Earth and Sun, and variation in the Sun's luminosity.

STONY TESTIMONY

Early Paleozoic: Cambrian-Ordovician Periods

Rocks of Glacial Origin in the Sahara Region

One of the Late Ordovician stand-out events is the Andean-Saharan glaciation between 440–460 Ma. Similar glacial rocks occur extensively in far off places—Morocco, Libya, South Africa, Saudi Arabia, and elsewhere. The ocean temperature was 5°C cooler than today and about 60% of marine life was affected by it.[5a] The distinctive glacial pavement striations, grooves and esker ridges are numerous in South Africa, Sahara desert, and the Arabian desert in Saudi Arabia.[5] (Fig.6.4)

Great American Carbonate Bank—Prolific Hydrocarbon Reservoir

One of the best examples of the Cambrian-Ordovician rock sequence is the Great American Carbonate Bank (GACB). The annular, tropical, shallow sea teeming with fauna, developed encircling a large swath of exposed land fringed by a coastal belt. The elongated Cambro-Ordovician shallow marine carbonate and sandy rocks extend >7000 km in the continental USA, from Texas in the south to Ohio, Michigan, New York, and to Ontario, Quebec, and Newfoundland, Canada; it continues even into Europe—Greenland, Scotland, and Svalbard in Norway.[4a]

6-4 Evidence of Permo-Carboniferous glaciation in the Arabian desert (Rub' al-Khali), Sarah Formation, Saudi Arabia. **left** A long series of glacial striations (arrow) pointing westward. **right** Deep groove marks (arrows) made by large glacially entrained boulders in the Paleozoic. *Photo credit: Bernabe Aguado.*

In Ontario, the Cambro-Ordovician limestone, shale, sandstone, conglomerate, etc., of the GACB, border the flanks of the NW-SE trending Frontenac Arch of Grenville rocks. The arch is like a geological bridge between the northerly Precambrian Canadian Shield, and the Adirondack Mountains to the south in upstate New York. It cuts across the eastern extremity of Lake Ontario and the St. Lawrence River where the Precambrian Grenville rocks show up as numerous, picturesque, tiny islands in the Thousand Islands area at the U.S.-Canada border region.

The Cambrian rocks of the GACB rest on the Precambrian Grenville gneiss, quartzite etc., separated by the 'Great Unconformity' in Ontario. Kingston, the original capital of Canada on the western edge of the Frontenac Arch, got its nickname 'The Limestone City' because the Cambro-Ordovician limestones abound in the area. Many historic buildings and monuments in the city are made of local GACB limestone.

Many oilfields in the Cambro-Ordovician Great American Carbonate Bank underlies the prominent, middle Ordovician karstic unconformity we mentioned before. The GACB has produced 7.7 billion barrels oil-equivalent from 28,000 wells in >3600 fields in North America.[4]

Formation of Subsurface Hydrocarbon Accumulations

We will keep mentioning oil or gas in underground reservoirs in many chapters. In a nutshell, hydrocarbon accumulation depends on several essential elements (Fig. 6.5).

Geoscientists and specialists talk of the petroleum system of five fundamental features. There will be no oil or gas accumulation if one of the five is missing. These are 1) a porous reservoir (sandstone, limestone, etc.) where oil can migrate into, and stay; 2) a trap or a structure that houses the reservoir; 3) a seal (a shale, or salt) that envelopes the reservoir, sealing the oil from leaking out; 4) an organic carbon-rich source rock (i.e., a black shale) that will produce oil after thermal maturation of carbonaceous detritus in the subsurface under heat and pressure; and finally, 5) a migration route (a porous carrier bed, and/or fractures/faults) that will facilitate the generated hydrocarbon in the source rock to migrate into the reservoir in a trap, protected by a seal above. There are two types of traps, and both can be effective—structural and stratigraphic (Fig. 6.5 - top and bottom).

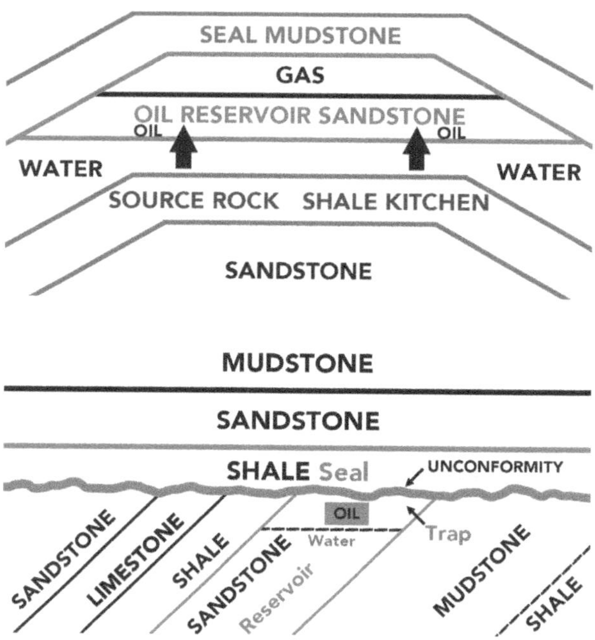

6-5 An illustration of petroleum system—trap (anticline or unconformity seal), reservoir (in this case, a sandstone), migration route (heavy arrows), seal (mudstone or shale), and source rock (shale). The structural trap is in reality a smooth anticlinal structure, whereas the stratigraphic trap may occur below an unconformity, or an impermeable rock. Water, oil, and gas separate in the reservoir because of density difference.

An Allochthonous Assemblage, Newfoundland, Canada

An out-of-place sequence of the Cow Head Group of the ancient Iapetus Ocean origin, lies superbly exposed in several areas in Newfoundland, Canada. The outcrops at the Green Point locality and nearby areas, show a Cambrian-Ordovician sequence of deep-water (base of continental slope) limestone, shale, and conglomerate. The area is part of the Gros Morne National Park by the Atlantic Ocean. A diagnostic conodont fossil demarcates the Cambrian-Ordovician boundary at Green Point.[34] Conodont is an extinct, venomous microfossil (0.2–5 mm), an eel-like, jawless vertebrate with sharp teeth-like elements. In 2000, the International Union of Geological Sciences declared the Green Point locality as a global Cambrian-Ordovician boundary stratotype (type section).

Middle Paleozoic—Silurian—Devonian Periods

The Rocks of the Majestic Niagara Falls, Canada, and USA

The Middle Paleozoic carbonate rocks occur extensively in many places in the world. The rocks of the world-famous Niagara waterfalls, gorge, and escarpment, expose Silurian limestone and clastic rocks (Fig. 6.6). Also, the pinnacle reef mounds of the Canadian Arctic, Michigan basin, USA, and Sweden, among others, are noteworthy. Silurian reef carbonate rocks occur extensively in the Indian Himalayas.

6-6 The Niagara Falls on the Niagara escarpment, ON, Canada, and New York, U.S. The bare Silurian scarp separates the American Falls (left) from the Canadian Horseshoe Falls.

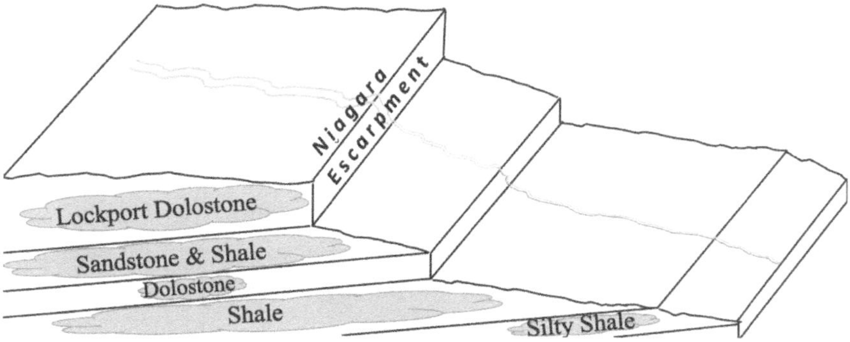

6-7 top The sketch shows how the top hard Lockport (and Amabel) dolostone formed the escarpment, giving rise to many waterfalls along the ridge. **bottom** The prominent Niagara escarpment near Elora, Ontario, continues as a ridge in southern Ontario, extending into the southerly upstate New York, USA.

The Niagara escarpment formed because of uneven erosion of softer rocks of a Paleozoic extensive carbonate platform of shallow marine origin. At the time, oceans were pulsating with new life after the stressful Precambrian Eons. The escarpment (>1000-km-long), formed over a long time, especially in the Late Quaternary, is a prominent geomorphic feature including the Niagara Falls.[35] It straddles the Lower and Middle Paleozoic as it includes limestone, shale, and sandstone of the upper Ordovician to Silurian age (Fig. 6.7). The Silurian dolostones of Lockport and Amabel formations at the top, form a protective hard cover on the underlying Rochester shale and upper Ordovician Queenston shale. The prominent ridge of the top dolostone has given rise to many waterfalls of great tourist interest along the Niagara escarpment. The Niagara River at the U.S./Canada border cascades down from the resistant Lockport dolostone to the plunge pool level of the soft Cabot Head Shale and Queenston Shale, creating the famous waterfalls. The 58-km-long, north flowing Niagara River connects Lake Erie with Lake Ontario downstream.

Fig. 6.8 illustrates how the melt water of the last ice sheet about 12

6-8 left The Grand River gorge in Elora, Ontario, traversing the resistant Silurian Amabel dolostone strata. **right** The right bank shows karstic cave development in the dolostone.

ka ago has carved out the narrow, steep gorge of the Grand River, near Elora, Ontario, in the resistant Silurian Amabel dolostone.

The Catskill Delta Produced Landfill of Orogenic Detritus

The Paleozoic orogenies gave rise to tall mountains that shed sediments forming large deltas at foothills. The Devonian Old Red Sandstone of northwestern Europe and eastern Canada is closely-linked with the uplift and erosion of the Caledonian Mountains. The iron oxide tainted red rocks are of fluvial, eolian dune, and lacustrine origin.

The Acadian orogeny produced the large, 3,000-m-thick Catskill delta sequence of sandstone, conglomerate, and mudrock. Widespread outcrops occur in the eastern United States mainly in New York State. Westerly flowing rivers from the rising mountains produced the Catskill delta in the marshy terrain between the ancient Kaskaskia Sea to the west and the Acadian highlands to the east. Western Pennsylvania's petroleum reservoirs occur in the delta facies of red sandstone, siltstone, and shale.

The deltaic Catskill of North America and the Old Red Sandstone of Britain are similar, yet geographically separate. Likewise, the Antler orogeny on the western United States also generated widespread coarse fluvial and deltaic deposits on both its flanks.

Devonian coal deposits occur in Canada, the U.S., and especially, in China. The Devonian petrified forest of silicified tree trunks in Arizona, United States, is a tourist attraction where many agatized tree trunks dot the park area. Earlier, the indigenous tribes routinely used agatized material to build homes. In upstate New York, the Gilboa petrified coal forest with radiating root structure is also impressive. Devonian fossiliferous limestones occur in many places in Canada and elsewhere. In 1858, the first commercial oil discovery of North America took place in the Devonian carbonate rocks of Oil Springs, Ontario. The oil keeps flowing from this area even today.[16a] The dolomitized stromatoporoid (extinct sponge) reef limestone of the Devonian Leduc Fm. is a proven petroleum reservoir in Alberta, Canada.

Upper Paleozoic: Carboniferous-Permian Periods

Carboniferous Period—Limestone, Coal Forests, and Glacial Tillite

The Carboniferous shows two distinct types of rocks in Laurasia, i.e., ancestral North America and Europe. The End-Mississippian tectonic uplift created deltas and river floodplains in the younger Pennsylvanian. The abundant marine Mississippian limestone sequence transitioned into swamps and forests in the Pennsylvanian time. The riverine floodplain and swamps were sites of exuberant growth of trees and vegetation. Eventually these formed thick commercial coal seams, which are now mined in Pennsylvania, West Virginia, and Kentucky and other states in the United States, and the Province of Nova Scotia in eastern Canada. The Sydney coalfield, Nova Scotia, now mostly submarine, is the largest in eastern Canada (Fig. 6.9).

6-9 Coastal outcrop of the Point Aconi bituminous coal seam of Pennsylvanian age in the Sydney coalfield, Nova Scotia, the largest in eastern Canada. *photo: Rygel, M. C., CC BY-SA 3.0.*

The prolific Upper Carboniferous Coal Measures of the U.K. are famous. Coal-forming terrain in the northern hemisphere was so vast and long-lasting that it has left behind a reserve of 312 billion tons of bituminous and anthracite variety of coal in the Pennsylvania State, USA, alone. The coal reserve was the motor of industrial revolution in the United States and Europe in the 19th century and later. Carboniferous Gondwana coal deposits occur in Brazil and Argentina.

When Gondwana Ruled the South Pole—Glaciers Covered the Land

Things evolved a bit differently in Antarctica, Australia, Africa, South America, and India, because of their geographic location near the South Pole. Glacial moraine with large boulders, cobbles, and pebbles, is a signature facies of the Gondwana. Another characteristic rock type is 'varve'—a millimeter scale, finely laminated claystone and siltstone. Varve developed seasonally in periglacial lakes in all Gondwana continents.

Carboniferous glacial tillites, diagnostic striations and groove marks occur in Australia and several Andean countries—Argentina and Brazil.

The Permian-Carboniferous glacial sandstone and conglomerate are known to be important heavy oil reservoirs in 20 fields in Oman.[18] The Unayzah Group, a Permo-Carboniferous glacial-eolian unit, produces superlight oil (up to 52^0 API) in several central Arabia fields (B. Aguado, personal communication, June 6, 2024). In South America, hydrocarbon reservoirs exist in Bolivia, Argentina and potentially in Brazil.[19] Many fields in the Australian Cooper Basin, and a few in the Canning Basin, produce mostly gas.[20]

The Permian Period

El Capitan Reef, and the Permian Basin, USA, and Salt, Coal, and Flood Basalt Elsewhere

The 1,000-m-thick sedimentary sequence of the Permian Basin of southwest United States, hosts many important oil and gas reservoirs.

The rocks include marine carbonate, shallow and deep-water sandstone, and organic-rich shale. The basin is a major source of commercial evaporitic potassium salts.

Freshwater leaching of the late Permian El Capitan Reef limestone led to the Carlsbad Caverns, New Mexico, USA, and the secondary gypsum deposits therein. Gypsum originated from H_2S and H_2S0_4, sourced from pyrite, aided by microbes. The acids dissolved the limestone and precipitated $CaSO_4$.

Gypsum in the White Sands National Park, New Mexico, USA, came from the drying up of a shallow Permian Sea. Tourists enjoy today the rare, extensive, white gypsum sand dunes of eolian origin of the Holocene age.

Indeed, the later part of the Permian was generally dry in the northern hemisphere. The numerous red beds with paleo-soils developed in ancient river flood plains and deltas in Texas, Oklahoma, West Virginia (Fig. 6.2), and elsewhere, including Canada.

Zechstein Salts, Europe—When Permian Oceans Dried Up

The vast Upper Permian Zechstein salt column occurs in Germany, UK, Poland, and the North Sea; it developed in an arid climate that climaxed in the Triassic. The principal lithologies are halite, anhydrite (anhydrous gypsum), dolomite, and shale.

Permian Coal—A Key Energy Source for Former Gondwana Nations

An important natural resource—coal, was the product of a benign, humid paleoclimate in the southerly Gondwana mega-continent. Thick commercial coal deposits are widespread in India, Australia, South Africa, and Argentina. The 'Jhingurda' coal seam in the Singrauli coalfield, India, is 131 m thick. Gondwana coals happen to be the main energy source today for several countries including India and Australia.

Siberian Flood Basalt—The Largest Glowing Lava Show Ever

The End-Permian to Triassic Siberian flood basalt province is the largest continental eruption in Earth history. It covers an area of 7 million km^2, comparable to Australia in size. Siberian traps are deep mantle-plume outpourings, like the flood basalts of LIPS (Large Igneous Provinces). The lava and gases would have likely contributed to the Permian mass extinction.

MIRACLE OF LIFE

Early Paleozoic: Cambrian-Ordovician Periods

The Singular Faunal Exuberance

The 'Cambrian Explosion,' an important evolutionary milestone in Earth history, heralded the sudden appearance of abundant, diverse, and complex organisms with shell—mineralized skeleton. The abundance of atmospheric oxygen and calcium ions in seawater led to the diversity of the Cambrian fauna. Brimming marine arthropod, bivalve, trilobite, graptolite, brachiopod, etc., suddenly thrived all over the oceans. Trilobite and graptolite fossils proved very useful in correlating numerous sequences of the Great American Carbonate Bank in the USA, Canada and elsewhere. The trilobite represents a unique Cambrian evolutionary jump when the multicellular creature first appeared with a well-developed, compound pair of eyes on the head. Its thorax separated both the head, and the tail section (Fig. 6.10). Fig. 6.11 shows abundant Paleozoic brachiopod fossils on a bedding plane of limestone in Canada. Graptolite fossils, typical of the black anoxic shale of the Ordovician-Silurian, resemble crayon drawings on a dark rock surface, hence the name *graptolite*, meaning 'writing on the rocks.' No one knows for sure if these were benthonic or planktonic (floating) fauna. The Cambrian *Anomalocarids* or 'abnormal shrimp' is the earliest mega-predator, up to 2 m in length but lacked teeth and jaws.[6] Another Cambrian first is the toothy flesh-eating 'penis-worms' with tooth-lined throats belonging to the *priapulid* phylum. The name priapulid comes from the Greek fertility god '*Priapus*, supposed son of Aphrodite. The famous Cambrian Burgess Shale in British Colum-

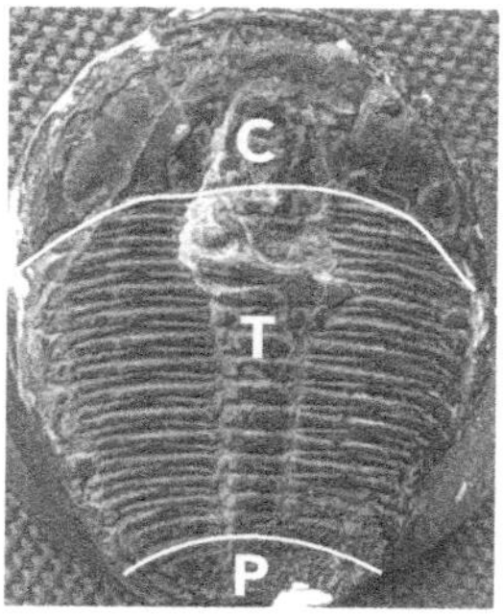

6-10 The 2.5-cm-long trilobite shows three body segments—*cephalon* (C) or head with two large eyes on either side, a ribbed *thorax* (T) in the middle, and a tiny tail or *pygidium* (P).

6-11 A veritable cemetery of bottom dwelling brachiopods in the tropical shallow marine environment of the Paleozoic ocean. Southern Ontario, Canada.

bia, Canada, contains a priapulid fossil '*Ottoia*,' in addition to other new forms like brachiopod, trilobite, etc.

North America remained close to the equator during the Paleozoic when shallow seas with teeming marine life flooded much of Canada, including Ontario. Thick Cambrian to Ordovician, mainly carbonate and shale sequence occurs widely in southern Ontario including the greater Toronto and adjacent areas. Abundant warm water fauna including organic tracks and trails on ancient seafloor testify the great biodiversity and abundance.

Elsewhere in the Georgian Bay region, Ontario, middle Ordovician fossiliferous limestones host abundant shallow marine fauna including some giant specimens. The vertical 3-m-long cephalopod fossil—*Orthoceras*, akin to squids and octopus, occurs here.[9]

The Ordovician saw the earliest colonization of land by primitive terrestrial arthropods and plant microfossils including spores and cuticles. Many more marine animals like brachiopod, coral, bivalve, gastropod, bryozoa, crinoid, etc., thrived in the Ordovician seas.

The 443.8-Ma-old, Ordovician Mass Extinction: First of the Phanerozoic Five

The Late Ordovician glacial episode is unique in that it likely caused a major extinction of marine life, the first one of the Phanerozoic Eon. Advected metal-rich, toxic deep water likely caused the extinction of graptolite. A marked sea level drop of 50–100 m and a lower ocean temperature could have contributed to mass mortality.[5] It is the second largest mass extinction in terms of biota loss, impacting 85% of all marine species.

The drift of the Gondwana continent toward the South Pole during the Ordovician, ushered in the icehouse condition affecting marine biota. The likely secondary causes were volcanic activity, anoxia, etc.[7] The Ordovician mass extinction may also be related to asteroid impacts. A 455-Ma-old, 4-km-wide buried crater in Glassford structure exists in Illinois, USA.[8] Another impact evidence comes from the 470-Ma-old, 7.5-km-across Lockne crater in Sweden.[9a] The impact crater is related to an Ordovician meteor shower because of the break-up of the L. Chondrite parent body in the Main Asteroid Belt.

Middle Paleozoic: the Silurian—Devonian Periods

Marine Animals Crawled Up to Dry Land

The Silurian high sea level covered much of the continents leading to the maximum expansion of marine organisms. The climate was very warm and generally dry. The Silurian can boast of many firsts in the animal world. A notable Silurian first was '*Agnatha*'—the jawless earliest fish,[12] its present-day examples being lampreys. A predator arthropod, famously nicknamed 'scorpion of the sea' also debuted, as documented in the Eramosa Formation, Ontario, Canada.

The first convincing vertebrate fossil occurred in the Lower Silurian. An interesting fact about predators is that vertebrate jaws first appeared in the very late Silurian, 420 Ma ago.[10] Other organisms were terrestrial arthropods including insects and early arachnids (spi-

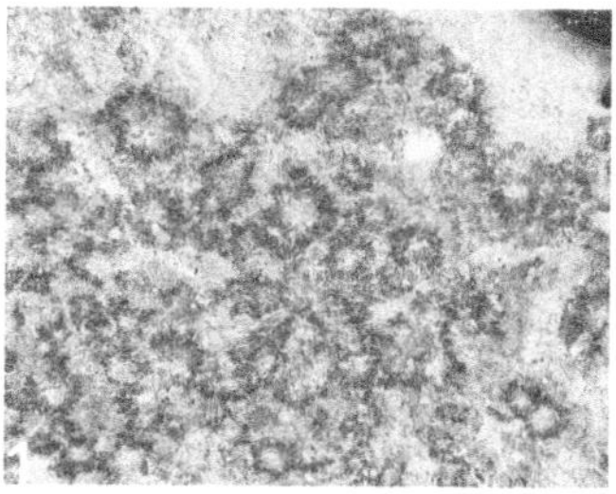

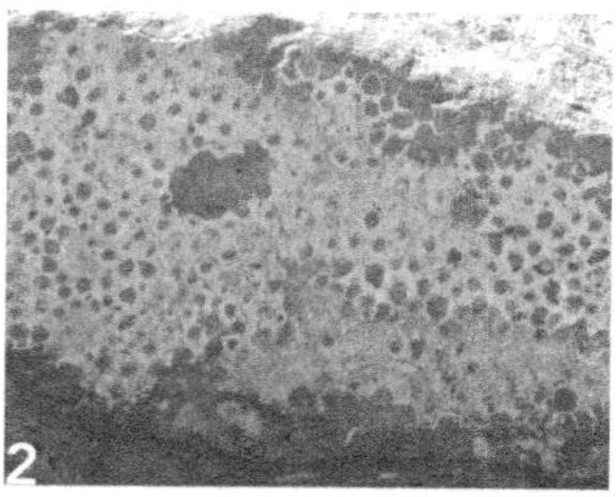

6-12 Middle Paleozoic corals **1** Silurian rugose corals of Amabel dolostone, Elora, Ontario. The black stain is likely bitumen impregnation. The surrounding pale mineral may be siderite **2** Tabulate coral colony showing a honeycomb texture. **3** Rugose coral polyps with radial fossil septa highlighted by likely black dead oil. **2,3** Extinct coral species from the Devonian Dundee Fm., Pelee Island on Lake Erie, Ontario.

der, scorpion family). Plants and fungi began to colonize land, and the first air-breathing creatures like centipedes, millipedes were crawling on Earth for the first time.

Devonian biota includes prolific reef builders like stromatoporoid (extinct sea sponges), colonial tabulate and rugose corals (Fig. 6.12), and other reef dwellers like bryozoa, brachiopod, gastropod, cephalopod, pelecypod, crinoid (sea lily) and trilobite. The first tetrapod or land vertebrate developed from the '*tetrapodomorph*' fish in the Early Devonian. The earliest bony fish, and later shark, are two very important paleontological markers. The *Acanthodian* (Greek 'acanthus'—'spines') was the first fish to develop jaws.[13] They evolved and greatly diversified in the Middle Paleozoic in China, the Old Red Sandstone, U.K., and the Man of the Hill locality (MOTH) in northwestern Canada.

The Devonian was indeed the 'Age of the fish,' a really good time for pristine seafood lovers. The fearsome, 6-m-long *Dunkleosteus* was a fish with thick armor. It had two pairs of sharp bony plates in its mouth serving as teeth with supposedly great bite force. It was omnivorous—diet included fish, shark and even had a cannibalistic streak—devouring specimen of its own kind.

The Period is famous for extensive reef buildups like the Canning basin reefs, W. Australia. It is an exhumed barrier reef today, a good analog of the more famous modern one Down Under. They aptly call it the 'Devonian Great Barrier Reef' because of its size.[14] The 350 -km-long reef, along the north margin of the Canning basin, is full of reef building coral, alga, and stromatoporoid. Other fauna are trilobite, ammonite, and characteristic placoderm fish.

The 380-Ma-old, Gogo Formation, W. Australia, is another Devonian fossil reef site.[15] It is part of a 1,400-km-long barrier reef with exceptional preservation of entombed fossils. The extinct tabular and rugose corals were the prime reef builders. Fossils of 45 types of fish are preserved in perfect, uncrushed pristine condition. The fossils mostly occur in calcareous nodules within dark anoxic shale. Acid etching of headshield of placoderm fish revealed fine intricate details of brain structures, including 'nerve pathways and olfactory system.' Paleontologists studied the first recorded Devonian shark fossil '*Doliodus problematicus*' discovered in 2003.[16] A new study confirmed that sharks likely evolved from the *Acanthodian.*

Devonian coral bioherms are present in much of western Ontario, and Alberta, Canada, and New York, USA, and elsewhere, showing well-developed buildups.

Devonian Mass Extinction(s)—
Second Oldest of the Phanerozoic Five

It is debatable whether the Devonian extinction was one solitary event, or it happened more than once between the Givetian and Famennian Stages. Some argue the whole 375–360-Ma interval is one mega-extinction event, while others disagree. The Kellwasser event is 375 Ma old, while the 360-Ma-old Hangenberg event occurred at the Devonian-Carboniferous boundary.

What Caused Mass Extinction? Many Suspects

Unfortunately, there are no smoking guns to rely on, as experts cite many probable causes. The coincidence of more than one natural event would have probably intensified the toxic physicochemical milieu leading to mass mortality. The main cited causes are algal bloom, CO_2 deficiency in the air, meteorite impact, and magma eruption.

One idea is that Devonian vascular land plant, i.e., seed-bearing tree and fern, is a major turning point in Earth history. The plant roots used to penetrate the ground, breaking apart rocks below and creating soil for the first time. The plants utilized the leached nutrients and minerals for nourishment. But some of the nutrients made their way into oceans via rivers, triggering algal bloom. When algae died, lowly bacteria would have feasted on the remains, using up ambient oxygen. The resulting environment might have asphyxiated the bottom-feeding marine life. The exquisite preservation of fossils implies that the animals died quickly and perhaps there were no scavengers to feast on the dead fauna in the oxygen-poor environment.[17] Many species survived because the ocean's upper layer had oxygen from air.

Carbon dioxide deficiency may be another possible cause. The newly arrived plants and trees had picked up plenty of CO_2 from the atmosphere for photosynthesis, drastically depleting the atmospheric CO_2 content; as a result, the environment became too cold for organic survival.

An apparent positive correlation exists between meteorite impact and mass extinction in two distant places. The Frasnian Kellwasser extinction event in Germany is synchronous with the impact of the 376.8-Ma-old, 52-km-across, Siljang meteorite in Sweden. The Australian 120-60-km diameter Woodleigh 2A crater is a 359-Ma-old impact structure. Its end-Devonian age coincides with the 360-Ma-old Hangenberg extinction event.

Some authors proposed a probable connection between Devonian mass extinction and magma eruption. The 360–370-Ma-old, Viluy traps of eastern Siberia, was practically simultaneous with the Late Devonian Kellwasser extinction.

In summary, Devonian mass extinction was a long-drawn affair lasting 360–375 Ma, where several natural events could have played out simultaneously to cause the disaster.

Upper Paleozoic: The Carboniferous-Permian Periods

THE CARBONIFEROUS PERIOD

Abundant Creatures in the Mississippian (Lower Carboniferous) Seas

The Mississippian warm shallow seas hosted many creatures with shelly skeleton. The abundance of crinoids was conspicuous, along with brachiopod, bryozoa and foraminifer. Some animals like trilobite, hitherto abundant in older periods, became scarce in the Carboniferous on the way to extinction later. Many modern-looking fish groups appeared, while the Devonian armored fish *Placoderm* became extinct. Sharks also thrived in the upper Paleozoic.

Carboniferous Reptiles Ventured on to Land

A major evolutionary breakthrough of the Carboniferous was when reptiles became the first land vertebrates, living independently off water. Another biological marvel, amniote eggs, was a Carboniferous first. The eggs donned a protective outer shell, common in birds, reptiles, and rare mammals. The innovation of nature gave animals freedom to lay eggs on land, where the outer shell kept the ovule soft, secure, and moist inside the egg, even in dry climates.

Amphibians normally laid their eggs in water. Reptiles evolved from advanced amphibians like the tetrapod *labyrinthodont*. The early Carboniferous crown tetrapod (nearest common ancestor) probably evolved from the *coelacanth* 'fish' (Latimeria—its 'living fossil'). All animals attained great size and height because of abundant nutrients and copious oxygen supply. *Meganeura*, an ancestor of the modern dragonfly, had a wingspan of 60–75 cm. Arthropleura, a millipede, was 1.5 m long.[22]

The rocks of the southerly Gondwana continent revealed a major End-Carboniferous/Early Permian global climate change event. It coincided with demise of many invertebrates.

Happy Time for Plants, and Later, Coal Forests

Many new plants struck root, huge ferns and tall trees resembling modern ones began thriving. The freshwater clam made its debut. The flourishing, dense, tropical to subtropical forests produced bark-bearing trees with woody tissues for the first time. Over time, the dead plants and trees were preserved in stagnant, reducing environment, soon covered by sediments, in oxygen-depleted swamps and bogs. The decaying vegetal matter quickly turned into peat and then into immature coal—lignite; later bituminous coal formed from peat/lignite under optimum pressure-temperature conditions. It takes nearly two thousand years to produce 1 m of peat from vegetal matter, and it takes 10 m of peat to produce 1 m of coal. It takes 20 ka to form 1 m of coal. One of the by-products of lush forest growth was an atmosphere with abundant oxygen from plant photosynthesis reactions. This, in turn, allowed the animal kingdom to thrive and expand in size.

Mainly vascular plants, like large lycopods, and giant ferns flourished and formed coal forests. Lycopods, ancestors of club mosses soon became extinct. *Lepidodendron*, a typical Carboniferous lycopod flora, was abundant. It was the first scale tree (with bark) with a trunk more than 1-meter-across, and up to 30-m-high.[21] The coal forests extended from the Appalachians in the United States to Poland and up to the Ural Mountains in the ancient Laurasia continent.

THE PERMIAN PERIOD

Fauna—Marine Life Thrived Until It Perished

Abundant marine biota—brachiopod (productid type), mollusk, echinoderm like crinoid or sea lily, etc., were dominant shallow water fauna. The ammonoid, reaching its acme in the Mesozoic, made its mark in the Permian. Modern nautilus is a distant relative of the same family. The 'Fusilinid' foraminifer '*Parafusulina bosei*,' is a diagnostic Middle Permian marker fossil. Insects and worms were very common in the upper Paleozoic. The Dunkard basin, USA, shows abundant fossilized calcite tubes of tiny *Microconchid* worm of lagoonal origin.[24]

Permian Tetrapod Explosion

Many species of tetrapod burst onto the scene in the ideal hot, arid climate. The drier expansive land surface was just the right setting for the amniote tetrapod to fan out. *Eryops*, a giant amphibian, *Cynodont*, a mammal-like reptile, and the carnivorous synapsid—*Dimetrodon*, with large sail-like spines on its back, all reigned in the Permian. Earlier, a farmer in PEI State, Atlantic Canada, found the intact upper jaw of a 4-m-long *Dimetrodon* fossil (*Bathygnathus borealis*) with curved teeth.[25]

In summary, the Paleozoic Era saw the blooming of diverse complex life after the lean Proterozoic. The amazing fauna started in the Cambrian seas and finally ended up broadening their habitat to yet unexplored vast dry lands during the upper Paleozoic. Indeed, the fauna showed a great evolutionary stride from the Cambrian to the Permian.

Flora—Naked Seeds, An Evolutionary Leap Forward

The *Gymnosperm* group (Greek—'naked seeds') is a vascular plant first appeared in the Devonian. It had a reproductive edge as a seed plant and greatly flourished in the dry climate.[23] The cycads (palm trees), gingko, yew, and conifer (pines) are gymnosperms. The seeds are not enclosed in an ovary or capsule. The conifers were particularly abundant then. The author remembers the ubiquitous *Glossopteris* (a gymnosperm) fossil leaf impressions in shaly rocks in India; it is the posterchild of Permian flora all over the Gondwana mega-continent. *Glossopteris* was replaced bv *Dicroidium* in the dry, early Triassic climate.

Another distinctive Upper Carboniferous-Lower Permian floral feature is the phylloid algal build-ups, found in many places in the United States and Europe (Fig. 6.13). These are skeletal-microbial reef mounds built by leaf-like phylloid algal mass in shallow marine environments. Sediment trapping and carbonate cement helped the algae in the forming of impressive reef mounds.

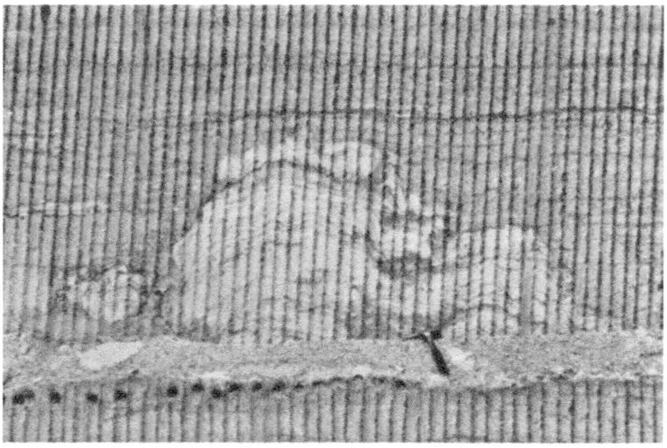

6-13 A phylloid algal mound about 2-m-high, exposed on an old quarry wall in upstate New York, USA. The hammer provides the scale.

The 'Great Dying': The End-Permian, Worst-Ever Mass Extinction

Earth had experienced five major mass extinction events in its 4.57-Ga history, the most noteworthy being the 'Great Dying' that straddles the Permian-Triassic (P/T) boundary. Such was its severity that 96% of all marine species and 70% of terrestrial vertebrates perished. About 57% of all biological families went extinct at the End-Permian. All the remaining trilobite, some corals, two groups of echinoderms—in the ocean, two-third of amphibian families, and all large herbivores disappeared.[28] The crinoid or sea lily and horn coral were also on their way to extinction at the End-Permian.

Probable Cause(s) of Extinction

Geologists have long debated about the earth-shattering cause or causes of mass extinction. A curious thing to note—all Phanerozoic mass extinctions occurred at the boundary between two successive periods. As we can see, our Permian-Triassic (P-T) mass extinction is no exception. There is no tell-tale evidence to pinpoint the cause or causes that had obliterated the greater part of Permian life. The commonly cited reasons for mass extinction are meteorite or asteroid impact, large scale volcanic eruption, drastic sea-level drop, global warming from magmatism-induced climate change, CO_2 increase in the atmosphere/oceans causing high ambient temperature, oxygen depletion in the environment, etc. It may also be tied to external factors such as the Sun's ultraviolet rays.[27, 29]

During the End-Permian, a 10-km-across asteroid created the 200-km-wide 'Bedout' crater, in the northwestern continental margin of Australia about 250 Ma ago. The event coincided with the End-Permian extinction.[31]

Large scale volcanic eruption may also be a plausible cause. Flood basalt eruption in Siberia might have contributed to the marked global warming. It likely left marine animals unable to breathe as warmer waters could not hold enough breathable oxygen. The trap eruption caused ocean temperatures to rise by 11°C, causing oxygen depletion. The oceans lost 80% of their oxygen content, and half of the seafloor area became anoxic, i.e., oxygen-free.[33] The ocean surface temperature rose to ~30°C, compared to modern day 14°C. Toxic gases of the Siberian lava could have damaged the ozone layer.[32] Earth's atmosphere was CO_2-rich and abnormally hot. To note, the End-Permian to Early Triassic interval is largely devoid of coal in many places.

The marked sea level fall on Earth at the Permian-Triassic boundary may be another possible cause. Both the coming together of the Pangea supercontinent, and a lack of seafloor spreading would have caused the sea level to drop. There was a 5-Ma spell after the Permian mass extinction, known as the 'Dead Zone.' Very poor biotic diversity and scarce recovery of life characterized the period.[30]

The main causes of End-Permian worst mass extinction would have been global warming, oxygen depletion in warm oceans, and toxic gases due to volcanic eruption.

CHAPTER HIGHLIGHTS

6 Paleozoic Periods in ascending order

Cambrian-Ordovician: Explosion of Hard-Shelled Life

- Highest sea levels, increased oxygen, and extensive marine carbonate platforms like the oil-rich Great American Carbonate Bank.
- The Sahara Desert, Morocco, South Africa, showing evidence of Ordovician glaciation.
- 1st mass extinction in Earth history.

Silurian-Devonian: Acadian and Caledonian Orogeny

- The mobile Gondwana and Laurentia mega-continents active.
- Disappearance of the Atlantic precursor—The Iapetus Ocean.
- Rocks of the Niagara escarpment, North America, its formation later due to erosion of softer Silurian rocks.
- Extensive fossil reefs preserved in Arctic Canada, USA, Sweden, and the Himalayas.
- Fauna: sharks, jawless and jawed fish, reef coral, sponge. Flora: plant, and coal forest.
- First tetrapod transitioning from the sea to land.
- The 2nd mass extinction.

Carboniferous-Permian: Orogenesis shaping the Appalachians

- Gondwana-Laurasia collision creating the Supercontinent Pangea.
- The largest lava eruption ever in Siberia.
- Glaciers on Gondwana near the South Pole, first icehouse-greenhouse cycle, and glaciation evidence in the Arabian Desert.
- Abundant marine fauna—brachiopod, echinoderm, and more.
- Marine reptiles move to dry land. Tetrapod explosion, amniote eggs notable.
- Lush flora—giant ferns, lycopods, first naked seeds, extensive coal forests.
- The 'Great Dying'– the third and worst mass extinction.

Aerial view of Pamir Mountains of Permian, Triassic and Jurassic rocks, in Tajikistan, uplifted during the Eurasia-India collision in the Eocene.

WHAT'S NEXT?

Chapter 7 starts with the Triassic Period, announcing the dawn of the dinosaur era. Initially, it was a rather trying time for life on Earth. However, the later Triassic, and especially the Jurassic Period, made ample amends with climatic normalcy when big reptiles were in their elements, and oceans were brimming with life.

7

Mesozoic Era—Triassic and Jurassic Periods: The Dinosaur Cometh

THE ERA 252 Ma—66 Ma:
TRIASSIC, JURASSIC AND CRETACEOUS PERIODS

The chapter narrates the story of the lower Mesozoic–Triassic and Jurassic Periods. Chapter 8 will discuss the Cretaceous Period (145 Ma to 66 Ma).

PALEOGEOGRAPHIC HIGHLIGHTS

The Triassic Period

Hot and Arid Pangaea: A Challenging Era for Flora and Fauna

Several natural events conspired to set the stage for the Triassic Period (252–201 Ma), the lowest rung of the Mesozoic Era. It ushers in a new tectonic scenario of the Pangea, the sole supercontinent holding together all the landmasses of Earth (Fig. 8.1). Mantle convection and tectonic plate movements were the key to Pangea's existence for about 100 Ma as a single entity in the Permian-Triassic. An animal could have trudged on solid ground from pole to pole, yet Pangea did not last forever. The same forces that made it were again at play unmaking it. This time, the Pangea started showing early signs of disintegration into two mega-continents—Laurasia and Gondwana during the Late Triassic.

Rift Basins of North America

On the west coast of North America, subduction-related volcanic island arcs collided and accreted to the stable continent. Several fault-bounded rift basins formed in eastern America, extending from the Carolina's, United States, up to Canada along the coast, and then continuing all the way to western Europe. The Triassic rift basins formed when the North American continent started separating from the African plate[32] (Fig. 7.1). In Canada, the Fundy basin basalt lava flows, and fluvial red bed rocks crop out continuously along the Bay of Fundy shore and elsewhere.

The 1,000-km by 350-km Sverdrup basin in Arctic Canada is a major rift sag, filled with 8 km of sediments.[2]

Pangea was largely intact in the Triassic. The Mid-Atlantic Ridge developed fully about 180 Ma ago in the Early Jurassic. New lava from a mantle source along a deep fissure in the Mid-Atlantic Ridge

7-1 Trend of exposed Triassic rifts along the North American eastern seaboard. The dashed line shows the general trend of the rift basins in the United States and Canada.[32]

started welling up and kept pushing the plates apart. Eurasia and Africa drifted eastward while the Americas sailed westward.

Triassic Climate—Toasty and Dry Initially, After the End-Permian Sea Level Fall

The Permian lowest sea level coincided with the worst mass extinction at the End-Permian. The basal Triassic was not a happy time for the surviving animals that scraped through the crunch of the worst extinction. The climate was hot and arid, a semidesert-like condition prevailed in many parts of Pangea. The mean temperature was 25°C and higher, compared to today's 14°C. The atmospheric CO_2 content was much higher, up to 10 times that of today, causing the high ambient temperature. The polar regions were moist and humid, with no snow in sight. Conditions became normal again in the upper Triassic when the 'Carnian Pluvial Episode,' made life surge anew amid a humid and moist congenial environment.[30] Soon the sea level began rising again.

The Jurassic Period

Pangea Goes Asunder—Dinosaurs in Their Elements

The Pangea supercontinent was in the process of breaking up by the Early to Middle Jurassic (201–165 Ma), widening the oceans around Laurentia (N. America), and the southerly Gondwanaland. The Proto-Atlantic rift basins continued forming in the Early Jurassic.[1] The Mid Atlantic Ridge was gradually taking shape, and North America and Africa/Europe were steadily separating (Fig. 8.1). The disassembly took many millions of years to complete. For example, crustal separation began off the coast of the United States in the mid-Jurassic and extended north toward the Labrador Sea in the Early Cretaceous. Greenland could extricate itself from both Canada and Europe only in the Early Cenozoic.[2]

Sea Level Rise and New Seaways Appear

The Jurassic climate, except for the earliest part, was mostly benign—warm, wet, allowing profuse vegetation to reestablish on land. The sea level rise caused marine waters to inundate coastal areas more frequently. The sea even reached areas now occupied by the Rocky Mountains along a shallow seaway from the Pacific Ocean. At the south end, the Gulf of Mexico opened and was connected all the way to the Atlantic and the Tethys Ocean beyond. Marine shelves surrounded Southeast Asia, China, and ancestral European regions.[14] Likewise, a shallow sea was lapping the northeastern flank of the southerly Gondwanaland. The ancestral South America and Africa remained welded together, but an arm of the Tethys Sea began to separate India and Africa. Plate tectonics was active again.

STONY TESTIMONY

The Triassic Period

The Pangea Story in Rocks

German geologists, first used the name Triassic, meaning a three-fold division. The lithological formations are Buntsandstein (Bunter) at base, Muschelkalk in the middle and the Keuper above. The Bunter is essentially continental red beds and conglomerates, the Muschelkalk is marine limestone, common in the alpine countries, and the Keuper is a mixture of claystone, dolomite, evaporite and sandstone. Except for the Muschelkalk marine limestone, the rocks in Germany are like the Triassic red beds and salts in the rest of Europe and North America. Evaporite precipitation was copious during the early phase of the Atlantic opening. The Permian-Triassic Zechstein salt basin extends from the U.K. to Poland, and halite is mined extensively in England, France, Germany, Hungary, Austria, and elsewhere in Europe.[4]

The Newark rift basin of thick red beds and salt deposits of eastern USA extends northward into the Fundy basin first, and then into the Jean-de-Arc basin, Canada, the site of the oil-rich offshore Hibernia oilfield. The rift basins continue farther north into the New Red Sandstone of the U.K., and the 'Buntsandstein' red beds, Germany. Similar Triassic rift-related red beds are present in China, Siberia, and Alaska, and in several Gondwana continents—Antarctica, Australia, South Africa, South America, and India.[3] Earlier, the author had mapped Triassic red beds in India.

In the Fundy basin, Canada, Triassic red beds and volcanic lavas outcrop in Parrsboro and surrounding areas, Nova Scotia (Fig. 7.2).

7-2 Triassic rocks in the Fundy basin, Nova Scotia, Canada. **1** The Five Islands Provincial Park in the background, about 20 kms from Parrsboro, has lava flows of Mid-Atlantic rift origin within Triassic sandstones. It was here that one of the earliest dinosaur footprints was discovered. The Pinnacle Island occurs to the far right. **2** At low tide, steep-dipping sandstone, siltstone, etc., lie exposed by the side of the Parrsboro tidal inlet. **3** Dipping siltstone and shale beds discordantly underlie the shaly strata. Note the angular contact (dashed line) between the thin-bedded rock below (white arrows) and the overlying strata.

The Jurassic Period

European and Gulf of Mexico Salts, and Other Rocks

The name 'Jurassic' comes from the Jura Mountains, part of the alpine belt from Nice, France, to Vienna, Austria, with the Swiss Alps making up the central part. The local geology consists of three shallow marine limestone and marl formations—Lias, Dogger, and Malm of the Early, Middle and Late Jurassic age respectively. The area boasts of a lot of underground river-like channels, streams, and outlets in the karstic limestone terrain. What is karst? It is a partly dissolved/corroded, pock-marked limestone with large cavities, even caverns. The creation of karst happened much later when acidic freshwater dissolved the exposed marine limestones.

The Early Jurassic break-up of the Pangea and the incipient rift basins caused marine salt precipitation. The nascent basins were shallow, leading to abundant evaporation of seawater and accumulation of huge volumes of salt on the ocean floor. Subsequent sediments—sand, mud, etc., shielded the salt from dissolution. The Jurassic Louann salt, Gulf of Mexico is 1,524 m thick.[15]

The Early Jurassic Global Rifting Phase and Signature Red Beds

A noteworthy rock type is the lower Jurassic continental red beds widely distributed in the Americas and elsewhere in the world (Fig.

7-3 Andean red shale and white sandstone beds of fluvial origin—Jurassic La Quinta Fm., Mérida State, Venezuela

7-4 **1** Polygonal sun cracks in the ~200-Ma-old red beds, La Quinta Fm., Táchira, Venezuela. **2** Laminated eolian Navajo sandstone, Colorado, USA (width 10 cm). **3** Dinosaur footprints (12–15-cm-long) in the 150–160-Ma-old Morrison Fm., Colorado, USA.

7.3). The red beds of Mongolia, India, USA, Venezuela, etc., yielded dinosaur remains among other fossils. These are nonmarine rocks where leached Fe-rich minerals produce the reddish iron hydroxide tint. The mud rocks may show signs of ancient sun cracks on the paleo-flood plain surfaces (Fig. 7.4). Dinosaur bones and footprints are common in the lower Jurassic of North America, as in Parrsboro, Nova Scotia, Canada.

The Jurassic red beds, sometimes cut by igneous dykes, developed in rift grabens. In North America, the rifts extended along the east-

7-5 The two outcrops occur on either side of the Cabo (Cape) Espichel where the mythologically-rich sanctuary of the Ermida Da Memória chapel is located. The outcrops are dissimilar—the left photo is more steep shale, marl and less sandy rocks, while the dipping Cretaceous Lagosteiro bay outcrop (right) consists of cross-stratified sandstone, shale, and marl. Geologists noticed a 20 Ma age difference in the two adjacent outcrops.

ern seaboard from Newfoundland, Canada, to North Carolina, USA. The red formations of Latin America are widespread across many countries—Mexico, Guatemala, Venezuela, Colombia, Peru, Chile, and Argentina etc.[16]

The Late Jurassic to Lower Cretaceous sequence by the Atlantic Ocean near Setubal, Portugal, is significant in that it yielded dinosaur memorabilia and footprints (Fig. 7.5). The host rocks from Jurassic Tithonian (left outcrop) to Early Cretaceous Hauterivian are of ancient lagoonal-estuarine-coastal origin. The strata show visible signs of karstification. We will discuss the historic dinosaur details in the 'Miracle of Life' section.

Dark Source Rocks—A Deep Underground 'Kitchen' Cooked Plenty of Hydrocarbons

Climate and sea level change may control the nature of sediments. After the arid, and drier lowermost Jurassic, gradual sea level rise and mild humid climate produced shallow water limestones with a cache of spectacular fossils. Highly organic-rich dark marine limestone and shale developed in the oxygen-poor deeper oceans.

Petroleum geoscientists prize anoxic *source rocks* because the organic matter therein produces oil and gas in the 'oil kitchen,' under optimum pressure-temperature conditions. The uppermost Jurassic (Tithonian) organic-rich, deep-water, marly limestone in the Gulf of Mexico, is a very coveted source rock. It proved to be the wellspring of numerous Jurassic/Cretaceous oil reservoirs. Similarly, the Jurassic Hanifa Formation is the source rock of the biggest oil field in the world: Ghawar, in Saudi Arabia (Pers. Comm. June 3, 2024, Bernabe Aguado).

The Kimmeridgian oolite limestones host many famous Jurassic oil reservoirs in the Gulf of Mexico, the Persian Gulf countries, and elsewhere. The grainy limestone reservoirs consist of ooids of shallow marine origin, where the pores between ooids store the precious oil.

MIRACLE OF LIFE

Triassic Period

Early Dinosaurs Take Timid Steps—The Archosaur Arrives

The famous Mesozoic dinosaurs likely descended from the Triassic *archosaur* (tetrapod ancestor) and attained great development between the Late Triassic to Early Jurassic. They were reptiles, air breathing vertebrates and had an edge over amphibians. The reptilian amniote eggshells were impermeable to external liquid, while letting the embryo breathe oxygen from the air outside.

The *Archosaur* was important as its different strands gave rise to the dinosaur, bird, flying reptile (pterodactyl), marine crocodile and alligator in the Mesozoic.[5] The dinosaur debuted ~235 Ma ago, coinciding with the Carnian pluvial episode. A total of 15 Triassic dinosaurs are known including Chindesaurus, Eoraptor and Coloradisaurus.[31]

True mammals appeared in the Triassic; they were small in size. The mammal-ancestor *therapsid* reptile, and *archosaur* were the main vertebrate animals. Birds and crocodilians are the only two living archosaurs today.

The word dinosaur comes from Greek—*dino* (terrible) + *sauros* (lizard). Dinosaurs are broadly divided into two categories—*Ornithischian* and *Saurischian*.[5, 6] *Ornithischians* are bird-hipped, plant-eating, bipedal, running dinosaurs. Examples are the Cretaceous, duck-billed hadrosaur *Edmontosaurus*, *Iguanodon*, horned *Triceratops*, and armored *Ankylosaurs* among many others. They did not fly like the *Pterosaur*, which is not a dinosaur but a flying reptile.[7]

The *Saurischian*, a lizard-hipped dinosaur class, includes the *Sauropod* and *Theropod*. The *Sauropods* are the largest herbivores, with lizard-like five toes, a very long neck, a tail, a small head, and pillar-like legs. The Theropods were four-toed, bipedal, master carnivores; they were normally smaller, but more ferocious—the Cretaceous *Tyrannosaurus*, *Giganotosaurus*, and *Velociraptor*, and the Jurassic *Al-*

losaurus. Both meat-eating Saurischians and vegetarian Ornithischians made their debut in the Triassic, flourishing greatly in the Jurassic—Cretaceous.

Dinosaur footprints exist in many places in the world, including the Triassic red beds of the Newark basin, United States, and the Fundy basin, Nova Scotia, Canada. The Fundy basin footprints are tiny, probably because the Triassic dinosaurs were still evolving. At the end of the Triassic, a restricted number of classic large dinosaurs were roaming around everywhere.

The Tiki Formation of India yielded a cache of Late Triassic *cynodont*—a mammal-like *therapsid* reptile, and other primitive dinosaur fossils.[8] The important Indian fossils include *Rhynchosaurs*, and saurischian dinosaur bones in the Tiki beds. The *Rhynchosaurs* are Triassic herbivorous reptiles related to the *archosaur*, and the likely ancestor of crocodiles and birds. The Indian fossil horizons correlate with other Gondwana strata in Brazil, Argentina, Madagascar, and coeval strata in the USA. The author had worked on the red mottled shale and paleo-soils of the Tiki Formation (400-m-thick) that had developed on ancient flood plains, in a warm, semi-arid paleoclimate with seasonal rainfall.

Marine Triassic Fauna

Many species went extinct at the End-Permian, yet the lucky survivors lost no time and thrived again in the warmer Triassic Sea. The giant reptiles, *Ichthyosaur* and *Plesiosaur*, were feeling at home preying on the plentiful fish population. Initially corals were fewer in the Triassic, but later came back strong. Echinoderm, ammonite, mollusk, conodont, etc., also successfully reestablished themselves.

The 'Dolomiti' Mountains of Italy contain Triassic marine fossils including reptiles. Leonardo Da Vinci used the famous Carrara Marble for his sculptures.

End-Triassic Mass Extinction—History's Fourth Repetition

The extinction event, lasting about 10 Ma during the uppermost Tri-

7-6 A 6-cm-long, ammonite section shows coil and individual chambers, lined with white calcite. Note the embedded snail fossil top right.

assic (Rhaetian), took a severe toll as 70% of all land and marine species disappeared. Many aquatic animals—brachiopod, gastropod, pelecypod, conodont (relative of modern hagfish with 4 hearts), marine reptiles, etc., did not survive. Vertebrates like marine *ichthyosaurs*, *plesiosaurs*, and some dinosaurs survived this extinction. Ammonite (Fig. 7.6) thrived, though many *ceratids* perished.

In the floral kingdom, club moss and horsetail died, but conifers and cycads thrived. The *Gymnosperm* survived better at higher latitudes. The thick Permian forests were gone in the Triassic, leaving scarce coal seams in the lower and middle Triassic for posterity. Mass extinction of many land vertebrate species was ecological good news for the dinosaurs who had a field day. They dominated the scene for 165 Ma, freely roaming around the riverine, coastal plains of the succeeding Jurassic-Cretaceous.

Cause(s) of Mass Extinction

One may ask what caused the extinction event at the Triassic-Jurassic boundary. No one knows for sure though a reasonable explanation exists.[9] The flood basalt lavas from the 200-Ma-old Central Atlantic Magmatic Province (CAMP) could have been the main cause of extinction. The lava erupted due to tectonic rifting between North

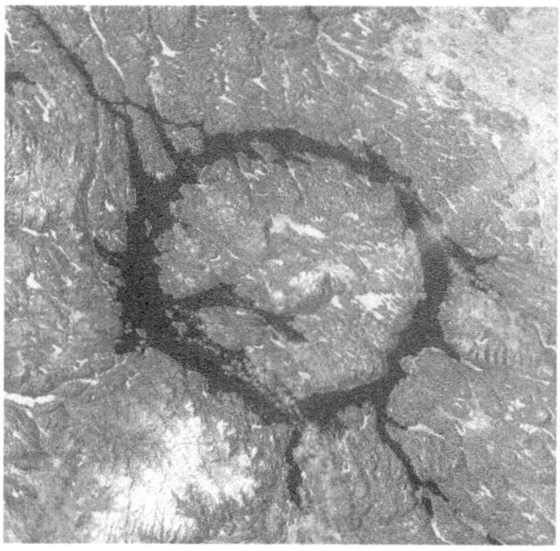

7-7 Satellite image—The 100-km-across, Manicouagan Crater and Reservoir, Quebec, Canada.

America and Africa. Extensive mercury concentration in End-Triassic rocks from around the world confirms a link between volcanism and mass extinction. The CAMP magmatism triggered sea level rise and released enormous amounts of atmospheric carbon dioxide.

In 2013, the U-Pb based zircon geochronology data from Morocco and eastern North America, proved a close synchronism between the Upper Triassic mass extinction and the CAMP volcanism that lasted 0.6 Ma. Palynological samples at the base of CAMP lavas in Morocco, yielded a Late Triassic age, further narrowing the onset of volcanism.[10, 11] The release of methane from permafrost is another probable cause, arising from CO_2-related global warming.[12] Some suggested that escape of atmospheric oxygen due to magnetic pole reversal could have been the cause of mass extinction.

The Early Jurassic rifting and drifting of the plates through inhospitable, desert-like climatic zone also could have been a factor in the extinction at the Triassic-Jurassic boundary.

The Navajo sandstone of eolian origin, Colorado Plateau, USA, is a colorful geologic testimony of the aridity.

Some speculated that the Manicouagan meteorite impact in Quebec, Canada, might have been the culprit (Fig. 7.7). A 5-km-diameter meteorite created the huge impact crater. It is the second largest in the Mesozoic, and the most visible crater on Earth, with a large ring lake inside. Radiometric dating of impact ejecta gave an age of 214 Ma.[13] The 13 Ma time difference between the extinction and impact age, makes Manicouagan an unlikely cause of extinction. However, the impact killed many Triassic Norian age marine species—sponge, radiolarian, conodont (eel-like jaw-less vertebrate) and ammonoid.

The tridactyl (3-toed) footprints of *Anchisauripus* (1–3-m-long), an upper Triassic-basal Jurassic dinosaur from the Bay of Fundy, Canada, proves it survived the extinction event. We recall dinosaur footprints also occur in the Five Islands Provincial Park near Parrsboro, Canada, and in Connecticut, and Massachusetts, United States. Some think that the End-Triassic may not be an extinction per se, rather a turnover of species.

The Jurassic Period

Romping Dinosaurs Chomping Around

While talking about the dwellers of the real Jurassic Period, one cannot but recall the famous 1993 science fiction movie 'Jurassic Park,' about the adventures of a billionaire and his closest people on an island in a wildlife theme park. A miraculously cloned/rejuvenated bunch of Mesozoic dinosaurs, produced from amber-sourced DNA, were freely roaming around in the wild. A good story but the timeframe does not match the fleshy giants. Neither the *Tyrannosaurus rex* nor the *Triceratops* appeared on the scene till the Late Cretaceous, many million years after the Jurassic.

The *Sauropod* genus examples of *Diplodocus*, *Brontosaurus*, *Apatosaurus*, *Titanosaurus*, etc., all appeared in the early Jurassic. The avian incarnation of *theropod* dinosaurs are birds arriving later in the Jurassic. One of the most notable places for dinosaur fossils and footprints of the Jurassic *Stegosaurus* and *Brontosaurus*, is the Tithonian Morrison Formation, Colorado, USA (Fig. 7.4). In India, the 18-m-long *Barapasauraus tagorei* was a Jurassic herbivorous dinosaur discovered in the Telangana State.[29]

The primitive *archosaur* reptiles came to dominate a wide variety of habitats during the Triassic and Jurassic.[5] Both the plant-and meat-eating dinosaurs from the archosaur lineages reigned in the Jurassic. We recall dinosaurs are grouped into herbivorous ornithischian, (example: *stegosaur*), and saurischian (both vegetarian *sauropod*, and carnivorous *theropod* like *T. rex*, and the birds). Oddly, birds are lizard-hipped, unlike the bird-hipped *Ornithischians*.

Another *Archosaur* lineage was the progenitor of the modern incarnation of dinosaurs—the avian class. The Jurassic *Archaeopteryx* is the first true primitive bird with mixed features—reptilian teeth, claws, yet with avian features like feathers.

A 200-Ma-old Dinosaur Family Near the Paleo-Equator, Venezuelan Andes

Across the Atlantic to the new world, paleontologists discovered in 2014, a new genus of *ornithischian* dinosaur—*Laquintassaura venezuelae* in the La Quinta Formation, Venezuela.[17] It is a dog-sized (1-m-long, 25-cm-high), ground-dwelling, bipedal herbivore, a relative of the famous *Stegosaurus* and *Triceratops*. The jumbled bones of four juvenile and adult dinosaurs make researchers think that the animals used to live in herds, a social trait unknown among early dinosaurs. The first-ever discovery of dinosaurs in northern South America shattered the myth that dinosaurs could not exist in the arid early Jurassic environment of equatorial South America. Radiometric dating of tiny crystals indicated that dinosaurs were living there only 0.5 Ma after the End-Triassic mass extinction event, proving that dinosaurs bounced back soon after the catastrophe. The discovery of such early species may be crucial in understanding not only the *Ornithischians*, but also the general dinosaur evolution.

Amphibians like frogs, toads, salamanders, as well as short-legged caecilians (large worms/small snake-like creatures) thrived in the Jurassic.[18] The Triassic mammals also continued to evolve in the earliest Jurassic.

Insects were abundant in the Jurassic—the '*Kalligrammatid*' looked like a butterfly and used to pollinate non-flowering plants.[19] The plant kingdom prospered in the congenial humid Jurassic climate. Conifers were dominant, while other gymnosperms—palm tree-like cycad and gingko, also thrived.

7-8 The historic sanctuary Ermida da Memória (Memory chapel) by the blue Atlantic Ocean is wrapped in mystery since the 15th century. Decorative tiles inside narrate the 1410 folklore of the apparition of Our Lady of the Cape, with the later construction of the chapel in the 18th century.

The Cabo da Espichel by the Atlantic—Where Myth and Geology Went Separate Ways

We mentioned the discovery of dinosaur footprints in the upper Jurassic and Early Cretaceous strata in the Cape of Espichel area on the southwestern tip of the Setubal Peninsula, Portugal (Fig. 7.5). The Pedra da Mua ('Mule footprints on rock') site of Late Jurassic age occurs on a steep rock face not far from the lighthouse. The younger Lagosteiro Bay footprints (Early Cretaceous), occur near the headland of the right photo of Fig. 7.5.

The serene place is wrapped in enchanting tales of faith and folklore. Myth has it that once two village elders had the same dream, i.e., Saint Mary rode a mule up the high cliff face to the top and then disappeared in the night, leaving behind unmistakable mule hoof marks on the rock surface. Indeed, there are real animal footprints going up the hill for all to see. The vision was so momentous that a chapel, a church and ancillary buildings were built at the site with the support of the then King of Portugal, and hordes of devotees regularly used to converge in the area (Fig. 7.8). Fast forward to 1976 when geoscientist M.T. Antunes, working in the Lagosteiro Bay area, dispelled the long-cherished myth with an inconvenient truth. His verdict—it was a vegetarian sauropod dinosaur that left behind the footprints. It was perhaps strolling or foraging in an estuarine lagoonal setting during the Late Jurassic time.

7-9 The meter-long skull of an *Ichthyosaur* fossil (Nemocon, Colombia). The dolphin-like, but much larger 'fish-lizard,' lived during the Mesozoic Era. It preyed on ammonite and other fish with numerous sharp teeth.

Further work and research during the following decades revealed about 600 tracks of dinosaurs of all stripes. Both sauropod and theropod dinosaurs had roamed in the area for millions of years from the Late Jurassic to the Early Cretaceous. In the Avelino quarry nearby, the Jurassic dinosaur footprints (30–100-cm-across) are of *Parabrontopodus* ichnogenus (footprints). Amazingly, geoscientists have come across 10 trackways on a bedding plane that showed that some dinosaurs used to move in herds—in this case 3 adults and 7 juveniles moving in the same general direction. The hip height of juveniles and adults at this Pedra de Mua section varied from 1.5–1.8 m, and 2.8–3.2 m respectively.[32, 33] The site also yielded data on hand and foot size, strolling speed, and claw details of the *Parabrontopodus*.

Jurassic Sea Rulers—Ichthyosaurs, Plesiosaurs, and More

Marine reptiles—*ichthyosaur*, *plesiosaur*, *sauropterygian*, *mosasaur* and turtle, came to dominate the Jurassic marine environment. True marine crocodiles existed only in the Jurassic.

Ichthyosaur fossils occur in the famous Jurassic Coast of England, the Colombian Andes (Fig. 7.9) and elsewhere. The recently discovered, largest recorded ichthyosaur from England is estimated to be 25 m in length, based on the size of the giant lower jawbone (more than 2-m).[35]

Marine invertebrates came back strong after the End-Triassic demise of many. Coral, rudist bivalve, sponge, and microbial reef builders widely flourished again.

Glass Sponges Ruled the Deep as Dinosaurs Roamed

Tiny glass sponges punched above their weight during the time. They built extensive reef edifices, especially across the ancient European continent, along the northern margin of the bygone Tethys Sea. It was one of the largest reef structures ever—about 3,000-km-long, winding through Portugal, Spain, France, Germany, up to Romania.[28] The extinct '*hexactinellid*' glass sponges had their skeletons built out of hard silica, unlike corals. A high dissolved silica content in the deep, very cold, dark oceans helped the sponge to erect the impressive glassy reef structure.

Jurassic Wonderland—Solenhofen, Germany's Outdoor Museum

A famous site of Jurassic fossils is the Solenhofen Limestone, Germany. Plant, vertebrate, invertebrate, and protist (eukaryotic, mainly unicellular organism unrelated to animals, plants, or fungi) fossils occur exquisitely entombed here. It is a 'lagerstatte' with exceptional preservation of biota. The area afforded scientists an unparalleled glimpse of the past geologic environment and the upper Jurassic organic evolution.[20]

A series of coral and sponge reef knolls developed in the Jurassic Tithonian Sea, near the present Solenhofen village, Bavaria region. Many coastal lagoons formed, landward of the reefs, disconnected from the open ocean and the surrounding land areas. The Solenhofen rocks are part of this deep lagoonal basin storing astonishing fauna and flora.[21] The host rock is a fine platy dark limestone, earlier quarried as lithographic stone for artwork by sculptors. The restricted ancient lagoon extended discontinuously for about 50 km, and experienced increased salinity from time to time because of evaporation. The lack of currents would have rendered some layers of sediments oxygen-poor. Many marine organisms, vertebrate and invertebrate, had drifted from the open ocean during periodic storm surges, settling into the stagnant, inhospitable lagoon. Later the animals got buried in the oxygen-free mud layers. The flora included cycad and conifer plants from adjacent lands. Geoscientists have discovered more than 700 species of plants and animals in the limestone.

The quiet milieu preserved many foreign and exotic flora and fauna, including delicate soft-bodied dragonfly, beetle, and jellyfish. Oceanic pelagic fauna from the outer reef include crinoid or 'sea lily,' ammonite, pelecypod (bivalve), fish, squid, jellyfish, star fish, crocodile, shrimp, horseshoe crab, etc. The algal coccolith and single-celled protist (foraminifer) from the open marine realm also ended up there. Researchers discovered 180 species of fossil insects here.[22] The stagnant bottom environment was toxic for benthic scavengers. The absence of bottom feeders also meant zero disturbance and churning by animal tracks on the muddy floor.

Archaeopteryx—A Unique Reptile-Bird Link

To date, geoscientists have collected 12 fossils of *Archaeopteryx*, the only genus in the world that shows that birds evolved from reptiles. It was like a modern raven about 0.5-m-long with a 0.5-m-long tail. It had a jaw with sharp teeth, three fingers with claws and distinctive feathers. The earlier discovery of a single fossilized feather here in 1860 gave scientists a clue to its existence. Darwin's 'The Origin of Species' predates this discovery by a year.[23] He felt vindicated upon the momentous discovery of *Archaeopteryx* body fossils subsequently. Darwin duly mentioned the significance of the find in the fourth edition of his book (*Wikipedia*).

Other than *Archaeopteryx*, fossils of another extremely large (wingspan 12-m) flying reptile—*Pterodactyl*, are also preserved in the Solenhofen limestone.

Jurassic Coast Fauna—Mary Anning's Remarkable Saga

The fossil treasure of the Jurassic Coast by the English Channel, a UNESCO World Heritage Site, is worth a special mention. It is a

155-km-long coastal strip, only a few hours' drive from London.[24] Here the Wessex basin holds the world's most complete Jurassic limestone sequence overlying the Triassic red beds. Anecdotes of fossil discovery here involved a country girl—Mary Anning (1799–1847). She grew up in the town of Lyme Regis, and without any formal training, taught herself geology and anatomy. An avid 'fossil hunter'—she was passionate about her hobby from a young age.[25] At age 12, she discovered the famous dolphin-like, 5.2-m-long, *Ichthyosaurus* or 'fish lizard.' Later in 1823, she found a complete skeleton of the marine reptile *Plesiosaurus giganteus* with a long curving neck and paddle-like flippers. Her third discovery was the *Scalidosaurus* ('Charmouth Dinosaur'), a heavily armored herbivore. She also discovered coprolites—fossilized animal droppings.

Legend has it that her story was the inspiration for Terry Sullivan's tongue-twister song in 1908 "she sells seashells…".[26]

The Jurassic site also yielded fossils of an extinct marine crocodile. The coastal cliffs yielded a fossil forest of great significance, where each of the 71 beds supposedly show a distinct Jurassic ammonite species.[27] The area's famous scenic wonders of white Cretaceous chalk cliffs attract hordes of tourists every year. The spectacular sculpted coastline, limestone arches, stacks, and pinnacles—all resulted from ages of erosion by wind, rain, and wave action.

The Triassic Period:

- The Pangea Supercontinent—A reality.
- Low initial sea level, an arid climate, and extensive red beds.
- The 'coal gap,' i.e., lack of coal seams, in the early and middle Triassic.
- The Carnian Pluvial Event ameliorated the dry Triassic landscape.
- Revival of flora and fauna after the End-Permian 'Great Dying.'
- Baby steps of dinosaurs, flying reptiles and marine reptiles.
- The 4th mass extinction at the Triassic-Jurassic transition.

The Jurassic Period:

- The Pangea Supercontinent starting to break up.
- The separation of Laurentia (N. America) and the southern Gondwana mega-continent.
- Formation of >1-km-thick Louann Salt strata in the Gulf of Mexico.
- Early aridity-related red beds, later a congenial climate favoring diverse fauna and flora.
- Fortuitous proximity of famous source rock and oil reservoir—the Jurassic Kimmeridgian carbonate reservoir, and Tithonian organic-rich shale in Mexico and the U.S.
- Diverse and abundant dinosaurs, marine reptiles, flying reptile *Pterodactyl*, and *Archeopteryx*—the first reptile-bird transition.
- Famous fossil sites—Solenhofen, Germany, the Jurassic Coast of England, and the Morrison Fm., Colorado, USA.

WHAT'S NEXT?

The Cretaceous Period of the Mesozoic Era represents the acme of dinosaur development in a congenial climate. However, their luck finally ran out. Some incredible Cretaceous tales await us in Chapter 8.

8

Mesozoic Era—Cretaceous Period: What Killed the Dinosaur?

MESOZOIC ERA 252 Ma—66 Ma

The Cretaceous Period 145–66 Ma—Geological Turning Point

The Cretaceous is the longest Phanerozoic interval. The Latin word 'Creta' means chalk, abundant in the white sea cliffs of Dover, England, and Normandy, France. Chalk consists of calcareous body plates of *coccolithophores*—tiny single-celled algae.

The Cretaceous marks an inflection point in Earth history at the transition of the Mesozoic-Cenozoic Era when the Supercontinent Pangea was no more. Many geologically important events transpired—the warm climate (av. 20–25°C), near all-time high sea level, marine flooding of a large swath of coastal marine areas, and snow-free Poles. The period saw the peak development of the dinosaur, ammonite and rudist bivalve. Finally, the twin earth-shattering events—the Chixulub meteorite impact, and the Deccan lava eruption changed the world deeply.

PALEOGEOGRAPHIC HIGHLIGHTS

The Unmaking of the Pangea—The Divorcing Continents

One major tectonic change in the Early Cretaceous is the clear dismemberment of the Pangea Supercontinent (Fig. 8.1). Nature never stands still; the same plate movements that made Pangea a reality, were active again in dismantling it in the Jurassic/Cretaceous. The two Pangea mega-continents, Laurasia and Gondwana, were clearly apart. The South Atlantic continued rifting in the Early Cretaceous, as did its northerly extension toward Labrador, Canada.

Indeed, the Cretaceous marked the crescendo of the Pangea demise. It set the stage for what modern continents would look like in the Quaternary. Two divorcing landmasses, South America and Africa, began to yield space for the South Atlantic. The Gondwana break-up coincided with the Jurassic Karoo-Ferrer Igneous Province (2 million km^3) magmatism that continued in the Cretaceous. The Americas and Eurasia/Africa became separate entities. Greenland started drifting away via Baffin Bay and Davis Strait, but not until later did it become detached from North America. The India plate extricated itself about 120 Ma back from the huge cluster of Africa, Madagascar, Antarctica, South America, and Australia. It began its long northward odyssey towards Asia, sailing on the Tethys for its tryst with its yet unknown Eurasian counterpart. Madagascar became free from Africa. Antarctica and Australia started separating but did not disunite totally till the Late Cretaceous. After the total breakup of Pangea, other new oceans appeared. The Atlantic Ocean began connecting with the Panthalassa and the Tethys Seas. The Panthalassa Sea soon morphed into the now familiar Pacific Ocean. A part of the Tethys Sea became the Indian Ocean. The last of the Tethys disappeared in the Eocene after the tectonic plates of India and Arabia, the latter a broken block of the African plate, each collided headlong against Eurasia about 40–50 Ma ago. The present-day Caspian, Black, Aral, and Mediterranean Seas remind us about the past glory of the Tethys.

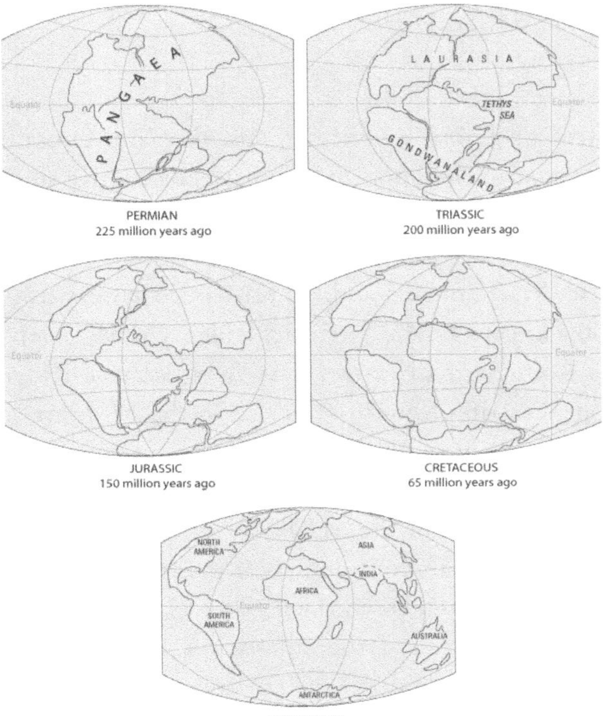

8-1 Phanerozoic plate movement from Permian to Present Day. The Pangea supercontinent broke up in the Mesozoic to transition into the modern continents and oceans.

The Laramide Orogeny— Rise of the Rockies and the Continental Divide

The Laramide Orogeny, another major tectonic event, affected the western part of North America and northern Mexico during the Late Cretaceous and Paleocene (80–55 Ma). The low angle subduction of the Pacific Kula and Farallon plates underneath western North America gave rise to the majestic Rocky Mountains. It runs for 3,000 km from British Colombia, Canada, down along the western United States and terminates in the southwestern State of New Mexico. Its maximum width is nearly 500 km. Laramide-related deep faults produced several large inter-montane basins within the Rockies—Uinta basin in Utah, and Powder River, Wind River and Big Horn basins, Wyoming. The Sierra Madre Oriental of Mexico may be the distal extension of the Rocky Mountains, across the Rio Grande River.

Mid-Oceanic Ridges and Mantle Hot Spots—Secrets of Unending Lava Flows

Other notable features of global significance are the Mid-Oceanic Ridge complex and submarine hotspots, which have been active since the Cretaceous. We broached the subject earlier in Chapter 4. The total length of the mid-oceanic ridges spread over all oceans is about 80,000 km. In 1872, C. W. Thomson and his HMS Challenger team discovered the 16,000-km-long Mid-Atlantic Ridge (MAR), separating the Americas from Africa/Europe. The Mid-Atlantic-Ridge extends from the southern tip of South America to the Mid-Arctic Ridge at extreme north, via Iceland. The 7,758-m-deep, Romanche Trench (an E-W transform, i.e., a strike-slip fault) has offset the northern and southern segments of the MAR. The nascent East African Rift System on the African continent will eventually mature and morph into a Mid-Oceanic-Ridge in the geological future.

The rift valley along the crest of the suboceanic Mid Atlantic Ridge is 80–120 km wide and is 1–3-km deep. The two plates on either side are moving away laterally at the rate of 2–5 cm/year.[1] The subsea mountain system occurs at a depth of 2,600 m below sea level while the ridge crest rises about 2 km above the deepest ocean floor. At least 9 big islands sit atop the Mid Atlantic Ridge, the prominent ones being Iceland and Azores to the north, and Tristan de Cunha to the south. The Icelandic volcanoes spew lava regularly, literally splitting the island apart.

A geological nugget: The island of Corvo falls on the North American tectonic plate while the rest of the Azores islands ride the European plate.

Can Mantle Convection Explain Seafloor Spreading?

Why is the Mid-Atlantic Ridge pushing apart two tectonic plates? The proponents of the idea argued that the ridge owes its origin to a mantle plume that heats up because of the radioactive decay of elements deep inside. The magma then rises by deep-seated convection cell to the top of the mantle. Hot mantle magma finds its way up through weak zones at plate junction and spills on the ocean floor. The oldest lava relict on the Atlantic Ocean floor is lower Jurassic in age, about 200 Ma old. It occurs now next to the continental margins because spilled fresh lava is continuously pushing away old lava from the ridge crest outward due to plate motion. The convection mechanism alone, though reasonable, may not satisfactorily explain the geometry of plate boundaries. The 'Ridge Push' and 'Slab-Pull' mechanisms we recall from Chapter 4, may better explain plate motions.[2]

HOT SPOTS

Submarine Hot Spots—
The Birth Story of Hawaiian and Other Island Chains

Some 50 to 100 hot spot volcanoes occur globally, totally unrelated to tectonic plate movements. Rather, they occur in the plate interior,[2] away from the mid-ocean ridges or subduction zones.

Some notable modern hot spot volcanoes are the Islands of Hawaii, Samoa, Galapagos, the Yellowstone Caldera, and the Ethiopia-Afar hot spots. Starting in the latest Cretaceous, the Hawaiian hot spot created a long string of volcanoes and sea mounts along the Emperor Seamounts—Hawaiian Ridge chain. It is 6,000 km long from the Aleutian Trench off Alaska to the Big Island of Hawaii.[3]

Prof. Tuzo Wilson first noticed the phenomenon in 1963 and interpreted the distribution as due to a stationary heat source or 'hot spot' deep inside Earth. Following Wilson, geophysicist W. Morgan proposed that the heat source was fueled by a hot mantle plume deep inside at the core-mantle boundary.[4] We will revisit the Hawaiian chain in the Neogene section.

Indian Ocean Hot Spots—
Basalt Lava Flooded the Rafting Indian Plate

Two spectacular Cretaceous hot spots lie under the Kerguelen and Reunion Islands in the southern Indian Ocean. The hotspots left their indelible mark on the storied stratigraphy of Mesozoic Indian plate when it was sailing along on its long journey northward for its eventual encounter with Eurasia.

As the Indian landmass broke away from Antarctica, Africa, and Australia, etc., during the Gondwana breakup, its northward route first passed over the Kerguelen hotspot. The mantle plume-related flood basalt eruption produced the early Cretaceous (Albian) lavas known as the Sylhet and Raajmahal traps in India.[5]

The story of India's lava shower did not stop in the Albian, rather it was an appetizer. The Deccan flood-basalt of Maastrichtian-Danian age (~66 Ma ago) was the last geologic event of the final Gondwana disassembly. The >2,000-m-thick Deccan trap lava owes its origin to another mantle plume hot spot now deep below the Reunion Island in the Indian Ocean, about 500 km east of Madagascar. The Indian plate was floating like an island-raft on the ocean above the hot spot, when it had the spectacular lava bath, thousands of miles away from its present Eurasian location. The details of the Deccan trap follow in the subsection 'Stony Testimony.'

Ninety East Ridge—
Milestones of the Indian Plate's Northward Odyssey

India's northerly journey had likely left its permanent trail on the Indian Ocean floor as the Ninety East Ridge. It is important to note that the India-Australia lithospheric plate is much larger than the Indian tectonic plate itself. The Ninety East Ridge (NER) of basalt lava separates the Indian Ocean into two west-east halves. The NER is a straight, almost north-south trending mid oceanic ridge running along the 90E Meridian from 34^0 S to 17^0 N. It is 200 km wide and 5,000 km long, from the southerly Broken Ridge near Australia to the northerly Bay of Bengal.[6]

Radiometric dating of samples along the ridge suggest a middle Eocene age at the southern tip, to an upper Cretaceous age at the north end. The temporal trend may mean that a Hawaii-like mantle hot spot beneath the Indo-Australian lithospheric plate created the ridge. However, some studies debate this conclusion.[7]

Cretaceous Passive Margins— The Vanished Western Interior Seaway, USA

Passive margins lack tectonic activity, and they form in Atlantic-like Ocean margins on flanks of continents. Multi-km-thick sedimentary strata develop along such basin margins. The Early Cretaceous saw the formation of wide marine platforms around all the splintered Pangea landmasses. The Western Interior Seaway—a vast inland sea, appeared where the Rocky Mountains stand now. It was more than 3,000 km long, and nearly 1,000 km wide. It disappeared about 70 Ma ago with the tectonic uplift of the Rockies. The west coast of North America was another story. Thrusting and accretion of island arc systems added new land areas to the continental landmass.

The Rising Andes, South America's Backbone, and the Nazca Plate Subduction

In South America, the early stage of ancestral Andes Mountain chain began to take shape in the Cretaceous. It involved deformation and uplift of older sedimentary and metamorphic rocks. The subduction of the Pacific Nazca Plate under South America started about 140 Ma ago. The high Central Andes started forming in the Oligocene and volcanism started near the End-Oligocene. We will read many fascinating details of the Andes in Chapter 10.

The Fateful Asteroid Impact at the K/Pg Boundary— Where is the Smoking Gun?

The End-Cretaceous (K/Pg) mass extinction of giant dinosaurs and other fauna intrigued many geoscientists over the years, who offered multiple probable causes of the disaster. Finally, Alvarez et al. (1980), a father-son duo proposed that an asteroid impact at the End-Cretaceous (K/Pg) juncture was the culprit.[66] However, geologists needed a smoking gun—a physical impact crater somewhere. Luckily, two Mexican oil company geologists (A. Camargo and G. Penfield) were carrying out exploration work then in the Mexican Yucatan Peninsula by the Caribbean. Earlier, through seismic imaging they identified a crater-like structure below the peninsula. Initially they thought it was just a volcanic cone with rocks in it, not an impact crater. But soon thereafter, they announced in 1981 that the crater was indeed of impact origin.[21a] Later in 1991, the team found more evidence of the impact structure, and named it after the nearby, tiny coastal fishing village 'Chixulub Puerto' (Fig. 8.2). The 180-km-across crater is sitting on a metamorphic basement 3–6-km-deep. Presently it lies buried partly under land and partly under the Caribbean Sea to the north. Recent studies show the original crater was entirely on land and that since then, it got partly submerged with sea level rise. The crater lies buried now under 1,000 m of Cenozoic sediments, mostly reef-forming coral limestone.[22] In 2016, a scientific project team drilled a borehole into the peak ring of the impact crater below the ocean floor. The peak ring is a circular rim around the center of a crater, formed by ejected breccia and melt rock on impact.[21] The borehole core evidence is convincing: high iridium anomaly, melt rock, shocked quartz, glass, and breccia of ejecta origin—all tell-tale signs of the Chixulub impact.

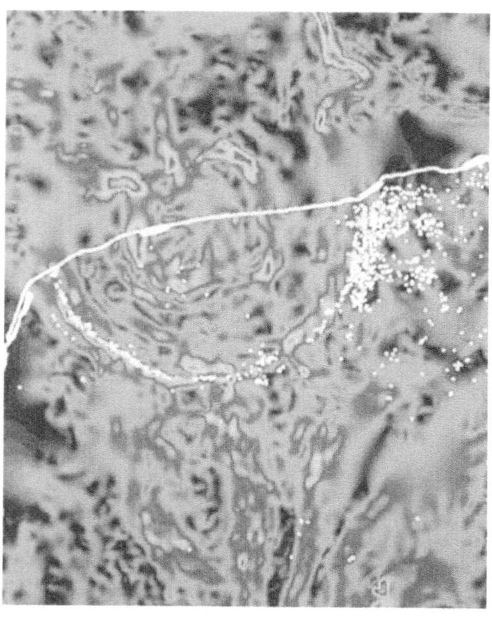

8-2 Chixulub crater outline from horizontal gradient map of Bouguer gravity anomaly. White line—modern Yucatan peninsula coastline, 25 km north of Mérida City, Mexico. White dots—Cenotes (sinkholes).

The Chixulub Crater, Mexico—
The Epicenter of the K/Pg Global Cataclysm?

The sheer effect of the 11-km-diameter Chixulub asteroid is revealing. It punched a crater 20-km-deep and had a mass of $1.0x10^{15}$ to $4.6x10^{17}$ kg. One marvels about the colossal power of the impact event. A stony/metallic body of that size upon hitting Earth at a speed of 20 km/sec, would throw up an enormous volume of dust and debris into the atmosphere. It would have blocked sunlight for a long period and plunged Earth in total darkness in a numbing cold spell. The heat of the impact fused rocks and soil into glassy matter. The impact had sent millions of tons of ejecta in the air. Drill core samples (1,300-m-deep) from the crater provided evidence of a mega-tsunami several hundred meters high flooding the crater soon after the impact.[23]

STONY TESTIMONY

Chalk, Reef, and Basalt

The Cretaceous global sea level was higher than at any other time in the geologic past, barring the Ordovician. The sedimentary strata faithfully registered the extraordinary sea level rise. For example, the Late Cretaceous Cenomanian transgression was one of the most extensive events in western Canada, where the shoreline migrated 150–200 km landward during each sea-level cycle.[8] Later during the Turonian, sea level reached a record height of about 220 m above the present sea level, flooding greater part of the Canadian Shield and elsewhere. An estimated one third of Earth's land area was under water at the peak inundation. One cannot imagine the consequences today if the sea level rises another 200 m soon. At the time, forests used to adorn the polar neighborhood, with no ice caps in sight anywhere. Dinosaurs roamed in areas only 15^0 off the South Pole, and in northerly Alaska.

The abundant, fossiliferous limestone—chalks, marl, shallow water grainy rock including oolite, is the hallmark of the Cretaceous;

deep-water anoxic shale, and abundant basaltic magma are also distinctive rocks. Silicic sandy rocks, shale, and conglomerate were products of shallow marine, river, delta, coastal, and deep environments.

Towering Chalk Cliffs Along the English Channel Coastline

The chalk beds developed worldwide because of the high sea level and warm climate. Thick chalk units occur in the North Sea, Greenland, China, United States, Canada, etc. The Jurassic coast of southern England, and the Normandy coast of France, boast of majestic chalk cliffs of great tourist attraction. The Cretaceous saw more chalk deposits than at any other time.

The *Coccolithophore*, a submicroscopic (2-to-75-micron cell diameter), floating marine, calcareous, unicellular alga, produces white chalk—it is porous, permeable, commonly friable, and soft to the touch. The high fluid flow capacity makes water quickly vanish from the surface of a wet specimen. The abundance of *coccolithophores* implies that the Cretaceous Ocean was saturated with calcium carbonate (Fig. 8.3). The algae live in warm, well-lit, clear surface waters. However, the dead ones accumulate on the deep shelf floor or beyond in the open ocean environment, hundreds of meters below sea level. The widespread chalk distribution proves that the ocean environment was largely uniform in many areas. Phosphatic chalks also occur.

Some chalks host highly productive oil and gas reservoirs, especially in the North Sea. The reservoirs show high reservoir quality (porosity

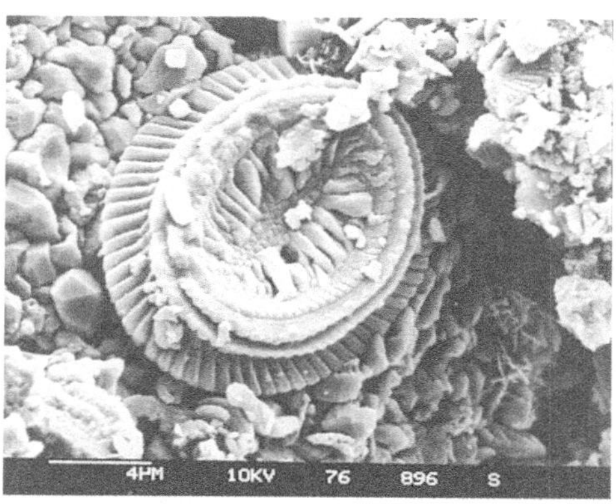

8-3 Scanning Electron Microscope image of a calcareous coccolithophore from the Upper Cretaceous Bornholm limestone, Denmark. The white bar is 4 micron long (1,000 micron = 1 mm).

of 40%), a reserve of 2.5 billion barrels of oil and 8 trillion ft^3 of gas.[9] The Austin Chalk, Texas, also produces hydrocarbons from chalks.

Rise of the Reef-Builder Rudists—Impressive, Yet Only for a While

We describe the rudists here because they built great reef edifices worldwide on Cretaceous carbonate platforms, though in places like Venezuela, rudist growth was more muted (Fig 8.5). The reefs form reservoirs of some of the largest oilfields in the world—the Middle East, Mexico, southern USA, etc. Several countries in Europe, the Mediterranean, and east African regions also show rudist reefs.[10] The massive Yucatan platform, Mexico, is a rudist-coral bioherm/reef complex. The main development took place in the upper Cretaceous, though they started appearing in the Late Jurassic. The warm, tropical-subtropical climate of the Cretaceous was a positive contributing factor (Fig. 8.4).

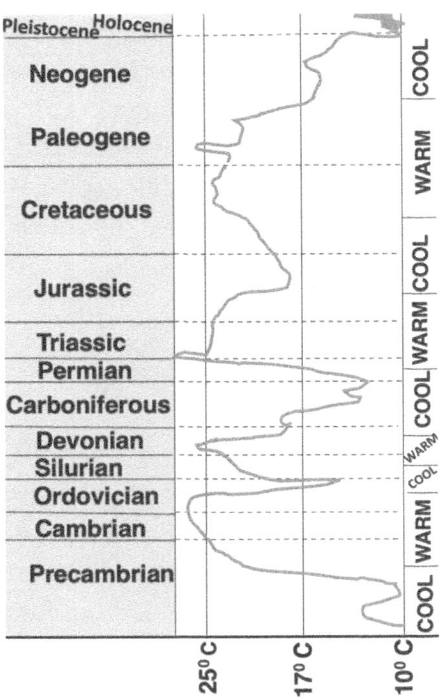

8-4 Global average geologic temperature variation during the last ~0.7 Ga. Note the highest temperature at the Permian-Triassic boundary.

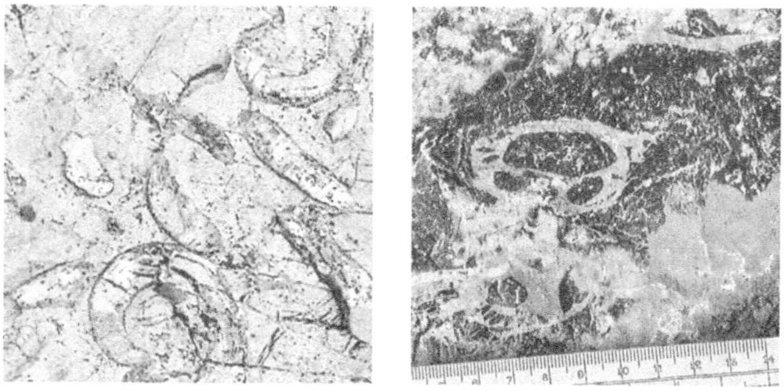

8-5. left Outcrop images of small juvenile rudists embedded in lime mud of backreef facies. **right** A lateral view of one valve seen in 3 rudists. Venezuelan Andes, Albian Peñas Altas Fm., Trujillo state.

Rudist-bearing limestone practically encircles the whole north and western margin of the Gulf of Mexico. Just the elliptic Actopan carbonate platform, Mexico, is 400 km long. It hosts about 2000 m of marine shelf limestone, bordered by fringing rudist reefs. Another example is the Stuart City Reef trend along the Cretaceous paleo-shelf margin in Texas.[11] The northeast-southwest belt, running from Texas to Louisiana, is maximum 10 km wide but extends for more than 1,000 km along the coast. Stuart City trend shows a 500-m-thick rudist-coralline reef facies. Rudist buildups in the Middle East host bountiful Cretaceous oil reservoirs in Iraq, Saudi Arabia, Qatar, the Trucial Coast of U.A.E., the Persian Gulf and Iran.[11]

Golden Lane, Mexico—
A Supersize Cretaceous 'Atoll' Hosts A Giant Oil Reservoir

The Cretaceous Faja de Oro (Golden Lane) Atoll, preserved partly on land and partly under sea, is an example of the prowess of rudist reef builders. The Mexican Golden Lane oilfields would always hold a pre-eminent place in the history of petroleum exploration.[12] It is an oval reservoir—a giant, rimmed carbonate 'paleo-atoll,' 150-km-long and 70-km-wide.[15] An abundance of rudists in the El Abra reef build-up gave rise to the concept of rudist reef early on.

The discovery of Golden Lane reservoir occurred in 1908. However, geologists thought of the 'atoll' concept only in 1963, when they struck oil in the Isla de Lobos-1 borehole on the marine half of the atoll. It produced 260,000 barrels of oil daily from the Cerro Azul—4 well, likely the world's most prolific oil well.[13] Such an atoll is hard to come by—modern or past.

The Golden Lane paleo-atoll had a complex geologic history, it occasionally emerged above water in the post-Cretaceous time due to tectonic movements. Undersaturated freshwater caused abundant solution in the limestone, leading to profuse karstic fractures, vugs, and cavern-like open spaces. The porous karstic limestone subsequently became buried, and later hosted a giant oil reservoir. From 1908 to 2001, the Golden Lane reservoir produced an estimated 3.9 billion barrels of oil equivalent.[14]

The paleo-atoll height was likely 1,000 m above the Cretaceous seafloor, and it now occurs half on land and the other half in the Gulf of Mexico. Huge amounts of reef debris from the high Golden Lane escarpment also slid down and accumulated as an apron of debris flow (Tamabra Fm.) in the marine-slope (>200-m depth) environment. The Tamabra Fm. also proved to be an important oil producer—2.3 billion barrels as of 1977.[15]

The famous Bikini is a large modern atoll in the Pacific Ocean; its inner lagoon measures 40 kmX24 km (Fig. 8.6). The gigantic Golden Lane paleo-atoll dwarfs the modern Bikini in size.

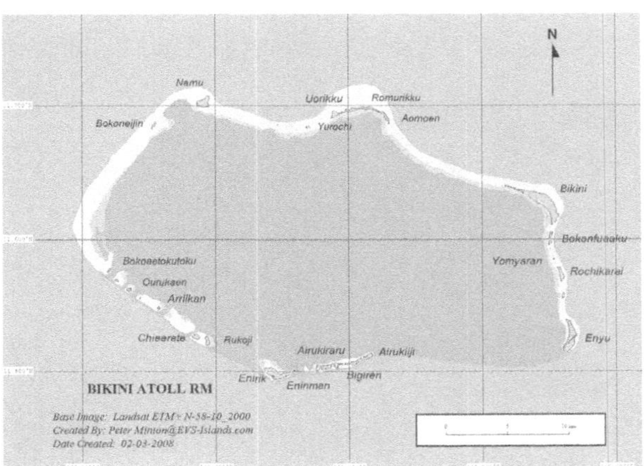

8-6 A view of the inner lagoon of the Bikini Atoll in the Pacific Ocean. Bikini and Enyu are the largest islands. The Cretaceous Golden Lane paleo-atoll is nearly three times larger than the Bikini Atoll.

Black Shale—The Main Ingredient in the Cretaceous Oil Kitchen

The Cretaceous bituminous shale and marly limestone are widespread—Venezuela, Colombia, Mexico, United States, Europe, Greenland, western Australia, east Africa, China, and other places. The Albian-Coniacian strata of deep marine origin are famous source rocks for oil and gas. Some 60% of the world's oil and gas occurrences originated from black shale, which is rich in organic matter.

Significance of Black Shale—
A Proxy for Sluggish Deep Ocean Currents

An abundance of disseminated, dark organic matter imparts black shale, and organic-rich dark limestone, the diagnostic color. The prime prerequisite is an oxygen-depleted, quiet, bottom environment to maintain settled organic matter in a reduced state (Fig. 8.7). How did the anoxic bottom waters develop? The Cretaceous forests thrived even near high latitudes with no snow at the poles. The temperature difference between the poles and the equator was rather minor. A thermal gradient between high latitudes and the equator is what triggers wind pattern and ocean currents. A weaker global wind pattern was unable to drive ocean circulation and mixing of surface and deep waters. The net effect was less upwelling and water turnover. This resulted in a more reducing, stagnant, euxinic ocean bottom environment, preserving the organic matter, free of any oxidative alteration. The Greeks used the word '*euxinia*' for the 'Black Sea' to

8-7 Oil-rich Cretaceous black shale and thin sandstone outcrop, Eastern Cordillera of the Andes, Colombia.

mean 'an inhospitable sea,' full of sulfidic matter including pyrite. The sulfur-rich bottom waters may contain sapropel (organic matter), with up to 20% carbon. A modern analog is the 1,470-m-deep Lake Tanganyika, its lower 1,200 m is biologically 'dead' due to high H_2S and oxygen-depletion.

The La Luna Formation, South America—A World-Class Oil Source Rock

The La Luna Formation, Maracaibo basin, is one of the most prolific petroleum source rocks in the world (Fig. 8.7.a);[16] it holds 53.6 billion barrels of oil equivalents. It is a dark, 100–300-m-thick, black shale-bituminous limestone assemblage of deep-water origin, with abundant planktic micro-foraminifers (Fig. 8.10). Many perfectly preserved ammonites (Figure 8.8) are present as nucleus of nodules (Fig.8.9). Black chert bands, pyrite nodules and bitumen impregnations are common. Coeval black shale and limestone occur extensively in the Querecual Fm., eastern Venezuela, and Trinidad and Tobago. The La Luna strata is capped by the phosphatic limestone

8-7a Several outcrop views of the La Luna Fm. in the Venezuelan Andes. **1** Parallel bedded dark limestone and shale. **2** Finely laminated, deformed folded/faulted rocks. **3** Close up of La Luna strata of black shale, limestone and a black chert band below the scale. **4** Chevron-folded, thinly bedded shale, limestone, and chert.

8-8 Upper Cretaceous ammonite (~10-cm-across) in the Andes, with prominent black suture lines, Chejende Fm., Trujillo State, Venezuela.

of the Tres Esquinas Member, up to 3-m-thick, with pyrite nodules, fish and shark-like selachian reptile remains with sharp teeth, and common bitumen impregnation.[17] The phosphatic limestone with abundant shiny, potato-sized pyrite concretions formed near an oxygen-poor paleo-ocean bottom. The high content of organic matter and ferrous iron characterize it. The newly formed mineral suite is: glauconite, chalcedony, apatite, dolomite, and pyrite replacing microfossils in a reducing environment. The author still remembers even after 40+ years, when he first saw the outcrop of marly limestone

8-9 Examples of nodules in the La Luna Fm., Trujillo, Venezuelan Andes. Some may grow up to a meter across. Often ammonite fossils, big or small, form the nucleus of such nodules.

8-10 Tres Esquinas Member, Mérida Andes, Venezuela. **1** A view of large shining pyrite nodules in a thin (~3 m) bioclastic limestone. Abundant stains are due to bleeding (oxidation) of Fe in pyrite. New cements of silica, carbonate and phosphate formed during prolonged marine still-stand period. The white lines are calcite in crisscrossing fractures. **2** Microscopic view (crossed nicols) of sub-millimetric microcrystalline glauconite, rhombic dolomite, black pyrite crystal, and bitumen. **3** Deep-water limestone of the La Luna Fm., Mérida Andes, with planktic micro-foraminifer *Globigerina* (~0.25 mm), pyrite, and bitumen.

in a remote Andean creek. Hundreds of fist-sized, gleaming pyrite nodules dazzled under a blazing midday Sun (Fig. 8.10, Image 1). Oxygen-depletion from algal bloom would have caused the sudden mass mortality of fish population and marine reptiles buried in these beds (Fig. 8.11). Excessive phytoplankton growth would have poisoned the water. Unusual 'bone beds' of exclusive fish remains attest to the occasional critical toxicity of the Cretaceous La Luna Sea.

Asteroid Impact Ejecta Forms the Largest Mexican Hydrocarbon Reservoir

Sometimes truth is stranger than fiction. The Chixulub asteroid impact created the Cantarell Field—the largest Mexican oil reservoir in the southern Gulf of Mexico. The glowing hot ejecta would have lit massive wildfires in faraway places during the impact.[21] The altered tektite glass, shocked minerals, and lapilli volcanic fragments in the breccia unit are telltale evidence of the impact, producing 10^4 km^3 of ejecta.[18] The outcrop and offshore borehole data confirm the presence of carbonate ejecta breccia over a vast area—200-km-long, 40-km-wide, and 300-m-deep.[24] The unit encircles the ancient Cretaceous platform edge.[19, 20]

8-11 1, 2 Plan view of thin bone beds, each ~40-cm-thick, in the Táchira Andes, Venezuela, illustrating sudden mass mortality of fish population due to periodic toxicity in the La Luna Sea. **3** Fossilized bones (up to 30-cm-long) of Cretaceous reptiles in the La Luna Sea. **4** A mandible with sharp shark-like teeth of selachian reptile. **3, 4** Tres Esquinas Member, Quebrada Sucia creek, Mérida Andes, Venezuela. Coin – 1.5-cm-across.

The chance discovery of the field is wrapped in a local anecdote. In 1972, Rudesindo Cantarell, a fisherman, noticed floating oil slicks in the Gulf of Campeche, about 100 km off the coast of the Yucatan Peninsula. He duly reported his findings to the Mexican authorities. From that lucky sighting of the oil slick, the giant Cantarell Field was born, happily named after Rudesindo Cantarell.[25] The discovery happened in 1976, it started producing in 1979, and oil production hit a record of 2.1 million barrels of oil/day in 2001. It was the world's second largest producing well in 2004. Who would have thought of this rare calamitous-propitious coincidence of a cosmic invader hurtling down to Earth near the little village of Chixulub, and that a giant oilfield would be entombed in the impact rubble?

The Deccan Traps—
Flood Basalt Impacting Continents, Oceans and Atmosphere

Another stellar Cretaceous geological event is the massive continental-scale outpourings of basalt lava on the already fragmented Gondwana landmasses including India. The catastrophic event marked one of the terminal stages of the breakup of the Gondwana continents, especially India, Australia, and Antarctica. After the 115–118-Ma-old initial eruptive phase of the Raajmahal and Sylhet Traps, the Deccan Trap flood basalt erupted in the Late Cretaceous. The 2200–2100-year-old, famous rock-cut caves of Ajanta and Ellora in western India, were carved out of the flat-lying Deccan lavas (Fig. 8.12). The word 'Deccan' comes from the Sanskrit 'Dakshin' meaning south. The word 'Trap' comes from the Swedish 'Trappa' or stairs, referring to the terraced tableland aspect of lava flows, topped by its rich black soil. The Maastrichtian-Danian, 68–63-Ma-old, Deccan lavas covered large part of the peninsular and offshore India. Its thickness ranges from >2,000 m near Bombay, to 600 m elsewhere in the Western Ghat range facing the Arabian Sea on the west. The erupted volume is about 512,000 km^3. It covers ~500,000 km^2 area, though the original area would have been 1.5 million km^2. Contemporaneous lava eruption had also occurred in Antarctica and Australia.[5]

Source of Deccan Flood Basalts—
The Reunion and Comoro Island Hotspots Far South

The Deccan flood basalt erupted, when the sailing Indian plate got temporarily perched atop a mantle plume, now under the Reunion Island in the south Atlantic. A newer study revealed that another plume, now hidden under the Comoro Island nearby, was active at about the same time.[26] The second plume also would have supplied

8-12 The 2.2-ka-old, Ajanta rock-cut caves with attractive Buddhist era sculptures and paintings in the visibly layered Deccan trap basalt, Maharashtra, India.

additional magma for the enormous Deccan effusive. Both the Reunion and Comoros hotspots are still active.

The Deccan eruption is noteworthy because of its continental scale and time frame. It is the second largest volcanic show lasting many thousands of years during the uppermost Cretaceous. Vesicles made by escaping gas bubbles in basalt, host many attractive minerals—zeolite, quartz, amethyst, calcite, etc. Radiometric age of zircon crystals in the volcanic ash reveals that lava erupted periodically between 68–63 Ma.[5] The peak eruption at 66 Ma (K/Pg boundary), coincided with the Chixulub asteroid impact.

MIRACLE OF LIFE

The Behemoths Thrived Everywhere

The sea level had reached an unearthly Late Cretaceous record—220 m higher than today, as the robust Mid Oceanic Ridge seafloor spreading displaced seawater towards land. As a result, the land areas occupied only 18% of Earth compared to today's nearly 30%. The climate was warmer and humid, great for dinosaurs and other fauna, and their food chain of plants and vegetation. It was also a congenial habitat for marine organisms. A tropical to subtropical climate prevailed extending as far as high latitudes. The CO_2 level would have reached 5 times the modern level (420 ppm) during the mid-Cretaceous 'hot house' state.[27] Marine fauna stayed in warm shallow seas, avoiding the large anoxic areas of the deep ocean. The flora and fauna began to show divergent evolutionary paths in the new, splintered Pangea continents.

The Dinosaur Dynasty—The Kingdom Lasted Long, ~165 Ma

A massive cache of dinosaur fossils occurs widely from every continent including Antarctica. In total, about 1,000 dinosaur species are known. However, the pride of place goes to North America, Argentina, Europe, Mongolia, and China, in terms of variety and number of fossils discovered. Both large and small dinosaurs dominated the scene. They became used to inhabiting even high latitudes where darkness reigned for half the year. The first-ever scientific mention of a dinosaur fossil was in the UK in 1676, probably a *Megalosaurus*, found in an Oxfordshire limestone quarry. The popular dinosaur sites are the Isle of Wight in southern England and the famous Jurassic coast of Dorset, already discussed.[36]

The Chicago Field Museum houses the renowned Maastrichtian *Tyrannosaurus rex* 'Sue,' discovered in 1990 in South Dakota, USA, by Sue Hendrickson. 'Sue' is the most complete fossil ever found (90% of bones intact). In 1991, a schoolteacher-paleontologist duo discovered 'Scotty,' the longest lived (~30 years) *T. rex.* from the bad lands in Saskatchewan, Canada.[28, 29] The 13-m-long, 4-m-tall dinosaur weighed an estimated 8,870 kg. Only 65% of its bones have been recovered, with evidence of bone injury. It appears the fellow toughed it out to make it to an old age.

New Late Cretaceous arrivals were herbivorous armored *Ankylosaurs*, horned *Triceratops* and a new duck-billed hadrosaur species. The Saharan desert of Morocco yielded the unusual theropod *Spinosaurus*—a giant, 15-m-long fish-eating river monster. It was more at home in water than on land, becoming extinct before the debut of the Titanosaurian sauropod family –*Argentinosaurus* and *Dreadnoughthus*, etc.

The Maastrichtian *Argentinosaurus*, discovered in 1987, was a giant dinosaur, the largest land animal ever—35-m-long, weighing up to 70,000 kg. The females laid 33-cm-long eggs.[30] The newer (2012), 98-Ma-old *Titanosaur* in Patagonia is even larger than the *Argentinosaurus*. '*Dreadnoughthus schrani*' was also a vegetarian giant. Remains of the big fellow from Patagonia, Argentina, measured 26 m long, weighing 58,967 kg.[31] Because of its size, it probably did not have to worry about predators, hence the name, meaning 'fear nothing.' Argentina's Desert Shield area is rightly a famous site for dinosaur fossils, comparable to the North American finds.

The Russian Titanosaur '*Volgatitan*' (1982) is a 130 Ma old, large, long-necked herbivore weighing 15,422 kg.[32]

In Canada, the world-famous Dinosaur National Park, Alberta, boasts of 40 distinct Cretaceous dinosaur species and over 500 individual specimens. Dinosaurs roamed all over the shoreline of the Western Interior Sea in ancestral Canada. Wide coastal plains by a warm sea were the ideal romping ground of the giants, nurtured by the abundant cycad, fern, and conifer. J. B. Tyrell discovered the first *T. rex* in Canada in the 1880s. Marine reptiles—*ichthyosaur*, *plesiosaur*, and flying reptile *pterosaur*, also called Alberta home.[37]

Beibeilong sinensis, and 'Baby Yingliang'—Dinosaur's Eastern Connection

A dinosaur 'hot spot' occurs in the Liaoning Province, northwest of Beijing. It's the home to China's famous feathered dinosaur variety, the *Sinosauropteryx*. In total, China reported 59 species of dinosaurs.[38] One famous Chinese dinosaur is the ~90-Ma-old 'Baby Louie,' a 38-cm-long embryo. It is a two-legged baby *oviraptorosaurian*, formally named '*Beibeilong sinensis*.' It had died prematurely surrounded by many unhatched eggs. Scientists think that the entire nest was in a late incubation stage when a probable flood would have abruptly buried it. Perinate (i.e., hatchling) skeletal fossils occur inside several eggs.[39] The unhatched eggs measured 40–45 cm long and 15 cm wide. Another perfectly preserved dinosaur embryo in China made big headlines in December 2021.[35] The intact embryo of the lovingly nicknamed 'Baby Yingliang' is 66 Ma old. Its vintage is significant, coinciding with the K/Pg extinction event.[40a]

The Pterosaur '*Quetzalcoatlus*'—A Flying Giant

Texas, USA, used to be home to the Maastrichtian, 250-kg monster *Pterosaur*—'*Quetzalcoatlus*,' tall as a giraffe. The 11-m-long creature with a wingspan of 10–12-m, reigned over the skies for 80 Ma, before perishing at the 66-Ma-old K/Pg boundary. The avian dinosaurs, i.e., birds, started sharing the skies with *pterosaurs* ('flying lizard') from the basal Cretaceous. Most birds perished at the End-Cretaceous.[33, 34]

A few survived the K/Pg inferno, diversified, and dominated the sky in the Cenozoic.

Why did the *pterosaurs* succumb in the Late Cretaceous? Probably their large size did them in. Research shows that few land animals with a body weight of more than 20 kg survived the disaster. Also, the challenging soaring conditions of toxic gases and clouds of dust in the air would have been lethal.

Dinosaur Egg Capital—India and China Strong Contenders

Recent finds put India on the list of important world dinosaur sites. In total, 20 native dinosaur species used to roam around, all with affinity to dinosaurs of its ex-Gondwana neighbors—Antarctica, South America, Madagascar, and Australia. The early mention of the Indian dinosaur in 1828, makes it the second oldest dinosaur discovery in the world. The giant carnivorous *Rajasaurus narmadensis* of Late Cretaceous, is the most famous one.[84]

India may be a place where the highest number of dinosaur eggs have been unearthed in one site. The discovery of nearly one thousand eggs in the Balasinor village site, Gujrat, and other states, is noteworthy. Many other places in Gujrat, Rajasthan, Madhya Pradesh, Maharashtra states, and elsewhere, reported the discovery of communal, or solitary nests with 10–24 eggs. Some eggs may weigh up to 5 kg. In addition to countless eggs, the finding of coprolites (fossilized droppings) throws light on the dinosaur's diet.

Unfortunately, many dinosaur bones and eggs are disappearing today because of a lack of adequate resources and protection. Mr. Mohaby, a geologist, lamented "in three decades we have destroyed what was left intact for 67 million years." The Dhar Museum alone, near Indore city, reported 27 stolen eggs.[84] In Jabalpur's Pat Baba Temple, priests protected the dinosaur eggs,[40] believing them to be a sacred symbol of Lord Shiva. They created a 'lingam' with dinosaur eggs.

In the Narmada valley, Madhya Pradesh, a cache of 256 *Titanosaur* eggs have been recently uncovered.[85] Tribal farmers in the area have been routinely worshipping the sacred 'stone balls' (dinosaur eggs) for generations as they found them while working in their fields and

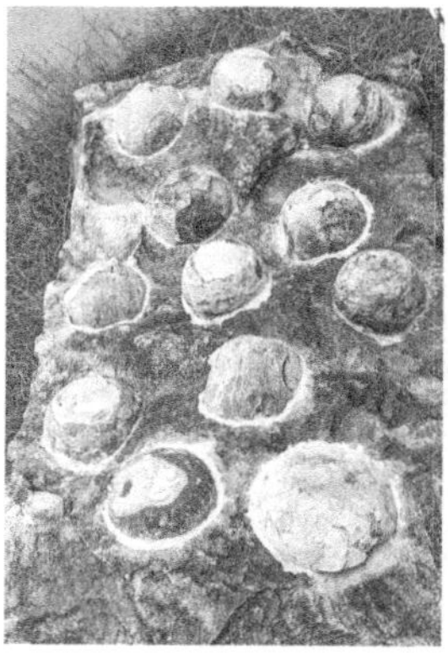

8-13 left A cluster of 12 cannon ball size Titanosaur eggs, now in the Indroda Fossil Park, Gandhinagar, Gujrat, India. **right** Paleontologist working in the field by the skeleton of a giant sauropod dinosaur—a *Titanosaur*, in Argentina.

open areas. After paleontologists examined the stone balls and told the villagers these are fossilized dinosaur eggs, the poor locals were in shock. However, a refreshing story comes from the impressive Indroda Fossil Park in Gandhinagar, Gujrat, India, by the Sabarmati river.[86] It is one of two dinosaur fossil parks in India, popularly known as India's Jurassic Park. The fossils come from the nearby Balasinor district in Gujrat. It showcases different dinosaur genus—*Megalosaurus*, *Tyrannosaurus* rex, *Titanosaurus*, *Brachiosaurus*, *Iguanodon*, *Barapasauraus*, *Stegosaurus*, etc. Also, dinosaur trackways and eggs feature here. Fig. 8.13 shows Titanosaur eggs now on display in the Indroda park, India, and a giant sauropod skeleton discovered in Argentina by paleontologist Bernardo Gonzalez Riga earlier in this millennium.

A Cretaceous Snake Caught Feasting on Dinosaur Eggs—Visit Gujrat, India

An amazing paleontological tale also comes from India, unusual but true. Researchers narrated a rare find of a 3.5-m-long 'macrostomatan snake' (large gape ingesting entire prey) fossil, at the Dholi Dungri area, Gujrat (Fig. 8.14, Image 1), in the vicinity of a sauropod dinosaur egg clutch. More unbelievable still, the snake appears coiled around one egg, with another intact egg, and a fossilized hatchling nearby. Experts think the snake was 'caught in the act' of predation—hunting and feasting upon dinosaur eggs and hatchlings.[41a] The snake nicknamed 'Sanajeh' apparently was buried alive, smothered by freak storm-flood debris. The discovery of three other snake remains in the vicinity strongly suggests that they purposely used to live and prowl around the nests, and habitually had their surreptitious but succulent meals of dinosaur eggs and hatchlings. Not all dinosaurs hatched their own eggs.

Young and Old Dinosaurs—Indelible Footprints Left Behind in the Bolivian High Andes

The dinosaur story will not be complete without highlighting the spectacular footprints they had left behind in soft clays of ancient inland coastal areas. Paleontologists learn a lot about dinosaur types, size and behavior from footprints or tracks. Many sets of tracks indicate they traveled in groups or herds, smaller footprints between larger ones imply some herds had their young ones in the center for protection. Footprints spaced apart mean walking gait, while closer

8-14 1 A fossil snake feasting on dinosaur eggs and hatchling, Dholi Dungri area, Gujrat, India. The snake head at top left corner. Its skeleton is coiled around a crushed egg, below the snake's head. To the top right of the snake skeleton, is an intact titanosaur fossil, and a 0.5-m-long hatchling within an egg below. Scale bar 5-cm. **2** Two prominent sets of dinosaur foot tracks on the 80-m-high and 1.2-km-long, vertical wall of Cretaceous rocks in the Cal Orcko Dinosaur Park in Sucre, Bolivia. Faint smaller tracks of a probable juvenile dinosaur to the right of the tracks.

footprints may mean a running activity.[41] The *Sauropod* and *Theropod* dinosaurs had distinctive footprints.

The Cal Orcko 'Parque Cretácico' near Sucre, Bolivia, may be the world's dinosaur footprint capital (Fig. 8.14, Image 2). The name means 'lime hill' in the Quechua language. Here 5,055 footprints, some 80-cm-long, adorn a steep bedding surface outcrop of limestone of Late Cretaceous age—1.2-km-long, 80-m-high.[42] The Park used to be a limestone quarry earlier. Geologically, the area was likely a sea inlet, that over time became a freshwater lake. Ancient freshwater fossils of catfish and crocodiles attest to this. The Andean orogeny uplifted and deformed the strata, forming vertical walls of the Cal Orcko site, 2,830-m above sea level. The wall shows at least seven layers of dinosaur footprints. Over 460 tracks made by both herbivores and carnivores of 15 different species are present. Some tracks are famous for their visible length.[43] The 'Sucre Life' municipal website proclaims that the most impressive one is a 347-m-long trail left by a baby *T. Rex*, fondly nicknamed 'Johnny Walker.'[44] Additionally, one *theropod* trackway continued for a never-documented length of 550 m. Sadly, many tectonic rock joints and fractures traverse the vertical wall and make it prone to rockfall and landslides. Rains damaged a large chunk of about 100-m of wall with the loss of 200 tracks.[45] However, when one layer disappears, a new one reveals exciting tracks again. The local authorities hope to receive funding to preserve the unique site.

Rare Andean *Pliosaur* in Colombia, also one Down-Under

The scenic Colombian town Villa de Leyva hosts a rare specimen of the *Kronosaurus* genus (Fig. 8.15). It is a marine, crocodile-like short-necked *Pliosaur* that lived about 110–115 Ma ago. The tail and one right flipper of the 10-m-long reptile is missing. It lived in the coastal waters of the Cretaceous Sea. It was a deadly marine reptile, came with a massive, large head, a short neck, and broad flippers. It is the star exhibit in a unique, single-fossil museum, built right at the site where the *Kronosaurus* breathed its last by an ancient sea inlet. A local farmer discovered the fossil in 1977, while tilling his land.[47] Geol-

Figure 8.15. Marine *Pliosaur* reptile—110-Ma-old *Kronosaurus*, 10-m-long, with a missing right flipper and tail. Museo Villa de Leyva, near Bogota, Colombia.

ogists in down-under Australia reported the only other fossil of a less preserved *Kronosaurus* earlier.

Mammals and Reptiles—Cold-Blooded vs Warm-Blooded

A significant Cretaceous arrival was our mammalian ancestors with unique traits: mammary or milk-generating glands, an enlarged brain, limbs under the body, a more complex physiology, etc.[48] Later, the mammals diversified into placentals (cows, dogs, humans), marsupials (kangaroo, opossum) and the uncommon monotreme (platypus). We may wonder what makes mammals different from a giant dinosaur. One difference is that dinosaurs are cold-blooded, like other reptiles. Cold-blooded animals need an external heat source to keep warm, whereas humans, cows or mice can generate heat internally. However, recent finds in Alaska suggest some dinosaurs were warm-blooded, like the avian dinosaurs—birds. Many think dinosaurs were generally warm-blooded creatures.

The other important feature is that a female dinosaur laid 3–20 or more eggs. Little embryos, snugly ensconced inside, would emerge from the shell after 3 to 6 months. In comparison, gestation of mammal embryos happens within the uterus for variable periods: 21 days for a mouse, up to 13 months for a Bactrian camel, and 20–21 months for an elephant.

End-Cretaceous—Some Survived, Many Didn't

Many marine reptiles were not lucky with the K-Pg extinction event. The *Plesiosaur* became extinct in the Late Cretaceous, so did the

4-m-long *Archelon* turtle, weighing 2,000 kg. The giant 'X-fish'—*Xiphactinus* used to swim both in the Western Interior Seaway of North America, and South American oceans (near Venezuela and Argentina) but perished at the End-Cretaceous.[54] Yet other marine animals like coral, and pelecypod were diversifying happily. Crustaceans—crab and lobster, reptiles—snake and lizard, and most fish varieties, began to look more like their modern successors.

We discussed the massive rudist reefs as oil reservoirs in the earlier section. What are the rudists? These are pelecypods like clam, mussel, scallop, and oyster. Each of the four rudist families lived in different parts of reefs. Some rudists show a weird evolutionary trait where the two valves are disproportionate in shape and size. In one variety, the larger conical valve grows attached to the ocean bottom ('sessile'), while the smaller valve acts as a lid on the other.[11] The rudists may grow up to 1.5 m long. The central cavity, and the thick vascular walls of the larger valve are porous. The rudists became extinct at the Cretaceous-Paleogene (K/Pg) boundary like the dinosaur, ammonite, and other organisms. The End-Cretaceous exit of rudists left an open field for other champion reef-builders—coral and algae, to thrive as Cenozoic reef makers.

The climate during the rudist clam era was tropical to subtropical and warm, 25°C (Fig. 8.4). A nearly all-time high global sea level inundated many inland seas with valuable nutrients for the Cretaceous rudist reef makers. A warm, tropical climate reigned at latitude 30^0 N in the Eduard's Plateau in Texas. The presence of dinosaur tracks and fossil cycad tree stumps in littoral sediments, close to the Cretaceous marine rudist reefs, proved the climate was benign. The reefs were generally 10–20 m high. But large reefs could grow up to 600 m.[10]

The Late Cretaceous witnessed the acme and demise of ammonite, a cephalopod. The Himalayan 'shaligram shila' (holy anthropomorphic symbol of Vishnu, the Hindu god) is a Jurassic ammonite fossil. It shows multiple shapes like coiled, straight, and even hairpin-like. Some grew to abnormal size, up to 1.8-m-across, the largest specimen ever, currently housed in a German museum. The author had the good fortune of enjoying the sight of a veritable sea of ammonite in Villa de Leyva, Colombia (Fig. 8.16). They are so plentiful that

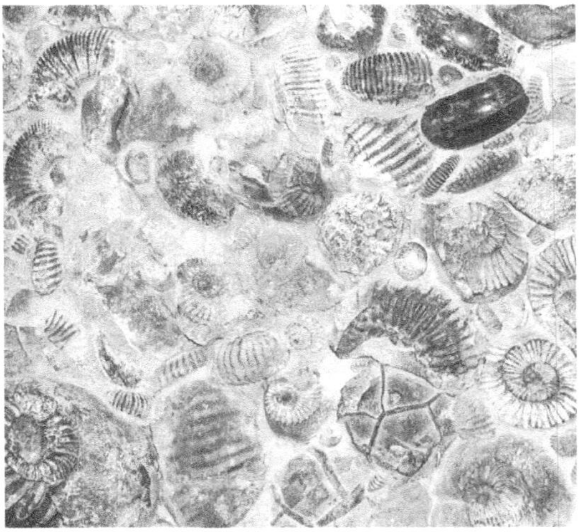

8-16 Cretaceous ammonites on the entrance wall, Villa de Leyva Museum, Colombia.

local churches, homes, restaurants, store fronts—all decorate their walls and floors with attractive examples of perfectly preserved ammonite fossils from a local geological site. The ammonites showed spectacular development, yet became extinct at the K/Pg cataclysm, like the dinosaurs.

Happily, all was not doomed at the End-Cretaceous. The amphibian salamander, frog, tortoise, and turtle thrived. The Late Cretaceous saw the first appearance of modern insects—ant, butterfly, termite, grasshopper, etc. It was a propitious time for the marine, unicellular, protist diatom (photosynthetic algae). Unlike the siliceous *diatom*, the calcareous *coccolithophores* (algae) dominated the tropical-subtropical milieu. Other protists—the single-celled *dinoflagellate*, *radiolaria*, *foraminifer* also survived.

The Cretaceous Flora—New Colors, Scents and Nectar Added

The Cretaceous flora included extensive pole to pole forests of cycads, ferns, and conifers. The most visible and attractive arrivals to dominate the floral kingdom later were the angiosperms—most trees with flowers, grasses, and shrubs. A female ovary of flowers ('carpel') encases angiosperm seeds. The pollinating bees came around to fill an ecological niche. The customary wind pollination aside, flowers learnt the trick to make pollination more fruitful by wearing eye-catching colors and scents to attract all pollinators. Flowers soon added nectar as an extra incentive, too sweet for the pollinators to want to fly away.[46]

What really caused the mass extinction at the Cretaceous-Paleogene juncture remains debatable. However, geoscientists now clearly know the damage and upheaval that Earth suffered at that time. We discuss here the major cited causes of the unprecedented havoc that occurred, that killed the Cretaceous behemoths and others. The earlier Ordovician, Devonian, Permian, and Triassic mass extinctions, had occurred at the transition between two geological periods. The Cretaceous-Paleogene (K/Pg) extinction event proved to be no exception.

THE CRETACEOUS-PALEOGENE (K/PG) MASS EXTINCTION

The 5th One in Earth History—Mundane or Cosmic Causes?

The causes of the extinction fall into two categories—Earthly and Cosmic. Earthly causes are—climate change, volcanism, etc., whereas 'Cosmic' means something out of the blue like any cosmic body impact, solar variations, etc.

PROBABLE EARTHLY CAUSES

Climate Change

A very probable scenario could be drastic global climate change. Any hypothesis to prop up the global climate change angle must explain two important things. First, what is it that unleashed the change so abruptly at the End-Cretaceous ? Second, whatever the root cause(s), it should have equally affected both the marine and land animals. Then, it should also explain why selectively 25% of all species of plants and animals could survive this infernal change, while 75% perished, in the blink of an eye, so to speak.

The Cretaceous Period was a time of ice-free poles with temperate climate. The entire globe enjoyed a warm humid, tropical to subtropical climate ~ 32°C in the North, while averaging 26°C at 53° latitude South.[49] However, the average Cretaceous temperature range was 20–25°C.[50] In either case, it was much higher than today's ~14°C. Both animals and plants, including planktons, thrived even at higher latitudes where forests extended up to the poles. If a mere temperature change was the culprit, then it had to be either a cold glacial climate, or a desert-like inhospitable climate. The problem is that geological evidence for either scenario is lacking. The succeeding early Paleocene climate was also normal, somewhat like the Cretaceous.

Volcanism and Plate Tectonics—Did They Play a Role in Mass Mortality?

Two of the potential earth-bound causes of mass extinction are volcanism and plate tectonic movements. Plate movements by itself cannot explain the worldwide K/Pg extinction event. However, the mantle plume hot spot volcanism may profoundly affect the atmosphere, harming extant fauna on land and water. Several authors invoked volcanism as the culprit for K/Pg mass extinction.[52]

The timing of K/Pg extinction and Deccan Trap eruption might be significant. Researchers noticed a convincing correlation between past extinctions and mass eruptions:

1. The End-Triassic extinction and eruption of the Central Atlantic Magmatic Province.
2. The End-Permian extinction and the 250-Ma-old Siberian Trap eruption.
3. The Permian (Guadalupian) extinction and the 260-Ma-old Emeishan Traps, China.

When the Deccan lava erupted, the India tectonic plate was thousands of kilometers out floating on the ocean, long before it rammed head-on with Eurasia. At End-Cretaceous, it was atop the mantle plumes of Reunion and Comoro Island hot spots in the Indian Ocean, near Antarctica.

Deccan Inter-Trappean Beds—What Do They Reveal?

Multiple dinosaur fossils and eggs occur in the inter-trappean thin sedimentary layers between individual lava flows. Paleontologists noted evidence that dinosaur diversity apparently declined later. A Maastrichtian assemblage of dinosaur fossils, the fish *Igbatis*, and spores/pollens, characterizes many inter-trappean beds. Two Paleocene beds also yielded fossils, but expectedly no dinosaur remains.[53]

Lava Layers Rose High—Deccan Traps Poisoned the Environment

The Deccan effusive event is important because of its sheer continental scale, and the time frame of eruption. It was the second largest volcanic show ever. The lavas yielded an age of 68–63 Ma, with a peak at 65.5–66 Ma.[55] The top Maastrichtian age (66 Ma) is coincident with the timing of the asteroid impact in the Yucatan Peninsula, Mexico. Radiometric dating of zircon crystals from interlayered volcanic ash beds suggests that the Deccan lava erupted in three distinct phases. U-Pb ages indicated that the phase 1 and 2 predated the extinction event. The second eruption phase, 'Poladpur flow,' was the major one, erupting 80% of lava and almost coincided with the K/Pg boundary.[57] It started tens of thousands of years before the Chixulub bolide impact and the K/Pg mass extinction.

The eruption of >2,000-m-thick lava flows also spewed enormous volumes of gases like CO_2, sulfur compounds and water. High CO_2, SO_2 levels would have impacted global climatic equilibrium—high continental weathering, acidic oceans, alternate global warming, and cooling, among others. CO_2 and SO_2 tend to have opposite effects on the atmosphere. A quick somewhat seasonal swing of hot and cold temperatures might have had a disruptive effect on the climate leading to food chain disruption and mass mortality. Particularly, the 'Poladpur' eruptive pulse could have been crucial.[56]

Micro-foraminifers—Witnesses to K/Pg Global Pollution

Micro-foraminiferal changes in 11 deep wells in offshore India, provided a reliable proxy for the catastrophic climate change.[57] After the main Poladpur eruption, ~75% of planktic foraminifers perished, leaving only 7–8 species that survived beyond the K/Pg boundary in India.

Another example comes from the El Kef site in Tunisia.[58] The change in foraminiferal abundance was most drastic at low latitudes, where almost total extinction of robust, complex, and tropical Cretaceous forms occurred at the K/Pg boundary. Some argue that a long-term oceanic instability might explain the observed foraminifer extinction. Another study posited that a high ocean acidity prevented foraminifers from creating calcareous shells in the post-K/Pg time.[51] Indeed, the Paleogene fossil shells were dwarf and less diverse.

Many Cretaceous spore and pollen varieties appeared visibly altered and showed significant diversity loss in the younger post-Chixulub-impact layers.[59] A clear trend was evident in the Paleocene: early appearance of opportunistic, robust 'crisis' taxa, next came a pioneer group of plants, and finally, a stabilization and recovery phase with new taxa.

Magnetic Pole Reversal—Any Effect on Extinction of Life?

Another process of potential impact is the reversal event of Earth's magnetic field.[60] We discussed in Chapter 4 that magnetic poles swapped position many times in the geologic past. The Cretaceous magnetic field remained stable and did not flip during the 118–81 Ma interval of the 'normal superchron' stage.[61] It was 'normal' because the magnet needle pointed toward the North Pole, same as today. A normal superchron may signal a long geological interval when Earth's internal geo-dynamo works at full efficiency.[62] To note, Earth had experienced several reversals in the remaining Cretaceous Period, finally coinciding with the K/Pg extinction.

As an aside, the most recent magnetic pole reversal happened 0.77 Ma ago. It did not happen in an instant, the whole process lasted about 22 ka. Recently the magnetic North Pole has been moving away ever so slowly since the 1830s, from the Nunavut Territory, Canada, towards Siberia in Russia, but it picked up speed in the 1990s.[63] The North Pole has traveled 2,240 km in 190 years.

Lava samples are excellent proxy of Earth's magnetic field and the relative pole orientation. The magnetic pole flips had a possible hand in the End-Triassic mass extinction event by stripping off oxygen from the atmosphere.[64] At that time, Earth went through 88 reversals in 30 Ma. The atmospheric oxygen level fell from 23% to 14%. The oxygen escape level had increased three orders of magnitude more than the present outflow rate. The polarity reversal stripped off 15 x 10^{18} mol (218 trillion tons) of oxygen from Earth's atmosphere during the End-Triassic when 76–84% of all species perished. However, researchers warn that the causal relation between polarity reversal and mass extinction is not infallible.

How and why does oxygen depletion happen? The dipole geomagnetic field weakens during Earth's polarity transition. It loses power to shield life from the harmful solar winds hitting Earth constantly. The unobstructed solar radiation may cause oxygen, so essential to life, to escape into outer space. Some opine that a cosmic body impact may also trigger geomagnetic field reversal. Scientists are unsure about the role of magnetic reversal in the End-Cretaceous mass extinction.

Cosmic Cause of Mass Extinction—A Father-Son Duo's Seminal Work

Earlier in the chapter, we learnt about the Chixulub asteroid impact in the Mexican Yucatan peninsula. Adherents to the cosmic origin hypothesis of mass extinction may outnumber the proponents of earthly causes. Here we briefly discuss how the cosmic impact idea gained such widespread currency in the geoscience community. The seed of the theory came from the thesis work of geologist Walter Alvarez. He was studying the Cretaceous-Tertiary boundary in Gubbio, an Italian village between Florence and Rome in 1978. An outcrop caught his attention, he zeroed in, sampled an unusual layer of red clay between the topmost Cretaceous strata, and the Paleocene limestone above.[66] Back at the University of California, Berkeley, Walter Alvarez discussed the samples with his father—Prof. Luis Alvarez, Nobel laureate (1968) in particle physics. Prof. Alvarez had the samples analyzed by chemist colleagues. Neutron Activation Analysis of samples gave some startling results to the father-son team. The red clay layer had 25 times more iridium than the Paleocene limestone above, and the Cretaceous below. Iridium (Ir), a transitional element to platinum, is the second densest (22.56 g/cm^3) after osmium (22.59 g/cm^3) in the periodic table. Compare density of iron—only 7.87 g/cm^3.

Iridium Anomaly at the K/Pg Holds the Clue

Very high-density metals like iridium and osmium are rare in Earth's crust because they had gravitated to the core of the primordial Earth a long time ago. The question is if the crust is so poor in iridium, where did it come from in Gubbio. Looking around, the Alvarez team found that many places worldwide—Denmark, New Zealand, the USA, and elsewhere, show the same phenomenon at the K/Pg boundary. In Canada, the boundary clay with high iridium value occurs in several sites, including the well-known Frenchman Valley in Saskatchewan.[65]

The Alvarez team proposed a novel theory in a seminal paper.[66] They already knew that iridium is abundant in cosmic bodies. They asserted that iridium anomaly everywhere at K/Pg was the handiwork of an extraterrestrial body—an asteroid impact. The team calculated that a 10–20-km-across body would have packed enough iridium to create the spike observed in all earthly samples. They argued that the cause of dinosaur demise would have been an earth-shattering cosmic impact that spilled rare iridium all over the globe. The impact-generated dust would later settle down as iridium-rich sedimentary layers, marking the Cretaceous—Paleogene contact (K/Pg boundary). The presence of soot confirms the effect of the impact and the generated heat. Happily, the smoking gun soon appeared with the discovery of the Chixulub crater in the Yucatan peninsula, Mexico.[21a] Geoscientists Glen Penfield and Antonio Camargo first identified the crater through seismic images and reported the discovery thereafter.[22]

How Big Was the Impact at Chicxulub Ground Zero?

It is not easy to assess what happens when a large, not so 'heavenly' body hit Earth at break-neck speed. What will be the real consequence of a large city-sized impactor on Earth? The 14-km-diameter Chixulub asteroid would have created an unprecedented shock wave. It had punched a 20–30-km-deep, and 180-km-across crater hole. The impacting body vaporized soon after hitting Earth. Impact melt rock shot out toward the stratosphere, before falling and scattering all over as glass spherule or tektite. Indeed, the effect of the Chixulub impact reached far and wide.

An asteroid creates gigantic volume of ejecta on hitting Earth. The Chixulub impact had dispersed 1.98×10^5 km^3 of sediment gulf-wide, and some of it traveled across what is now North America and rest of the world.[70] Such a cataclysmic event would have had many negative outcomes for the hapless fauna and flora—hundreds of meter high tsunamis, widespread infernal fire from hot debris raining down, lighting up the forests like tinderbox. Also, it created abundant CO_2, CO, SO_2 from the disintegration of limestone and sulfate minerals. The other effects were acid rain, acidity of ocean waters, short term intense global warming due to CO_2, shortage of day lights etc. The hydrous aerosols from sulfuric acid droplets, would have caused long-term blockage of sunlight in the southern hemisphere.[68] The shutdown of plant photosynthesis due to the protracted blockage of sunlight by dust and debris, would have led to a total food chain collapse. The disequilibrium led to an apocalyptic level of organic demise.

Several Caribbean Sites Confirm the Culprit Asteroid's Fateful Visit

Similar supporting evidence of impact ejecta came from places around the Caribbean and parts of Texas and North Dakota, United States. Voluminous evidence of the impact is the widespread limestone breccia, scattered all over Mexico as an apron around the Chixulub impact crater. The author had visited impact debris in Belize, Central America. The debris include beds of altered glass spherules or microtektite, and adjacent coarser dolomitic limestone breccia beds set in a clayey paste (Figs. 8.17 and 8.18). Belize to Chixulub impact crater distance is 400 km.

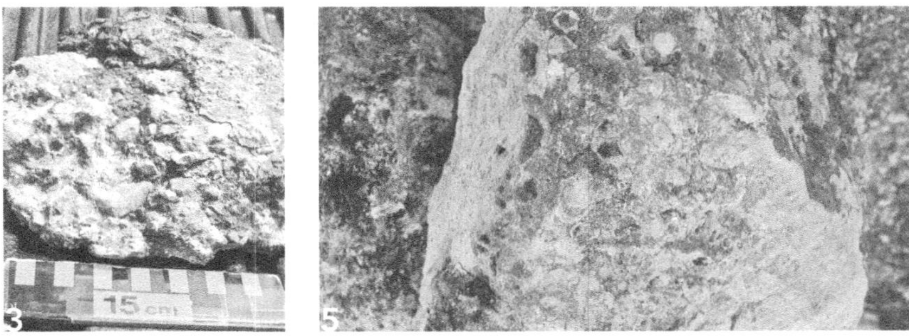

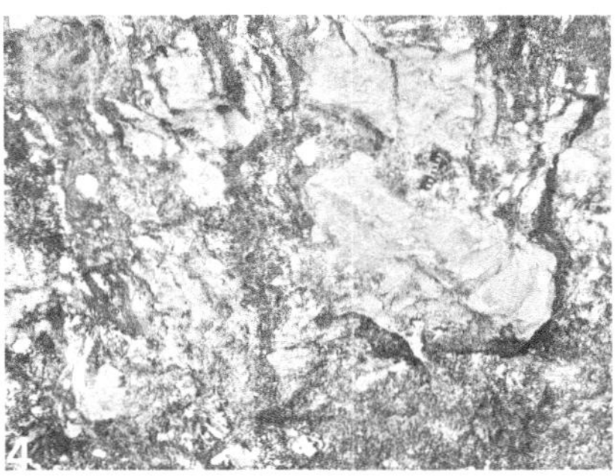

8-17 **1** The white beds are ejecta layers on a Belize road section. **2** The white beds of image 1 (30–40-cm-thick), contain impact-related tektite spherules of fused glass origin. Black arrows show spherules (~3- mm-across) and molds. **3** Close-up of Image 1 shows breccia clasts and tektite molds. **4** View of black, oil impregnated, poorly sorted, shattered angular dolomite clasts and boulders. **5** Partly dissolved breccia clasts in a Fe-rich matrix.

The Albion Island in Belize contains a 15-m-thick volcanic ejecta layer with altered impact glass, dolomite spherules, and accretionary lapilli (aggregated wet volcanic ash particles of 2–64-mm size).[67] Lapilli in Latin means 'little stone.' The limestone breccia near the K/Pg boundary in Guatemala, shows altered glass spherules and elevated amount of iridium.

The Pacific Gorgonilla Island of Colombia confirmed the wide dispersal of Chixulub asteroid impact ejecta.[69] A sedimentary outcrop of ancient deep marine origin displays a 2-cm-thick layer of mm-size spherules that rained down from above in the Cretaceous Ocean. The

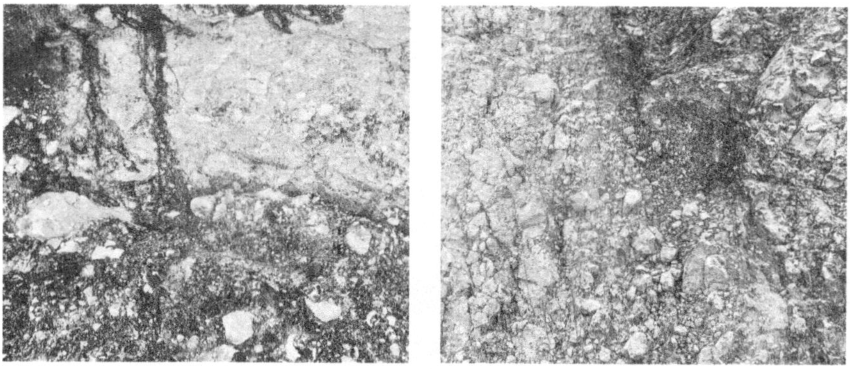

8-18 Two images of the Late Cretaceous Barton Creek Limestone, Belize, showing dead oil impregnated breccia. Oil would have dripped from above. **left** It highlights the difference between the oil-impregnated lower part, and light-colored upper half. **right** A gradual change occurs in breccia texture—fractured, crushed aspect on the two sides, and oil-stained smaller clasts at the center.

chemical composition of the spherules matches those in Beloc, Haiti and elsewhere.

The Tanis Site, North Dakota, USA—Havoc Written in Stone

Barnum Brown of the American Museum of Natural History discovered the first *T. Rex* fossil in 1902 in the Hell Creek Formation, N. Dakota, USA. The Late Cretaceous rocks here preserved in graphic detail one of the most remarkable far-reaching effect of the Chixulub impact.[71] The soft clays, mudstone, etc., of Maastrichtian age, straddle the K/Pg boundary, and continue over the Dakotas, Montana, and Wyoming. The area used to be a Cretaceous subtropical river flood plain near an estuarine coastline, where animals and fish were buried instantaneously.

Hell Descended on the Distant Hell Creek Formation—All Life Buried Alive

The Tanis site is about 5,000 km away from the ground zero in Chixulub, Mexico. However, it was only at a short distance from the ancient Western Interior Seaway. The unique site contained direct victims of Chixulub impact event. Robert DePalma, then a doctoral student, found a meter-and-half-thick strata of jumbled mass of everything from air, sea, and land. He noted abundant charcoal, and hexagonal minute impact diamond (‘*lonsdaleite*’). DePalma discovered countless tektite and micro-tektite cratering the surface, making funnel forms. The excellent three-dimensional preservation of fauna in soft, porous sediments was significant. It made researchers think that the Hell Creek event had recorded the devastating effect of the far-off asteroid impact in the first hour or so. About 50% of the fossilized fish here shows gills choked with tektite. Fossils are so perfectly preserved that even the scales are intact. Random to near vertical orientation of fossil fish attests to a rapid burial in a dense viscous slurry of clay, mud, and sand. They probably suffocated to death (‘tetany’ effect). The remains of burned fossilized tree trunks contained traces of amber enclosing tektite. There were *pterosaur* remains including egg, *Triceratops* carcass, sea weeds, ammonite, marine *dinoflagellate* cysts, insects, and all varieties of other dinosaur memorabilia—feathers, teeth, bones, hatchling remains, intact eggs with embryo, and a Mosasaur, an extinct marine reptile.[71] The coexistence of fully marine fauna and flora with terrestrial biota is significant.

Experts M. Richards and W. Alvarez visited the Tamis site, argued that the impact would have created a massive 10–11 magnitude earthquake; concurrent seismic waves likely would have arrived at the Tanis site within 10 minutes of cosmic impact at the ground zero.[72] Richards, a coauthor of DePalma, likened the Tanis site as an End-Cretaceous museum in a 1.5-m strata.

A Devastating ‘Seiche’ Wave, 3000-km Away

High seismic waves would have created a ‘seiche’ wave in all water bodies including inland sea, rivers, and ponds. The ‘Seiche’ is simply standing waves created by sloshing of water by seismic waves in any confined or semi-confined water body—a harbor, a lake, or a pond. Two seiche waves likely had left behind sediments covering the fossil bed in the Tanis site. An iridium-rich clay layer rests above the fossil bed. The site is unique because impact particles from the fireball in the sky settled on the same layer where animals and plants were smothered by the impact. We may gauge the ferocity of waves from a modern analog—The 9.2 magnitude Alaska temblor, 1964, created strong seiche waves, sinking several boats in Louisiana, USA.[73]

Misfortune Didn't Come Alone—
Other Accomplices of the Heavenly Kind

The End-Cretaceous appears be a juncture of more than one killer space objects hurtling down to Earth. The notion that Chixulub was a one-off impact that caused so much havoc may no longer be true, considering findings in the new millennium. At about the same time, a 24-km-across asteroid had hit the Boltysh village site in Ukraine. The impact happened just before the Chixulub disaster.[74] Researchers found a post-Boltysh-impact flora that formed on the ejecta layer. They also found that the new flora was decimated by the slightly later K/Pg Chixulub impact in Mexico. The Boltysh impact predated the Chixulub by only 2–5 ka.[75]

We have more surprises to digest. In 2002, an oil company consultant unexpectedly discovered a crater about 1 km beneath the sea in seismic images. He came upon the 'silver pit' crater, while exploring a North Sea gas field, about 140-km east of Britain.[76] A Late Cretaceous asteroid 200–500-m across, created the hole, 3-km-diameter and 300-m-deep in chalk layers. The crater appeared to be rather unique. It revealed 10 concentric fracture rings, the outermost one being 19-km-across. The subsea images revealed that the chalky rock at the center of the crater rebounded upwards forming a peak ring after the impact. The asteroid, though rather small, would have had spawned a huge tsunami, affecting the area and beyond.

DINOSAUR DEMISE—EARTHLY OR COSMIC CAUSE?

Role of Deccan Trap Eruption

The first phase of Deccan trap eruption was unimportant, it spewed only 6% of lava. The most powerful 'Poladpur' second phase ended at the K/Pg boundary, accounting for 80% of lava. It began only tens of thousands of years before the Chixulub impact. The third and last phase of volcanism (14%) was inconsequential, it happened after the K-Pg boundary (66 Ma), in the Paleocene Danian stage (64.7 Ma).[77]

An apparent decrease in dinosaur diversity in the Deccan trap country was palpable, though they survived through the early stage of volcanism.[53] Does it mean that dinosaurs were on a decline even before the meteorite impact in the Yucatan Peninsula? Does it mean that the lava itself was not the cause, rather the related gaseous outpourings proved to be too toxic for the dinosaur population and their food chain? Undoubtedly, successive lava flows emitted enormous volume of toxic gases including CO, CO_2 and sulfur compounds critically contaminating the atmosphere. The excessive volcanic heat would have scorched the environment over a period of several hundred thousand years. We recall the curious positive correlation between large-scale volcanic eruptions, and the last 4 known mass extinctions.

Role of Chixulub Asteroid Impact

Data on the Chixulub impact and its intimate temporal connection to the K/Pg boundary and mass extinction, convinced numerous experts about the weight of the impact hypothesis. In 2010, on the 30th anniversary of the cosmic impact theory, 41 experts in the field jointly reaffirmed in the journal 'Science' that the asteroid impact event was the cause of K/Pg mass extinction. They unanimously espoused the original idea of Alvarez et al, 1980,[66] i.e., the Chixulub asteroid was the culprit that killed the dinosaurs, not volcanism or other natural events.[78] However, a larger data base does not seem to support a direct link between asteroid impact and mass extinction on Earth. The Earth-impact database of extra-terrestrial bodies, run by the University of New Brunswick, Canada, documented 190 impact craters in the world of all geologic ages.[79] Apparently, no significant causal relation exists between meteorite impact and an extinction event.

Some geoscientists including paleontologists, doubt that a single cosmic event could have caused worldwide mass extinction at any time including K/Pg transition. After all, nearly 25% of species survived the cataclysm, and merrily went on to thrive and populate the world in the Cenozoic. Paleontologists may prefer a more mundane set of causes of slow changes and evolution, á la Charles Darwin.

The naysayers are a few, not only the fossil-guys: Two computer modelers showed that an asteroid impact has low probability of causing mass extinction. They opine that only a climatic debacle, such as a prolonged global cooling (8°C–11°C cooler) event and/or drought because of stratospheric soot and sulfate aerosols, can cause mass extinction. They postulated two preconditions—an asteroid must hit rocks rich in hydrocarbons and sulfate minerals like anhydrite, the two sources of soot and sulfur-aerosols. Their second condition is that the impacted site should cover at least 13% of Earth's surface area. The Chixulub impact,[80] arguably met none of the conditions. The probability of an asteroid causing mass extinction is quite low for them.

The critics of cosmic body impact hypothesis argued that the Chixulub impact predates the K/Pg boundary by about 300 ka, based on paleontological work in the Brazos River section in Texas, United States. Drill core data of a well within the Chixulub crater also confirms the early age of the impact event.[81] In other words, Chixulub impact itself did not kill the dinosaurs at the End-Cretaceous.

However, the vapor plume created by 600 km^3 of sedimentary rocks vaporized upon asteroid impact, would have had a lasting effect on the atmosphere.[59] Some think that prolonged sunlight blockage and significant cooling in the southern hemisphere, were mainly due to atmospheric hydrous aerosols from sulfuric acid droplets.

A Novel Proposal—
Chixulub Impact Triggered Deccan Volcanism

The debate about the cause(s) of K/Pg mass extinction went on for decades. Some specialists came up with a new proposal: the two events—asteroid impact and Deccan volcanism are inter-related. Chixulub impact would have had triggered the massive lava outpourings, dealing a one-two punch on all creatures.[82] The age of the impact rocks and Deccan lavas showed that the two events are essentially contemporaneous, especially the largest 'Poladpur' second flow of the

Deccan traps. They argued the asteroid impact could have generated enough seismic energy to trigger Deccan eruptions, especially the second one. Another study concluded that lava flows, and asteroid impact were equal drivers of the singular catastrophe leading to the mass extinction at K/Pg boundary.[55]

Points to Ponder—
Why Did the Cretaceous End the Way It Did?

We reviewed several probable or possible causes of mass extinction as we went through this topic. Any satisfactory hypothesis should address several observations geoscientists have made. First, the hypothesis should explain the global scale of the K/Pg boundary. The iridium-enriched clay bed at the Cretaceous-Paleogene boundary occurs in every continent, confirming the global fingerprint of the asteroid impact. Dinosaur fossils are conspicuously absent in the Paleogene rocks, above the K/Pg boundary. On the other hand, mass extinction is not unique in Earth history because four other mass extinctions happened at the turning point from one geological period to another. In that sense, the K/Pg extinction event is no exception. However, dinosaurs proved earlier that they were a hardy bunch. They survived the End-Triassic catastrophe successfully and proliferated in the Jurassic-Cretaceous. Studies confirm that dinosaur diversity was not declining in the latest Cretaceous except for two herbivore varieties.[83]

How could the snake, lizard, frog, turtle, crocodile, the avian-dinosaur, i.e., the bird, and more importantly, our mini-sized mammalian forebears, avert this calamity of apocalyptic proportion and come out swinging and stronger in the Cenozoic? At least 25% of all species escaped demise, even when large creatures like dinosaurs, otherwise sturdy, couldn't outlive this calamity. Why could they not withstand the blow of nature? Was it just luck for the survivors?

The widespread presence of literally thousands of intact dinosaur eggs in multiple nests in India and China, may imply that something happened very quickly that prevented dinosaurs from hatching their eggs. They perished along with their eggs in a geological blink of an eye, so to speak. Scientists agree that dinosaur eggs needed incubation of 3–6 months before hatchlings come out of the shells, and then likely they needed long-term care in their infancy. The impressive 347-m-long trail of baby dinosaur *T. rex* footprints between adult ones confirms this in the world-famous Bolivian site in the high Andes. So many intact eggs in India only validate that something of cataclysmic nature had happened at the K/Pg junction in a matter of days or months. Looking closely, the lucky survivors of the disaster were smaller, mainly omnivorous, more adaptive to new situations. They included amphibian reptiles like the crocodile, frog, and turtle that could get sustenance both from land and water. Things were stacked against the dinosaurs; they were at a disadvantage because of body size and specialized diet both in quality and quantity. The feathered avian dinosaurs—the birds, survived probably because they could cover larger territories for their omnivorous diet. They reproduced faster than the big behemoths, and likely could fly and stay away from hellfire more successfully because of body size.

It is likely that the Chixulub impact created a chain of natural disasters of global proportion—infernal heat and wildfire, seismic energy-related mega-magnitude earthquakes, unprecedented tsunamis and 'seiche' events. Other negative effects of Chixulub were impact-triggered volcanism, glassy droplets raining down from the sky arising out of vaporized molten rocks, toxic gases from vaporized anhydrite salts and carbonates. Blockage of sunlight and massive collapse of photosynthesis leading to almost total disruption of the food chain for animals proved critical. The list can go on.

The impact and its effects were compounded by another simultaneous natural event of global proportion—the long-drawn Deccan Trap volcanism on the Indian plate, and its wide-ranging chemical and physical impact on land, air, and ocean over a protracted span of time. Probably two global events of such magnitude had never converged so strikingly at one juncture in our geologic history.

CHAPTER HIGHLIGHTS

- Dismembered Pangea supercontinent transitioning to modern continents.
- The early phase of the rising Rockies, and the ancestral Andes.
- Mid-oceanic ridge, and Atlantic seafloor spreading in full swing. Review of mechanisms, drivers, and details of these natural phenomena.
- Mantle hot spot activity starting the creation of the Hawaiian and other island chains.
- Extensive 2-km-thick Deccan lava eruption on the Indian plate on the go.
- Discovery of global high iridium anomaly at the End-Cretaceous in many places.
- Discovery of the Chixulub asteroid impact crater, Yucatan peninsula, Mexico.
- Coeval Ukrainian Boltysh, and North Sea impacts magnifying the Chixulub effect.
- Dinosaur, ammonite and rudist bivalve, showing acme of development, yet finally becoming extinct at the K/Pg, i.e., Cretaceous-Paleogene boundary.
- Many dinosaur tales—unhatched dinosaur eggs, snake feasting on eggs, baby dinosaur embryo, and the 500-m-long trail of dinosaur footprints in the Bolivian Andes.
- Warm climate even at the poles. Flora—cycads, ferns, conifers, and debut of the flowering angiosperms.
- Flowers with the color, scent, and nectar to attract pollinators.
- Abundant carbonate reefs everywhere.
- Asteroid impact ejecta hosting the giant, Mexican Cantarell oilfield, Gulf of Mexico.
- Commentary on the two main competing theories about the 5th mass extinction event—Chixulub cosmic body impact vs. Deccan volcanic eruption. A novel theory—asteroid impact triggering the Deccan lava eruption.

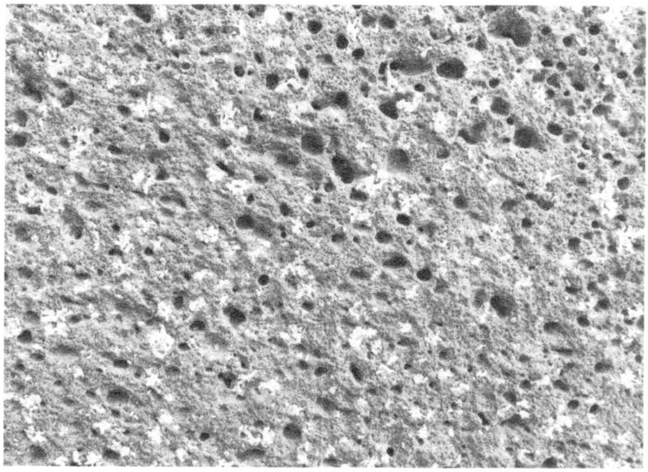

Vesicular Deccan basalt in the Ellora Archeological Site, Maharashtra, India. The off-white crystal patches are secondary zeolite mineral precipitated from hydrothermal solutions.

Paleontologist Dr. M. A. Lorente follows the giant footsteps (~1-m-across) of the Jurassic sauropod (vegetarian) dinosaurs near the 'La Playa de las Griegas' beach, Asturias, Spain. Photo Credit: Jose Luis Perez Sanchez.

9

Paleogene Period: Scorching to Chilly–A Boom for Mammals

THE CENOZOIC ERA: 66 Ma—PRESENT

The Paleogene Period (66–23 Ma)—How It Differed from the Cretaceous

The Paleogene Period consists of three epochs—Paleocene (66–56 Ma), Eocene (56–34 Ma), and Oligocene (34–23 Ma). It shows a gradual paleogeographic change from the preceding Cretaceous Period. Also, extreme temperature fluctuations marked the period. The snow-free Paleocene-Eocene poles contrasted with the Antarctic icehouse debut in the Oligocene.

A shallow North Sea appeared between Britain and continental Europe, while a large inland sea covered most of eastern Europe. The vast Western Interior Seaway of North America disappeared. Many natural events—dominant plate tectonic activity, rising peaks (i.e., Everest), extensive coal forests, mammalian dominance, are some of the signature Paleogene features.

PALEOGEOGRAPHIC HIGHLIGHTS

The Restless Tectonic Plates

By Paleocene, the old southern Gondwana mega-continent was no longer part of the world map as one entity. Antarctica and Australian landmasses became independent. The Atlantic Ocean from south to north was wide open. Laurasia had not yet broken into three landmasses. Greenland, still attached to Europe, was now separating from North America. Marine waters of the Western Interior Seaway had receded to the Gulf of Mexico by the early Paleocene. The Eocene Laramide orogeny caused the final uplift of the Rocky Mountains. Farther south, a pair of colliding plates proved to be transcendental. The east-moving Pacific Plate, subducting under the South American plate, gave rise to the Andes Mountains. The Andes runs north-south, hugging the west coast of South America. The Andean uplift, though mainly of Oligocene-Miocene age, continues today as the Pacific plate is actively subducting under the continent.

To the north, transform (strike-slip) plate motion marked the eastern Pacific shoreline (near present California, USA), 40 Ma ago. The Pacific and North America plates have been sliding past each other (Fig 3.16.1) along the N-S trending, 1,280-km-long and 20-km-

9-1 The Continental Divide at Loveland Pass, Colorado, and the backdrop of the snowy Rockies.

deep, San Andreas Fault System. The plates have slipped past 300–350 km since the Oligocene, ~30 Ma ago, along the western edge of North America.[1]

The earthquake-prone fault zone marks the boundary between continental North America and the oceanic Pacific plate. In recent times, several major earthquakes along the San Andreas Fault struck San Francisco in 1812, 1857 and 1906. The fault literally slices the State of California. While San Diego and Los Angeles are passengers of the north-going Pacific plate, the oblivious San Francisco, Sacramento, and Sierra Nevada Range are gently gliding by on the American continental plate.

The Rise of the Rockies—The Wall That Formed the Continental Divide

The peak Laramide orogeny of the Early Eocene put the finishing touches to the towering backbone of North America—The Rocky Mountains. The Rockies start at the northerly eastern British Columbia and Alberta (of Jasper and Banff fame), Canada, all the way through the western United States up to the state of New Mexico at the south end. The foothills of the Canadian Rockies separate the Rocky Mountain fold thrust belt from the Western Canada Sedimentary Basin. The lofty range rises far from the flat-slab (i.e., low angle) subduction of the Pacific Plate under the stable North American plate. The rocks range from Precambrian to Pleistocene.

The Rockies have a special distinction of forming a watershed between two oceans. The challenging Berthoud Pass Trailhead along the US Route 40 follows the crest of the continental divide and the watershed between the Pacific and Atlantic oceans. It is a singular wilderness area of breath-taking landscape, snowy peaks, pristine trees and forests, and unique wildlife. The Loveland Pass in Colorado, at 11,109 ft height, on Highway CO-6 also sits on the continental divide (Fig. 9.1). Similarly, the majestic Canadian Rockies Continental Divide separate two watersheds, and the Kootenays and Jasper national parks in British Columbia and Alberta respectively.

9-2 A great view of the rolling snow-clad peaks of the Rockies from the road outside Denver, Colorado. The peaks are sharp, while the broad symmetrical valleys are synformal.

About 100 Ma ago there were no Rocky Mountains, only an arm of the sea—the Western Interior Seaway (WIS) in the area. The sea accumulated thousands of meters of sediments during its lifetime. The WIS became a victim of the Laramide orogeny uplifting the older rocks and sediments and forming the Rockies. Later volcanism added igneous rocks to the mountain edifice. The final Neogene push shaped the Rocky Mountains as we know today (Fig. 9.2). Plio-Pleistocene glacial events stripped off much of the rock sequence.[61]

Many oil and gas basins occur along the length of the Rockies. The Canadian British Colombia Province is rich in natural gas (Horn River, Liard, and Cordova basins). In the US, the Denver basin in Colorado and Wyoming, and the Uinta and Paradox basins in Utah, store a wealth of hydrocarbons—oil shale, natural gas, oil, tar sands, etc. Other than energy resources, the Rockies also produce kimberlite, silver and other minerals and metals from the above mentioned, and other basins—Big Horn, Wind River, Green River, San Juan, Raton and Powder River, etc. The last two basins are rich in coal and coalbed methane resources.

The New Drake Passage Banished Antarctica Forever in Permanent Chill

The Late Eocene northward 5^0 drift of the South American Plate cre-

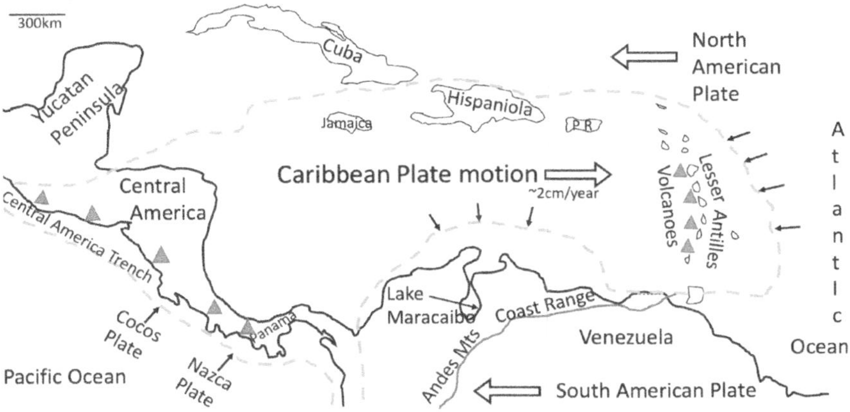

9-3 The Caribbean Plate moving east, and North and South American plates moving west. The Cocos and Nazca Plates dive under the Central American island arc dotted with volcanoes. Solid triangles—volcanoes above subduction zones.

ated the 'Drake Passage' at its trailing edge. The passage lies between the continent's southern tip—'Tierra del Fuego' in Argentina, and Antarctica.[2] The opening of the Drake Passage spawned the early Oligocene Circum-Antarctic-Polar Current, triggering the birth of the Antarctic ice cap.

Rafting Plates Go Places—The Caribbean Plate Traveled from Galapagos to the Tropics

The origin and evolution of the Caribbean plate is complex and probably some details are still missing. Most experts think that the Proto-Caribbean Plate had traveled from the vicinity of the far-off Galapagos Island in the Pacific Ocean.[3] It made the voyage after the Jurassic Pangea breakup, when Laurasia (including North America) and Gondwana mega-continents went their separate ways. The Caribbean plate conveniently inserted itself in the open space between North and South America and began its steady, eastward journey in the Paleogene (Fig. 9.3). The Caribbean plate kept colliding or side-swiping with the westerly moving North and South American plate margins. Some distinctive Caribbean plate fragments got stitched to the adjacent North and South American continents along

9-4 Poás volcanic crater in Costa Rica located on the Central America trench. It is an active volcano, still erupting in the new millennium.

suture zones. The intense tectonic activities gave rise to novel geographic features like the Greater and Lesser Antilles, Dutch Antilles, Venezuelan Coast Range, Barbados Prism, Yucatan basin, etc.

Two arc trenches border the east-west extremities of the Caribbean plate. The westerly Pacific Cocos and Nazca plates are diving under the lighter Caribbean plate at the Central America trench. A string of active volcanoes occurs in Nicaragua, Costa Rica, Guatemala, Honduras etc., situated on the adjacent island arc (Figs. 9.3 and 9.4).

At the east end, the subduction of the North American plate under the Caribbean created the arcuate Lesser Antilles Island Arc, and a chain of volcanoes of Miocene and younger age. These islands are tourist attractions today, famous for scenic beauty and turquoise blue waters—Montserrat, Martinique, Guadalupe, Saint Vincent, St. Lucia, Grenada, and others. The 'Kick 'em Jenny' is a submarine seamount, 10 km north of Grenada. Barbados is the only Caribbean Island of non-volcanic, biogenic (coral) origin. Older allochthonous suture zone rocks mark the islands of Cuba, Hispaniola (Haiti and Dominican Republic), Jamaica, and Puerto Rico, etc., on the northern strike-slip Caribbean margin.

The Oil-rich Maracaibo Basin, Venezuela— The Caribbean-South America Collision Shaped It

The South American plate's northerly edge tells an interesting story. Its Paleocene-Eocene oblique collision with the Caribbean plate gave rise to the 'foreland' Maracaibo basin.[4] A 'foreland' is a sag basin that forms parallel to a newly created mountain belt. The Rockies, Andes, etc., have their foreland sags on one of the flanks.

Volcanoes do not dot the south Caribbean coastline in the absence of a major subduction process. The orographic high of the Venezuelan Coast Range is mainly due to lateral and opposite, east-west sliding motion of the Caribbean and South American plates. The Villa de Cura nappe of metavolcanics and metasediments in Venezuela is a fragment of the Caribbean oceanic crust that obducted, i.e., jumped from the subducting Caribbean plate on to the South American continental plate above.

India-Eurasia Collision—
End of Indian Plate's Long Journey from Deep South

The Indian plate continued to head closer to its Eurasian destination in the Paleocene Epoch. Eurasia extended from Spain-Portugal on the west, all the way east through Russia and China up to the Bering Sea. It runs from Russia to India in a north-south direction. The intervening Tethys Ocean between Eurasia and India likewise had been gradually vanishing under the Eurasian continent. The African and Arabian plates were also marching in tandem with the Indian plate northward. The sailing plates eventually collided with Eurasia in the Eocene, shaping the Alpine-Himalayan Mountain belt, and many high peaks (Fig. 9.5).

9-5 View from the 3636-m-high Sandakphu summit, West Bengal, India. Mt. Kanchenjunga, at the India-Nepal border, the world's third highest peak, resembles the 'Sleeping Buddha.' The head (left) is 'Kumbhakarna' Peak, torso—Mt. Kanchenjunga, the feet is 'Pandim' (far right).

To note, India's original continental collision began 2,000 km south of its present position. The India-Eurasia rendezvous had occurred in 2 stages.[5,6] Apparently, a string of island arcs from the south first collided with Eurasia 50 Ma ago. Thereafter, India collided head-on with Eurasia ~40 Ma ago. The Kohistan-Ladakh Arc region in the high Himalayas represents island arc volcanic rocks sandwiched between the colliding Eurasia and India. Since that time, India had rammed farther into the belly of Eurasia, deforming both its own leading edge as well as the Eurasia margin.

Paleomagnetic data suggest that the Himalayan thrust sheets caused crustal shortening of 700 km, shaping the highest mountain chain in the world.[7] 'Crustal shortening' results from horizontal compression of plates; the crumpled rock units will create majestic 'highs' like the Himalayas. Imagine what happens if one pushes a rug from two ends. The rug will form a fold, i.e., an arch in the middle, while the rug length will be shortened because the ends come closer. When landmasses of same density collide, neither can go under. The only way to release stress is to buckle upward as the Himalayas did. The Everest rose >9 km in 50 Ma.[8]

Wonders of the Everest and the Tibetan Plateau

Mount Everest consists of metasedimentary rocks, piled high by thrusting. At the summit, a ~450-Ma-old Ordovician limestone-dolostone unit crops out with marine trilobite, crinoid, and ostracod fossils. Algal cyanobacterial stromatolites are also common lower down. The fossils and rocks were all dredged up from the 'Sea' that is no more—the Tethys Ocean. The Indian plate lost nearly 1,000 km of its leading edge in 20 Ma, at the rate of 5 cm/year.[9] Mass balance analysis reveals that the mantle had consumed about 50% of the mass of India and Eurasia in the last 60 Ma.[10]

The Tibetan Plateau lies north of the Himalayan Mountain chain, crowned by 13 peaks higher than 8,000 m, including the Everest. The India-Eurasia collision pushed up the vast Tibetan Plateau, measuring 1,000-km-wide, 2,500-km-long, and more than 4,500-m-high.

The plateau displays the thickest continental crust on Earth, about 70 km thick, rarely even 100 km.[11]

The whole region is tectonically still very active as India plate continues pushing into the Tibetan plateau. The underplating of the Indian plate is exerting pressure on the stable Eurasian landmass to the north. The result: an enormous amount of stress is slowly building up there. Numerous cross-cutting faults, earthquakes, and detachments release the stress from time to time, affecting countries like Afghanistan, Pakistan, India, Nepal, Tibet, China, and others.

The Tethys Saga—The Sea That Just Vanished

The idea of a disappeared Mesozoic equatorial sea separating a pair of continents dates back to the 1880s. Later, Edward Suess christened the sea as 'Tethys,' after the sister and consort of the Greek God of the Ocean—Poseidon. Incidentally, the sea experienced two incarnations—Paleo-Tethys, and Neo-Tethys. In the Late Paleozoic, the Paleo-Tethys existed between the northerly Laurasia (N. America/Eurasia) and E-W trending Cimmerian terrane that had just splintered from the Gondwana. The Cimmerian terrane was a string of northern Gondwana microcontinents now representing Turkey, Iran, Afghanistan, Tibetan Lhasa, and Malay Peninsula including the Sibamasu terrane.[12, 13] Glacial tillite and diagnostic fossils confirm the Gondwana origin of the microcontinents. The Eurasia-Cimmerian collision during the Triassic closed the intervening Paleo-Tethys Ocean. The Neo-Tethys formed behind the northward drifting Cimmerian terrane, remaining as the only sea between Eurasia and the Indian plate. However, the Neo-Tethys had the same fate as Paleo-Tethys, it vanished after India docked into Eurasia 40 Ma ago.

Lake Baikal—a Siberian Lake of Rift Origin, Deeper-Than-Ocean

Changing the subject while still in Asia, let us look at a unique place in the southern Siberian craton, about 3,000 km north of the Himalayas across the Mongolian territory. The 1,642-m-deep Lake Baikal is the largest freshwater lake in the world. It lies below sea level because of active subsidence. The lake water is so clear that visibility may reach a depth of 40 m below the surface. The rift-related fissure started in the Cretaceous, then continued into the Oligocene when active opening produced the half-graben sag. Soon the Baikal depression became a lake with moderate sedimentation. Baikal's renewed Plio-Pleistocene rifting may be a collateral effect of the Eocene India-Eurasia collision.[14]

The Berber Country, Atlas Mountains, Africa—Once Was So Close to North America

It embodies a complex geologic history dating back to the Alleghenian orogeny of the Carboniferous-Permian age. The Africa-North America collision had produced the Anti-Atlas Mountains of Morocco. Across the Atlantic, the 1,400-km-long, Piedmont escarpment 'Fall line region' of the eastern United States, is the other half of the Anti-Atlas.

The 2,500-km-long High Atlas Mountain chain extends from the Atlantic coast of Morocco to the Mediterranean coast of Tunisia along the northern border of the Sahara Desert. The Atlas likely developed in two stages.[15, 16] The High Atlas horst developed, starting in the Cretaceous from *inversion* of an older rift basin. Later, the Africa-Europe collision finally resulted in the Tell Atlas *inversion*, coinciding with the Oligo-Miocene Alpine orogeny.[17]

Talking about the *inversion* of basin-fill, what is it? A basin is a saucer-shaped depression where sediments accumulate. 'Inversion' simply means that sediment in a basin is turned upside down by tectonic deformation and uplift. A bowl of pasta may be an apt metaphor to imagine the process of inversion. When a cook inverts a bowl of pasta upside down, a little 'mountain' of the pasta forms on a plate resulting in a central high and sloping flanks.

The Fate of the Aegean-Anatolia Plate—It's in a Tough Place

The Paleogene collision of the northward moving African-Arabian tectonic plate and Eurasia triggered the Alpine orogeny, including the later Aegean-Greece-Turkey configuration. The Aegean-Anatolia region of southern Greece and Turkey forms a microplate, sandwiched between the stable Eurasia and the southerly mobile Africa-Arabia plates. The North Anatolian Fault, (NAF) of Miocene age separates the Aegean-Anatolia microplate from Eurasia that carries northern Greece and a sliver of Turkey including Istanbul.[18] It is one of the world's most seismically active zones, producing many historic earthquakes. The Istanbul City of 15 million people is only 20 km away from the North Anatolia Fault. The vise-like tectonic grip of the combined Eurasia and Africa-Arabia plates from north and south, has been pushing out the Aegean-Anatolia microplate westward along the North Anatolian Fault, creating periodic earthquakes. The latest, a devastating 7.8 magnitude (Feb 2023) temblor killed over 50,000 people and toppled >160,000 buildings in Turkey and Syria. Geologists compare the Anatolia fault with the earthquake-prone San Andreas Fault of Oligocene age, in California, USA, also a strike-slip fault between the Pacific and the N. America plate.

Eurasian Duel With Africa/Arabia— The Birth of the Alpine-Caucasus Mountains

The Alps, spread over many countries in Europe, is another fascinating tale of how plate tectonics constantly reshaped Earth. The Oligocene push by the African plate gave rise to the Pyrenees first, and then the majestic Alps. Between the crystalline rocks of Eurasia and the leading edge of the African plate, lay the western Tethys Ocean with its varied sediments and fauna. The Tethys gradually disappeared because of the northward push of the African Plate. Some of the vanished Tethyan deep-sea ophiolite and sediments ended up high on the mountains, along the Alps-Himalayan tectonic belt, extending from the Mediterranean to the Pacific.

Further east of the Alps, the Caucasus Mountain belt (~1,200-km-long) was the product of the Oligocene Arabia-Eurasia plate collision. The Caucasus region, shared by Russia, Georgia, Azerbaijan, Armenia, Turkey, and Iran, lies midway of the 16,000-km-long Alpine-Himalayan Belt. The Greater Caucasus belt lies north of the

Lesser Caucasus. The Russian Chechnya region falls in the Greater Caucasus, an uplifted horst block.[19] The Caucasus belt stretches from the westerly Black Sea to the easterly Caspian Sea, both are likely relics of the Neo-Tethys. Azerbaijan and the Caspian Sea host important oil reservoirs in the region.

Post-Cretaceous Global Climate—Drastic Ups and Downs

The marked change of sea level and climate sums up the main difference from the Cretaceous to the Paleogene. While the Cretaceous Sea level reached a near all-time high, the Paleogene ushered in a marked sea level fall, an overall cooler climate, and finally, the first Antarctic snow cap. The Cretaceous milieu favored the biotic eruption of lime secreting organisms leading to thick carbonate strata. In comparison, the Paleogene registered a progressive withdrawal of the sea, reaching its acme at the End-Oligocene (23 Ma ago). The sea level was 160 m below the peak in Cretaceous (still 60 m above the present sea level). Extensive land areas emerged as the sea receded. Also, tectonic rearrangement of plates triggered the uplift of basins. The net result was widespread non-marine, sand and mud sedimentation, and extensive swamps for lush coal forests in the Paleocene-Eocene.

Why Was the Paleocene-Eocene Transition So Hot? Large Igneous Province Events?

The Paleocene-Eocene witnessed a marked climate change, in tandem with the lower sea level. Paleocene pollens from a borehole core in the Arctic revealed that the region was warmer than today. But the short-lived Paleocene-Eocene Thermal Maximum (PETM) event was more extreme when high temperature was greatly upsetting global oceanic and atmospheric circulation patterns. The PETM caused extreme changes in Earth's carbon cycle along with a spike of 5–8°C from the earlier Paleocene time.[20] The mean temperature hit ~30°C in the lower Eocene. Earth's Poles were ice-free, pleasantly temperate. The Arctic was supposedly enjoying a balmy subtropical 13–14.6°C, like today's average global temperature.[21] The global climate cooled down later to ~20°C toward the End-Eocene.

What can satisfactorily explain the sudden thermal maximum and carbon release in the atmosphere at the Paleocene to Eocene transition? Apparently, thermogenic CO_2 and CH_4 release from large igneous province (LIP) events can explain the PETM warming. However, Marlboro Clay samples in Maryland, USA, suggest that CO_2 release from a cometary impact might probably better explain the ~5°C global warming at PETM.[22]

THE NEXT SURPRISE—THE OLIGOCENE DEEP FREEZE

The Antarctic Ice Cap

The Oligocene was much cooler (av.17–21°C) when the Antarctic ice cap formed ~34 Ma back. The sudden cooling sharply contrasted with the fleeting Eocene extreme heat wave. Geologists often refer to the Eocene-Oligocene transition as the "great divide" between the greenhouse and icehouse conditions. The climate was particularly cold in the Antarctic region, while Greenland and Pan-Arctic region saw ephemeral sea ice during Eocene-Oligocene transition.[23] Analysis of marine samples from many localities confirms appreciable cooling at high latitude (45^0–70^0) in both the hemispheres.[24] In the earliest Oligocene, both ice volume and atmospheric cooling event could have caused the observed 60 m sea level fall.

What was the cause of significant cooling at the Eocene-Oligocene transition? One theory is that a significant reduction of atmospheric CO_2 level would have triggered the temperature drop.[25] The marked increase of weathering and erosion of silicates, from the new Himalayas and Tibetan plateau, would have consumed much of the atmospheric CO_2. Carbonic acid from CO_2 + H_2O would have lowered the CO_2 budget in the atmosphere, thereby causing the Oligocene chill.[26]

Oligocene Ice Sheets—Why only at the South Pole?

The other puzzle is—why did the Antarctic alone experience an icehouse condition? We recall the northward drift of the South American plate in the Late Eocene opened the Drake Passage. This facilitated the east-west flow of cold Antarctic Circumpolar Current. As a result, Antarctica could not benefit from the caressing warmer currents of the Pacific, Atlantic and Indian Oceans, and glaciers started to appear. The Oligocene ushered in a new ice age, the first time since the Permian icehouse event. The cooling trend continued unabated in the Miocene, except for a minor blip of 21°C at the Mid-Miocene Climate Optimum (14–15 Ma ago). The Miocene average temperature was 15–17°C; the Pliocene average was 10–15°C (in general, not lower than today), when an ice cap finally crowned the Arctic Pole 3 Ma back. Glacial growth peaked in the Pleistocene when a less than 10°C average temperature was the norm.

STONY TESTIMONY

Rocks of Laramide Origin— The Basin and Range Province, Rocky Mountains

The rock stories of the Paleogene are too many to do full justice here. We can only highlight a few of these from the different continents and oceans to get a sense of the exciting developments in the post-dinosaur world.

We recall the Laramide orogenic event of North America due to the subduction of the Pacific Farallon plate under western USA, gave rise to the majestic Rocky Mountains. Several faulted basins developed in the Basin and Range Province of the Rockies. The Big Horn, Powder River, and Wind River basins in Wyoming, show the best development of these intermontane sags, filled with ~5,000 m of mostly Paleogene sandy sediments.[30] We mentioned the Powder River Basin is rich in energy resources—coal, natural gas, and oil. The region supplies 40% of coal demand in the United States. The Fort Union Formation of Wyoming and Montana, also provides economic resources like coal, uranium, etc.

9-6 The white line marks a great angular unconformity, Zulia State, Venezuela. A fluvial Neogene sequence lies above the inclined marine shaly Eocene (Paují Fm.) sequence below. Here the entire Oligocene rock record (11 Ma) is missing.

The Western Interior Seaway of North America—The 'Sea' No More

The seaway, lasting 70-Ma during the Cretaceous, finally receded towards the Gulf of Mexico. It still existed near the modern Mississippi embayment in the early Paleocene.[31] In its prime, the 1000-km-wide and 3200-km-long seaway had split North America in two parts, extending all the way from the Arctic Ocean to the Gulf of Mexico/Atlantic Ocean. The hallmark of the ancient seaway is a widespread carbonate sequence and teeming fauna. The Chalk cliffs of the 'Monument Rocks,' Kansas, USA, is a testament to the bygone ocean.

South America's Maracaibo Basin—Famous Reservoir and Oil-Kitchen

The basin resulted from the interaction of the Caribbean and the South American plates. It hosts a 4-km-thick wedge of deltaic-coastal marine sandstone and shale of the Eocene Misoa Formation—the main oil reservoir. By Late Eocene-Oligocene, the migrating Caribbean plate had moved away further east from the Maracaibo area.[27] As a result, tectonic loading ended, and the foreland basin became inverted. A major angular unconformity formed above the exposed Eocene strata (Fig. 9.6). Later, sedimentation again buried the Eocene reservoir rocks in the deep subsurface. Once the thick Misoa

9-7 Eocene outcrops of the Misoa Fm, the main reservoir rocks of the subsurface Maracaibo basin. **1** Large mega-ripples on a thin bed of near-shore tidal origin. **2** A knee-bend fold of the sandstone-shale sequence. **3** A 5-m-thick high-dipping sandstone in the center flanked by two shale beds. **4** Two thick sandstones alternating with two carbonaceous shales. Road sections, Zulia State, Venezuela.

sandstone reservoir was in place, it had to wait for the oil to trickle in. But where is the oil, and how will it reach the big sandy reservoirs? Maturation of organic matter generates hydrocarbon in an 'oil kitchen' in the deep underground. The 'kitchen' 'cooked' the carbon-rich source rock with heat and pressure to generate oil and/or gas. In the Maracaibo basin, the source rock and reservoir rock are fortuitously close by. The Cretaceous La Luna Formation, a world class petroleum source rock, was already generating oil right below the Eocene Misoa reservoir (Fig. 9.7). Deep faults and fractures acted as easy vertical plumbing pathways from the La Luna kitchen to the Eocene and younger Neogene reservoirs in the Maracaibo basin.[28] Nearly 44 billion barrels of oil are still left in the reservoirs for future exploitation.[29]

Coal, an Effective Carbon Sink—
A Needed Energy Source for Many Countries

Paleocene-Eocene coal and its cousin lignite of paralic (i.e., transitional coastal) swampy origin, is abundant and widespread. A warm, humid climate favored the growth of lush coal forests. Important coal and lignite reserves occur in China, India, United States, Canada, Venezuela, Colombia, Australia, Indonesia, and other places. China is the largest producer in the world. Wyoming in the US is the main producer of Paleogene coal, followed by Central Texas. In India, coal/lignite of Paleogene age occurs mainly in northeast Indian states—Assam, Meghalaya, and elsewhere,[47] where the author worked many years back. In Venezuela, commercial coal deposits occur in the Paleocene Marcelina Fm., in the Zulia State, and the author's familiar

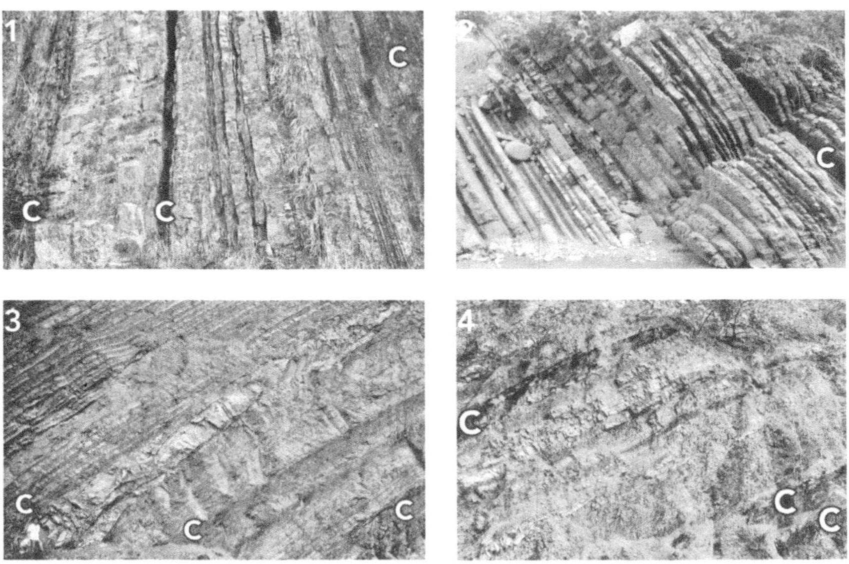

9-8 Images of some Venezuelan coal-bearing formations. The symbol 'C' denotes coal beds. **1** Vertical sequence of sandstone, shale, and coal beds, Valle Hondo Fm., Trujillo. **2** A sequence of sandstone, shale, carbonaceous shale and thin coals, Carbonera Fm., Táchira. **3** A lenticular sandstone in the middle flanked by thinly-bedded coal and shale, Valle Hondo Fm., Trujillo. **4** Thin coal seams interbedded with sandstone and shale, Los Cuervos Fm. Táchira.

Los Cuervos and Carbonera formations of Eocene age near San Cristobal in the Táchira state (Fig. 9.8).

The Colombian Cerrejón open-pit coal mine, tectonically related to the uplift of the Andean Perijá range, is the largest in Latin America and one of the largest in the world. It is located in the Guajira peninsula near the Venezuelan border. About 40 coal seams are part of the Paleocene Cerrejón Formation sequence. The seams range from 0.7–10 m thick, the average being 3 m.[62] In 2007, paleontologists discovered a never-before-seen fossil in the mine—a 58–60-Ma-old vertebra of the largest snake that ever lived, *Titanoboa cerrejónensis*. It was estimated to be 15–16 m long (45–50 ft), about a meter (3 ft) wide and weighing more than a ton.[63]

Other than coastal facies like the Misoa, and the widespread coal-bearing Paleogene formations, coeval deep-water facies (rocks)

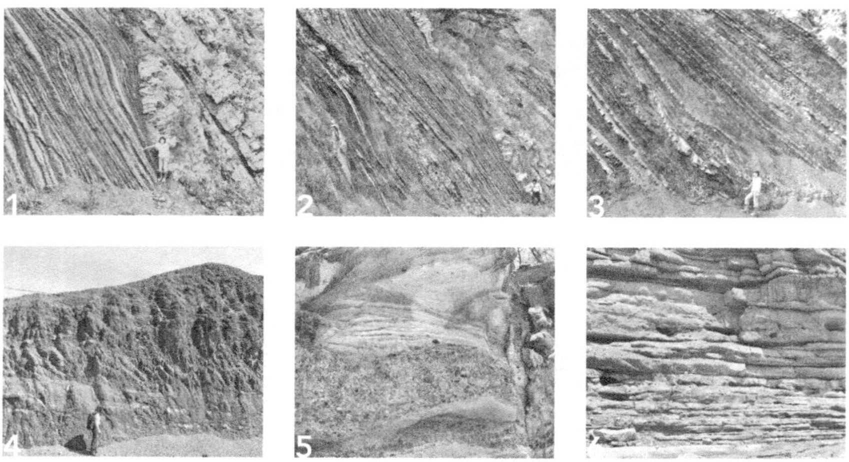

9-9 1 A partially exposed submarine turbidite channel with a scoured erosional base. It sits on an interval of interbedded thin sandstone and dark shale of overbank levee or inner fan origin. The channel is about 8 m thick. **2** A sequence of interbedded sandstone-shale and small turbidite channels. **3** A bathyal (>200-m-deep) dark shale-rich facies with thin, continuous, interstratified sandstone and siltstone beds. Images 1, 2 and 3 from Trujillo Fm. Valle Hondo, Venezuela. **4** Deep water organic-rich shale (source rock) with calcareous nodules, Oligo-Miocene Heath Fm. **5** A basal conglomerate and sandstone package in a submarine channel, Eocene Verdun Fm. **6** Stacked thin-bedded sandstone of turbidite fan origin, Verdun Fm. Note beds are thickening upward. Images 4, 5 and 6 from Tumbes forearc basin, northwest Peru.

9-10 A circular spiraling trace fossil about 10–12-cm-across of ichnogenus *Spirorhaphe*. It was made by a bottom-grazing Paleogene Sea creature from the allochthonous Caribbean deep-sea flysch rocks. Guárico Formation, Venezuela.

are present extensively in many places—North Sea, Andaman Islands (India), Trujillo State (Venezuela), Tumbes and Talara forearc basins in northwest Peru, Barbados in the Caribbean, the Greek Aegean Karpathos Island, the outer Carpathians of southern Poland, etc. In Peru, the deep-water facies formed in the forearc basin as a sequel to the subduction of the Pacific Nazca plate under South America.

The Paleocene-Eocene Trujillo Formation shows deep-water submarine channels and associated distal environments. In Peru, the Eocene Verdun Fm. consists of a basal conglomerate channel and sheet-like sandstone package of submarine fan origin (Fig. 9.9).

The Venezuelan Coast Range— The Caribbean Plate Was the Architect

After shaping the Maracaibo basin, the Caribbean plate moved farther eastward, scraping tangentially the northern margin of South America in the Oligocene. The encounter produced the Venezuelan Coast Range and adjoining foreland basin farther east in northeastern Venezuela, and Trinidad and Tobago.[2] The Venezuelan Coast Range, the surrounding mainland areas, the Margarita Island in the Caribbean Ocean, the Antilles Islands, host some transported Caribbean deep-sea rocks—ophiolite, slope turbidites, metavolcanics, and deep-water fossils (Fig. 9.10).

While we discuss rocks of deep-water origin, a brief mention of deep-water olistoliths is in order. What are olistoliths? These are chaotic mélange of older rocks of all sizes—blocks, boulders, cobbles/

9-11 Images of the allochthonous Paleogene Matatere Fm, Lara state, Venezuela. **1**, **2** and **3** show a gradual increase of sandy beds and laminae within the mud rock. Secondary calcite veins cut through **3** of inclined and deformed sand-shale rhythmites. **4** Vertical, deformed, thin-bedded sand-shale strata. **5**, **6** Two large (long axis 10–12 m) olistolith blocks of limestone (**5**) and interbedded sand-shale (**6**) are products of submarine sliding in the Paleogene deep-sea.

pebbles and sand, mud—all violently transported by natural agents like submarine gravity sliding and slumping. The Eocene Matatere Fm. is an allochthonous package of rhythmites, and spectacular olistoliths in northern Venezuela. The emplacement of the Matatere is related to the interaction between the east-moving Caribbean plate with the northern margin of South America (Fig. 9.11). However, similar jumbled Paleogene rocks are widespread, not only within Venezuela, but also in Colombia, Ecuador, Peru, and throughout the numerous Caribbean countries.[64] Traditionally, the Laramide orogeny has been invoked as the cause of this phenomenon. However, some geologists wonder if a Paleogene extraterrestrial visitor like an asteroid had a hand in the wide distribution of deep-sea rocks in northern South America, and especially, so widely in the Caribbean islands.[64] But where is the smoking gun?

The Alps: Partly African, as Told by European Rocks

The Alps is an Oligo-Miocene Mountain belt along the south margin of Europe. It is 200 km wide, 1,200 km long, spread over Italy, France, Switzerland, Germany, Austria, Slovenia, Croatia, and others. The Alps embodies rocks of three distinct source.

The highest point is the Matterhorn Peak—the name comes from German 'matte' for meadows, and 'horn' for a deer antler. The pyramidal, 4,478-m-high peak at the Swiss-Italian border, has its four faces aligned with the cardinal directions. The glacially carved sharp ridges mark the peak's four corners. A nearly 1,000-m-thick Paleozoic allochthonous granite gneiss adorns the summit 'Mont Cervin'—a memento of the African plate. It is standing tectonically tall, above a middle section—a mélange of transported Tethys assemblage of Jurassic marine limestone, sandstone, ophiolite, peridotite, etc. Autochthonous Eurasian marine limestone occurs at the base.[45, 46] Matterhorn is the finished, chiseled testimony of the hard collision between two plates—Africa and Eurasia.

Egyptian Pyramids Crowned with Eocene Limestone of Unicellular Critters

Nummulite, a noteworthy marine, bottom-dwelling foraminifer, is widely present in Middle Eocene limestones. The nummulitic limestone is extensive in southern Europe, Arabian Peninsula, Egypt, Turkey, Sahara, UAE, India, Philippines, Indonesia, Cuba, Venezuela, etc. The ancient Mediterranean people often used nummulitic rock as a building stone. *Nummulite* is a single-celled protozoa, usually 1–5-cm-across, rarely even 10-cm, as in Turkey. The genus is a critical extinct fauna ('index fossil') for precise age determination and geological correlation work. Dinosaurs are index fossils of the Mesozoic. The "Nummulitic Period" used to be the earlier moniker of the Paleogene, because of its importance in stratigraphy.[34]

The nummulite-bearing limestone was one of the building materials of the famous Egyptian pyramids (Fig. 9.12). It came from the nearby Tura quarry across the Nile, near Giza. The outer smooth cas-

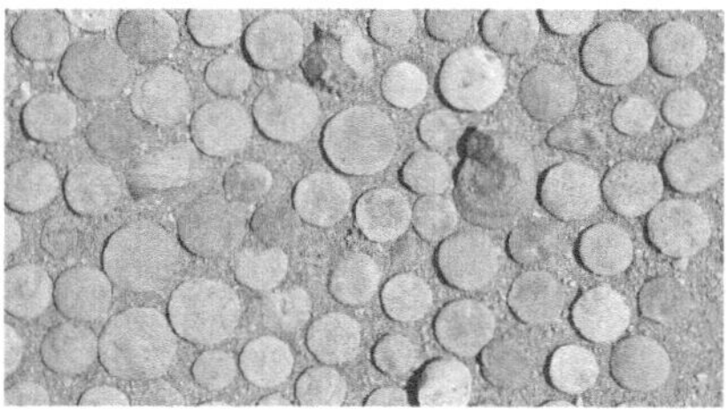

9-12 top Limestone building blocks (left) of the Giza Pyramid (right). **bottom** Platy, circular fossils of *Nummulite*, from Qarara (near Cairo), Egypt, adorn the top of the pyramid.

ings of the pyramids were of high-quality fine white limestone, now largely gone because of the ravages of time. The inner structure of some pyramids is made of sturdier, coarser limestone, supporting the entire edifice. Only the innermost burial chamber is made of a reddish granite brought ~4,500 years ago from the Aswan area—800 km away.

The ancient Egyptians also used the *nummulite* shells as coins or money for everyday barter or trade. No wonder, the Latin word '*Nummulus*' means 'little coin.' On a separate note, nummulitic limestone forms oil reservoirs in the Philippines and Indonesia.

Why does nummulitic limestone occur so extensively, but is only restricted temporally to the Middle Eocene? Data from Kutch, India, reveals that a higher strontium and calcium ratio coincides with the rapid growth of *nummulite* shells during the much warmer Middle Eocene. The foraminifers also showed a symbiotic relationship with photosynthetic algae. Favorable warm climate during the high PETM—Paleocene-Eocene Temperature Maximum, would have been crucial for the rapid growth of *nummulite*. On the other hand, a rapid enrichment of ^{18}O isotope, and a reduction in both shell size and diversity of the fossil at the Oligocene transition is meaningful. It speaks for a rapidly cooling climate, and a corresponding slower growth in the Oligocene. The cooling jolt would have killed most Eocene *nummulite* species.[35]

Everest: Anatomy of the World's Tallest Himalayan Peak

'Himalaya' means abode of snow (*Hima*, i.e., snow + *Alaya*, i.e., abode). It stretches for 2,400 km from the Indus River, Pakistan, to the Brahmaputra River, India, in South Asia. Everest stores rocks of multiple origin. It has autochthonous rocks of the Asian continent, smashed up fragments of colliding Indian landmass, and many slices of the bygone Tethys Ocean. The latter supplied deep ocean ophiolite, fossil-rich marine limestone, sandstone, etc. Slices of ophiolite got tectonically dredged up from the basement of the Tethys and now occur along the Himalayan belt. As India rammed farther into Asia, the collision pushed Tethys limestone layers high enough to crown the Everest.

Many geoscientists investigated the general geology of the Everest over the years. The British India Survey Company coined the name 'Mount Everest' for the peak straddling between Nepal and Tibet (China). R. N. Sikdar, an Indian mathematician and surveyor, first measured its height (29,002 ft or 8,924 m) as the world's highest peak in 1852. The Chinese name of Mt. Everest is Qomolangma. The Nepalese name 'Sagarmatha' or 'Peak in the Sky,' literally conveys the grandeur and staggering height of the peak. The 8,948.86-m-high (new data) Everest consists of four rock formations:[38, 39]

Bottom—Rongbuk Formation: Basement rocks of schist, gneiss, etc. intruded by leucogranite, near the lower part of Everest. A high-grade metamorphic sillimanite gneiss occurs in it.[41] A fault separates it from the Everest Series above.

Everest Series: Black schist, quartzite, and marble of original Tethys source (Cambrian or older).

Yellow Band Fm. (YB): A fault separates the top limestone from this eye-catching, fossiliferous, scarp-forming Cambrian marble (200–250-m-thick) of Tethys origin.

Top—'Gray limestone' or *"Qomolangma Limestone"* at the summit is a 225-m-thick, grainy rock. Many crinoid, trilobite, ostracod, brachiopod fossils (Ordovician age) from the Tethys Sea are now capping Everest. How did a Tethys limestone end up at the summit?

It is at the top because it was tectonically transported and placed above older rocks.[40]

Tibetan Plateau—'The Roof of the World'

Tibet is the 'Roof of the World' because of its vast expanse and height, with countless crisscrossing glaciers. It is also 'The Asian Water Tower' because many mighty rivers like Brahmaputra, Indus, Huang He, Yangtze, Mekong, Salween and Tarim originated on the plateau. Its other name 'The Third Pole' is meaningful—the many lakes, rivers, and snow make it the third largest repository of fresh water after the two poles.

The Tibetan Plateau is not just a giant single mass of rock, rather a collage of continental 'terranes' successively sutured to the Eurasian plate during the Phanerozoic.[42] A 'suture zone' is a linear belt where transported terranes from afar get glued together through intense tectonic stress, heat, and pressure in the deep subsurface. A *terrane* originated somewhere else, rather than where it is now. Major faults and dislocations characterize a suture zone, often lined with ocean bottom ophiolite. Paleomagnetic studies reveal that the 'terranes' were once part of the distant Gondwana continent. They successively said bye to the northern Gondwana during the Paleozoic/Mesozoic Eras, and progressively rammed into Eurasia.

The Himalayan belt lies south of all the Tibetan terranes, and in turn, separated from the Indian Plate by the Main Frontal Thrust. Terrane accretion had finished before the Cretaceous, ahead of the momentous India-Eurasia collision during the Eocene (Fig. 9.13).

The uplift of Tibet occurred in stages after the India-Eurasia collision. It began ~20 Ma ago and reached its present majestic stature about ~8 Ma ago in the Late Miocene.[42] However, Chinese scientists observe that the strongest uplift had jolted the plateau in the Pliocene, 3.6 Ma ago, and gained its present height only in the Quaternary period, ~1.1–0.15 Ma ago.[44] The crust became abnormally thick—about 70–100-km, due to tectonic pile-up of strata.

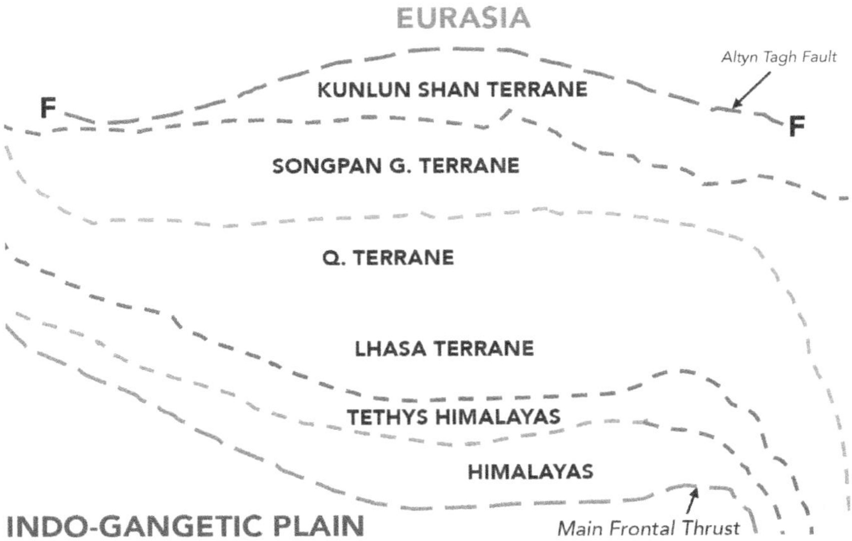

9-13 A schematic diagram (not to scale) of the Tibetan terranes (Kunlun Shan, Songpan Ganzi, Quiangtang (Q) and Lhasa) that accreted to Eurasia during the Mesozoic Era along suture zones (short, dashed lines) and faults (long dashed lines). India and Eurasia collided in the Eocene.

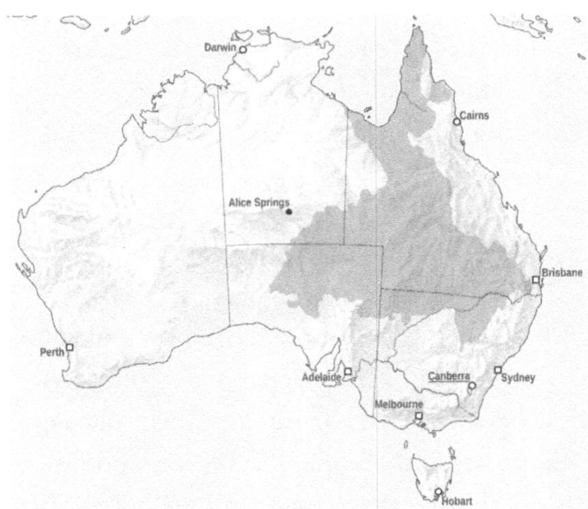

9-14 Location of the Paleogene Eromanga Sea, east-central Australia.

The Down-Under Eromanga Sea—
Upon Demise, It Gave Birth to the Opal Country

The Eromanga basin in Australia is another telling instance of sea level fall and disappearance of a marine basin. It was a shallow sea, occupying most of the eastern half of Australia during the Late Cretaceous (Fig. 9.14). It took shape 110 Ma ago when the Australian plate divorced Antarctica during the final Gondwana breakup and headed north. The plate separation caused east-central Australia to subside nearly 350 m, establishing a northerly marine connection. Gradually, seawater flooded the basin, converting it into a large inland sea.

Eastern Australia appeared like several emergent islands rising from the Eromanga Sea. The sea existed for about 20 Ma, and then it disappeared because of new plate rearrangements.

Soon Late Cretaceous and Paleogene fluvial-lacustrine sands and mud filled up the Eromanga depression and created the Great Artesian Basin of Australia. It's an invaluable source of groundwater nourishing 200,000 people in the area.

The marine Cretaceous Eromanga strata contain many important fossils and mineral sites. Many well-preserved reptile remains are entombed in the shaly host rock. The sequence is also the most important source of the famous Australian opals with unique play of colors, and rare opalized marine fossils.[32, 33, 65]

The 40,000-km-long Pacific Ring of Fire—
Vulcan's Fury and Earthquake

The Ring of Fire, a long chain of roughly 450 volcanoes, is part and parcel of the Pacific plate. The seismically sensitive Circum-Pacific-Volcanic Belt became active from the Cenozoic onward. It shows a horse-shoe shape around the Pacific Ocean and adjacent countries (Fig. 9.15). As the denser and heavier Pacific plate dives deep down below the continents, it heats up, melts in the mantle and resurfaces as basalt lava in the Ring of Fire. It encircles almost half the world and is also the subduction site of numerous smaller plates—Nazca, Cocos, and Philippine. Many eruptions occur every year along the

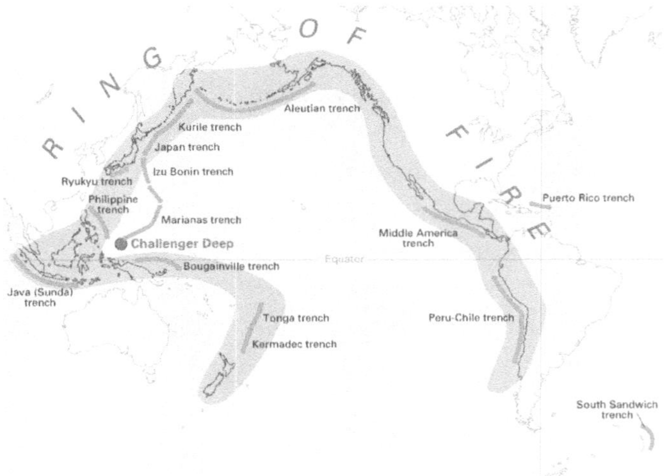

9-15 Pacific Ring of Fire girdling the Americas, Australia, and east Asia. The thin gray lines are trenches. The Challenger Deep is at the southern end of the Mariana trench.

belt. About 20 of the 25 largest lava eruptions of the past 11,700 years occurred on the Ring of Fire. It is also a highly earthquake prone zone—the site of nearly 90% of earthquakes in the world.[37] The subducting Pacific plate has formed numerous deep-sea trenches in the Pacific Ocean: Peru-Chile, Mid America, Aleutian, Japan, Mariana, Java-Sunda, Tonga, etc. The Mariana Trench is the deepest one, next to Mariana Island, about 2,285 km northeast of the Philippines (Fig. 9.15). The 11,034-m-deep trench ('Challenger Deep') dwarfs even Mount Everest.

Subduction activity continues everywhere around the Pacific Ocean except near offshore California, USA. The Pacific plate is sliding past the North American plate along the San Andres strike-slip fault, source of several high magnitude earthquakes, as in 1906.

MIRACLE OF LIFE

The "Age of the Mammals" – They Thrived in the Paleogene

After the End-Cretaceous upheaval, life came back slowly but steadily in the Paleocene and beyond. The exit of the leviathans made the field wide open, inviting in the diminutive Mesozoic mammals who were scurrying in and out of the hidey-holes or caverns to keep a safe distance from the daunting dinosaurs. Mammals diversified to all corners of the globe during this period.

Evidence from the USA reveals that all animal groups came back strong barely within 700,000 years after the asteroid impact at the K/Pg boundary. Some mammals flourished, becoming 100 times larger (from a tiny rodent to wolf size). Some 233 varieties of plants including legumes like beans also showed good recovery.[48]

The earliest Paleocene-Eocene primates are related to ancestral lemurs, originating in Africa and Mozambique. Primitive monkeys appeared in the Eocene.

Other Paleocene-Eocene Fauna and Flora—Life Beyond the K/Pg Apocalypse

Indeed, life was far from over at the K/Pg juncture. The lucky holdouts—snake, turtle, crocodile, etc., thrived in the Paleogene, after the traumatic experience of the K/Pg cataclysm. Whale, the giant of the ocean, evolved from land mammal *Pakicetus*. The skeleton of *Archaeoceti* in the Al-Hitan desert, Egypt, is one of the earliest forms of whale and included hind legs, feet, toes (Fig. 9.16). Other fossils of shark, crocodile, sawfish, etc., in the area, suggest a Paleogene coastal marine environment of the present Al-Hitan desert. The desert site, about 150 kms from Cairo, is a World heritage Site of UNESCO.

9-16 Fossilized skeleton of a giant whale ancestor (40–37-Ma-old)—*Archaeoceti*, in the Wadi Al-Hitan desert of Egypt.

The avian dinosaurs, birds, did beautifully, dominating the sky in the Paleogene. They diversified rapidly after the K/Pg mass extinction into all major groups present today. Paleontologists discovered the fossil of a tiny bird *'Tsidiiyazhi'* ('little morning bird' in *Navajo* dialect) in 62.5-Ma-old rocks in New Mexico, United States. It is the ancestral species of the modern mousebird, now restricted only to the African continent. The antiquity of this bird confirms that winged creatures did not take long to come back, likely only a few million years after the K/Pg.[52, 53] The flora did well in the humid and warm environment. Abundant coal forests of swamp origin are hallmarks of this period.

Evolution of Perissodactyls—Horse and Others

Some of the species we know today like camel, rhino, and horse, made their debut in the Eocene. The story of ancestral horse is an interesting Cenozoic evolutionary tale. The Early Eocene clan started with the ancestral horse—the *Eohippus*, a diminutive animal about the size of a dog. The Oligocene *Mesohippus* was slightly larger—61 cm high and weighed 45 kg, compared to a 500 kg modern horse.[49] The *Miohippus*, *Parahippus* and *Merychippus* were Miocene horses (102 cm high). The *Pliohippus* of the Pliocene age, looked more like a modern horse. The *Equus* evolved into the modern horse about 4.5 Ma ago.

Horses used to freely roam around North America until the recent past. They inexplicably became extinct (epidemic, hunting?) about 10 ka ago. Earlier, they had spread to South America, and to the rest of the old world via the Bering Bridge. The North American modern horse came from elsewhere, principally from Europe. Hernán Cortés and Spanish settlers reintroduced horses of Spanish origin to México in 1519. Thereafter, indigenous peoples transported them to the rest of North America along trade routes.

The discovery of the horse-like *Cambaytherium thewissi*, a 25–30-kg animal, was significant. The ~54.5-Ma-old fossil in the Cambay Shale Formation, Gujrat, India, is likely of the Perissodactyl order.[50] The Perissodactyls are non-ruminative hoofed mammals, the ancestors of horse, tapir, and rhino. Some argue that Perissodactyls originated in

India when it was still part of the Gondwana. Later they would have radiated out to Africa and other continents.[51] China is the site of one of the earliest Eocene Perissodactyls.

Oligocene Life—The New Asian Arrivals Compensated Extinction in Europe

The Eocene-Oligocene transition is significant as it heralded a cool period because of two potential culprits—volcanic eruption and meteorite impact. The Chesapeake Bay crater, United States (40-km-across), and the Popigai crater (100-km-across) in Siberia, bear testimony to the latest Eocene impacts.[54] These might have had a hand in the Paleogene turnover in flora, and particularly marine fauna during the Mid-Oligocene.

'Grande Coupure' and 'Mongolian Remodeling'—Two Cases of Faunal Turnover

The Circum-Antarctic-polar current in the Drake Passage suddenly cooled down Earth's environment, especially Antarctica. The general cooling caused significant demise of marine invertebrate and mammalian fauna in Europe and Asia. The European "*Grande Coupure*" or 'big break' at the Eocene-Oligocene transition, marked a turnover of mammalian fauna. About 17 Eocene endemic mammalian species became extinct. However, 20 new arrivals enriched the European population.[55] These included *Perissodactyls* (odd-toed), and even-toed *Artiodactyls* (camels, pigs, ruminants) from Asia.

The 'Mongolian Remodeling' is a similar episode of abrupt Eocene-Oligocene climate change, inducing a remarkable shift of flora and fauna in Asia. The large *Perissodactyls* of warm Eocene climate disappeared from the Jungaar Basin, China. The smaller *lagomorph* (Hare family) fauna dominated the forest-steppe dry climate.[56] The cow size '*Paleomastodon*,' an early elephant, appeared in the Oligocene. The Asian *Indricotherium*,' a rhinoceros, was 6 m tall at the shoulders. The largest land mammal ever, it weighed as much as 4 elephants.[36]

The Oligocene Flora—Some Retreated to Warmer Latitudes

The flowering plants, i.e., *angiosperms* were becoming dominant worldwide in the generally cool and dry Oligocene climate. So did grasses of different types and plants with non-woody stems (annuals). The *Gymnosperm* suffered some extinction in the cooler Oligocene onward and had to beat a retreat to warmer latitudes. The cooler open plains and savannas accelerated the spread of grasslands.[58, 59] The subtropical cashew and lychee trees, and temperate rose, pine, etc., developed, along with leguminous plants like peas and beans.

The Quadruped Primates in the Oligocene

Paleoanthropologists discovered old-world monkey fossils on a riverbed in Tanzania.[57] The Fayum locality, Egypt, yielded similar fossils of the genus *Apidium*, *Oligopithecus*, and others.[60] The apes appeared much later. The lesser ape gibbon debuted in the earliest Neogene, while the great apes in the middle Neogene to Quaternary.

An intact Eocene fossil fish *Knightia* (7-cm-long) from the ~50-Ma-old, Green River Fm., Wyoming, USA.

CHAPTER HIGHLIGHTS

- Plate tectonic-related dispersal of Gondwana landmasses from the south pole region, isolating Antarctica from the northerly warm currents.
- Plate movements initiating the earthquake-prone San Andreas fault, California.
- The Caribbean plate wedging into the empty space between the two Americas, during its northward travel from the Galapagos area.
- Collision of India-Eurasia, the Tethys demise, and rise of the Himalayas and the Everest.
- Common origin of the Atlas Mountain, Africa, and the Piedmont region, eastern USA.
- Eurasia's duel with Africa/Arabia plate leading to the Alpine-Caucasus mountains from the upper Oligocene onward.
- Rock details of the Pacific Ring of Fire, the Everest, Tibetan plateau, the Alpine Matterhorn Peak, etc.
- Appearance of early primates—lemur and old-world monkey. Also, the debut of horse, camel, rhino, snake, crocodile, etc. Debut of whale. Diversification of birds. Dominance of flowering plants and grasses.
- Paleogene climate swings—Hot (av. 30 ^{0}C) Eocene with lush vegetation and coal forest, preceding the cooler Oligocene with the new Antarctic snow cap.
- Significant Oligocene faunal turnover.

WHAT'S NEXT?

Chapter 10 reveals many new, fascinating aspects of the Neogene Earth, largely preserved intact. The Oligocene cooling trend continued unabated, as our early ancestors felt the first glacial chill at the End-Neogene.

10

Neogene Period: Mountain Chains Set—Icehouse Returning

The Neogene Period (23–2.58 Ma): Backstage of the Quaternary Period

The Neogene is momentous—our forebears finally descended from the trees and emerged on the plains. The Period offers one advantage—its rock record is largely intact. Weathering agents—rain, erosion, mass wasting, etc., did not get enough time to do the messy work, making the geologist's task somewhat easier.

Earth began to look somewhat like our modern world. The mountain building events continued in full force in the Neogene. The Himalayas, Alps, Andes, Rockies, Appalachians, Colorado Plateau—all were tectonically active in the Mio-Pliocene.

Plate tectonics created the towering mountains that spawned large rivers like Amazon, Indus, Ganges, Yangtze, Yellow and others. Salt diapirs and domes also crisscross the subsurface in some regions, including the Gulf of Mexico and North Sea. Regarding the salt, the temporary (geologically speaking) vanishing act of the Miocene Mediterranean Sea was an amazing event. It resulted in kilometers-thick salt deposits during the Messinian Salinity Crisis (7.25–5.33 Ma). Extensive deserts appeared on all continents due to climate change. The Oligocene cooling trend continued unabated in the Neogene. It finally culminated in the prolonged icehouse stage, though with transient, yet welcome, warmer interglacial breaks.

PALEOGEOGRAPHIC HIGHLIGHTS

Continents and Oceans in the Neogene

If we miraculously had satellite images of the Mio-Pliocene Earth, it would not have looked that much different. The continental plates virtually reached their present location in the Miocene (23–5.33 Ma) and more so in the Pliocene Epoch (5.33–2.58 Ma). Antarctica remained as an island continent, a permanent abode of snow. The Indian Ocean Islands, including Madagascar, and the Ninety East Ridge were in place. The firmly welded India had been still under-thrusting the Eurasian continent. Australia got a clean divorce from Antarctica, then traveled north, steadily readying for a rendezvous with the southeast Asian Archipelago.

The Gulf of Mexico on the west, became the theater of operation, where the North American plate has been actively subducting, i.e., diving under the Caribbean. Subduction created the Greater and Lesser Antilles volcanic arcs over time. About 19 active volcanoes dot the Lesser Antilles—Saba to St Kitts cluster, Montserrat, Guadalupe, Dominica, Martinique, St. Lucia, St Vincent, Grenada, etc. The only exception is Barbados, a non-volcanic coral island.

North America and Greenland finally became totally separate. Iceland appeared on the map. The Pacific Juan de Fuca Plate, a broken piece of the Farallon plate, has been still diving beneath the North American plate, creating the volcanic Cascade Ranges in California.

The Basin and Range Province, the site of the Las Vegas City, stretches from Oregon down to northwestern Mexico; we recall it is a major tectonic feature in the western United States. Alternate faulted blocks went up and down, sculpting the typical physiography. The escarpment faults separate the horsts (ranges) from the intermontane topographic lows (basins) of the 800-km-wide rift belt.[1]

Global Ocean Conveyer Belt— Finally the Arctic Polar Ice Cap Appeared

'The Global Ocean Conveyer' current carried warm tropical Gulf Stream moisture to northern latitudes where it came down as rain or ice in the Arctic Ocean (Fig. 10.1). Abundant fresh water hastened the formation of snow and ice in the Arctic,[2] coinciding with the general cooling in the Neogene. In the Pliocene, ice was somewhat ephemeral until about 2.8 Ma ago, when the North Pole donned its permanent icy crown. The Neogene cooling trend is long term, restarting after the brief warm interlude of the Middle Miocene Climatic Optimum

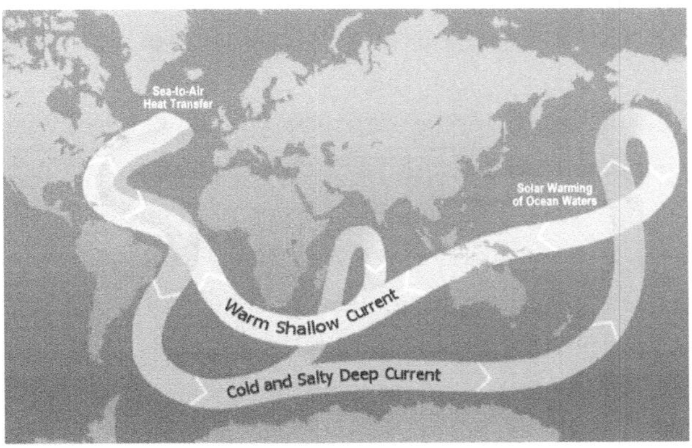

10-1 Deep Ocean current, 'The Global Conveyer belt' distributes heat uniformly between the warm shallow current, and the cold and salty deep current in the oceans.

interval (17–14.5 Ma). The Pliocene cooling event was 2–3°C warmer than today, and the sea level was also 25–30-m-higher. The Mio-Pliocene trend reached its icy climax in the Pleistocene.

The Isthmus of Panama—A Bridge Between the Americas, and a Wall Between Oceans

One singular End-Pliocene tectonic event affected Earth's climate, environment, and biodiversity. The large-scale migration and

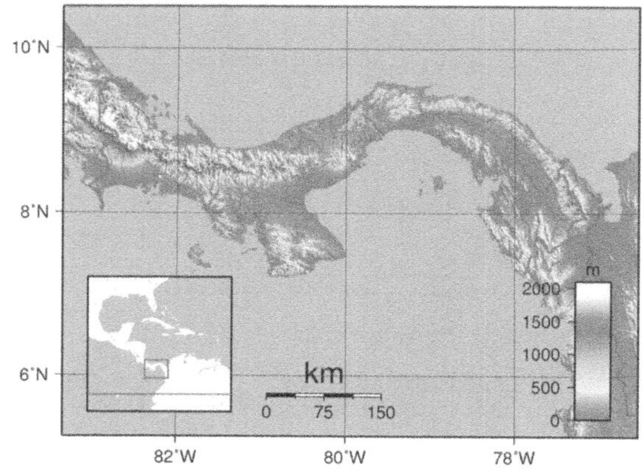

10-2 Panama extends from westerly Costa Rica to Colombia on the east. The Panama Canal extends from the southerly Gulf of Panama to the northerly Caribbean Sea near the 80 °W longitude.

flow of animal species between two continents became a reality. A ground-breaking first in Earth history—the Isthmus of Panama surfaced (Fig.10.2). It controlled maritime weather pattern and oceanic circulation, influenced climate in Europe and the northern part of North America. It helped to create the new Arctic ice cap and stood firm as a land bridge between two continents.

The Central American Volcanic Arc is a 1,100-km-long chain, with 15 active volcanoes, parallel to the Pacific coastline of Panama, Costa Rica, and Nicaragua (Fig. 9.3). The main arc volcanism ended in the Miocene (16–10 Ma ago). Soon after, the Costa Rica-Panama Island Arc of Pacific origin collided with the northwest corner of the South American plate, obstructing deep circulation between the Pacific and Caribbean.[3] However, the seaway was not permanently sealed until about 2.8 ka ago, when the Panama 'Isthmus' had appeared.[4] A narrow strip of land connected South, Central and North America with oceans on either side.

What triggered the isthmus formation? Data from old volcanoes and lavas of the Panama Canal region show that volcanism started 25 Ma ago, laying the foundation of the isthmus. The Pacific-Caribbean plate collision, as well as active volcanism epitomized the isthmian tectonic event. The submarine volcanoes rose to shallower level as collision progressed, while new eruption added more lavas over time, largely completing the edifice.[5]

The Ometepe Island of Lake Nicaragua and its hourglass-like pair of volcanoes, offer a small-scale modern isthmus analog. The conical Concepcion and Maderas stratovolcanoes were originally separate. Lavas slowly constructed a 3-km-long narrow bridge on the lake connecting the volcanoes. The still active Concepcion first erupted approximately 800 BCE, while Volcan Maderas last erupted 70 ka ago.

The Panama 'sluice gate' permanently stopped the intermixing of Atlantic and Pacific waters, greatly impacting Earth's later development. It created a permanent land bridge between the Americas, closing the Central American seaway.

Earlier, tropical waters used to flow freely between the oceans, balancing temperature and salinity in both. The circulation pattern in them had changed since then. The new isthmus deprived the Pacific of the warmer Atlantic waters that had veered northward toward Europe. The Gulf Stream warmed up southern Europe by about 10^{0}C. The cold, nutrient-rich upwelling Pacific waters produce a rich plankton and fish population, compared to the nutrient-poor, yet inviting, warm, turquoise Caribbean waters.[6]

Mantle Plumes Create Necklace of Islands—Hawaii and Similar Chains

The island chain of Hawaii is a prime proof of the incessant inner workings of the Earth engine. A deep, stationary, hot plume from a mantle convection cell, underneath the Hawaii Big Island, is rising and heating up the upper mantle, producing magma at shallow depths.[7] The mantle plume acts as a hot spot, erupting constantly on the ocean floor. The submarine eruptions created seamounts, atolls, volcanoes that eventually became emergent island volcanoes like the Hawaiian chain (Fig. 10.3).

Now, that is a simple enough story of how volcanic islands formed. But how and why do the island chains show a progressively younger age from the northwest to the southeast? Prof. Tujo Wilson (1963) first suspected the progressive age variation in all three Pacific Island chains (Hawaii, Tuamotu, and Austral Groups). He painstakingly plotted each island's radiometric age and observed the same trend in each island chain. The oldest island is at the northwest, and the youngest, presently erupting ones, at the southeast end. He visualized

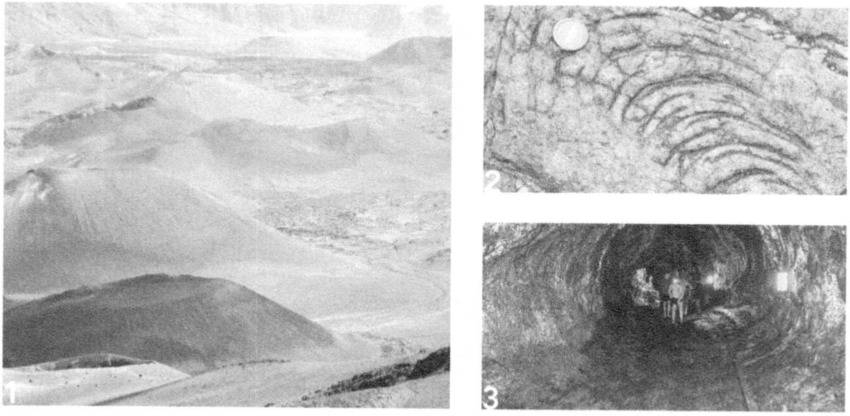

10-3 **1** Multicolored cinder cones, Haleakala Volcano, Maui **2** Solidified 'pahoehoe' lava, Maui **3** An old circular ~7-m-across lava tube, Big Island of Hawaii.

a fixed, deep-seated hot spot plume continuously erupting, while the sailing Pacific plate, like a raft, is slowly moving 8–9 cm/year northwestward above it.

The Hawaiian chain best exemplifies the antiquity trend. The Emperor Seamount chain at the northwest end varies progressively in age from 85 Ma to 39 Ma. The trend continues eastward—Midway Atoll 27.7 Ma, Necker Island 10 Ma, Kauai 5.1 Ma, Oahu 3.7–2.6 Ma, Molokai 1.9–1.76 Ma, Maui 1.3–0.75 Ma, and finally, the Hawaii Island at the extreme southeast 0.4 Ma–Present. Tomographic images showed that the mantle hot spot, now lying under the Big Island, is 2,000-km-deep and nearly 500-km-wide. The hot spots have been active from the upper Cretaceous onward. To note, older seamounts, islands, etc., are widely spaced, up to 200 km apart. The younger ones are closely spaced clusters, each with a shorter active life: example—The Hawaiian Islands.[8]

Fairy Chimney, Cappadocia—Seamless Blend of Geology, Culture, History and Religion

Nature took millions of years to shape the surreal scenery in middle Turkey. The marvels of 'Fairy Chimney' of Cappadocia, and the amazing labyrinthine maze of the world's largest underground rock-

10-4 1 View of the fairy chimneys from a hot air balloon. The hillocks on the horizon are full of rock-shelters, with a skirt of modern settlement below. *Photo: Lorenzo Farronato* **2** Lava, ignimbrite (tuffs), and sedimentary beds make up the plateau of Cappadocia, Goreme Yolu, Nevsehir. Note the dark layer behind the village forming the lower half of the plateau and showing typical fairy chimneys **3** A solitary fairy chimney in Nevsehir; it shows how the pointed cap of dark lava protects the parallel bedded softer (partly eroded) tuff layer with rock-cut dwellings.

hewn cities below, is unique enough that UNESCO declared it a World Heritage Site in 1985. The area falls on the ancient silk road in the Central Anatolian Volcanic Province (CAVP)—a sequel to the clash of the Afro-Arabian and Eurasian plates, and later extensional tectonics.[80, 81] Far away, Mt. Erciyes (3917 m) and Mt. Hasandag are two of the largest snowclad volcanoes in central Turkey. Radiometric dating assigns a Late Miocene (<10 Ma)—Quaternary (2.58–0.00 Ma) age to the Cappadocian ignimbrites. What is an ignimbrite?

It is simply a poorly sorted pyroclastic flow of hot volcanic ash and pumice, mixed with floating crystal, glass, and lithic fragments (ignimbrite = lithified volcanic ash, i.e., tuff). Fluvio-lacustrine sediments occasionally overlie the ignimbrites. Local stratovolcanoes like the Kardak caldera, long since eroded, produced the volcanic material.[80] Typically, the area shows plateaus crossed by valleys dotted with fairy chimneys with numerous rock-cut dwellings, and pigeon-holes, amid orchards and grape vines. The welded, hard, topmost ignimbrite layer or lava in the area forms a conspicuous plateau or table mountain (Fig. 10.4, image 2).

10-5 **1** Early-stage development of fairy chimney; the dark continuous cap (lava?) prevents pointy chimneys, though the soft ignimbrite below shows abundant signs of erosion **2** Tall towers with troglodytes **3** Near Urgup, a fairy chimney 'family' ('parents and a child'), with precariously hanging caps **4** Imposing multistoried, honey-combed, high-rise dwellings **5** Creamy cone-shaped fairy chimneys with dwellings, caps missing.

The unearthly fairy chimneys are pointy conical landforms dotting the magical valley scene (Figs. 10.4 and 10.5). How do they really form? On a gently sloping plateau, unequal erosion of ignimbrites along cooling fractures initiates the formation of chimneys. Harder interstratified lava, fluvio-lacustrine strata, and pyroclastic air-fall deposit may act as chimney caps. Fairy chimneys may show a spectrum of forms because the caps, though harder, also gradually wear down over time. Images 2 and 3 (Fig. 10.5) show vestige of caps, while total erosion of caps produced the pointed chimneys in photo 5. Experts opine that rainfall, temperature, and contrast in densities, accelerated the erosion of the fairy chimneys since the Last Glacial Maximum onward.[81]

The honeycombed troglodyte dwellings (Fig. 10.4 and 10.5) have a long history of human occupation over centuries. Ancient Turkey was the theater of many rivalries among Romans, Greeks, Persians, and Byzantine Christians; especially, the central Anatolia region was at the heart of it. People often took refuge in the fairy chimneys, because the soft, less dense, porous ignimbrite proved easy to carve cells, shelters, dwellings, etc., with simple tools—hammers and chisels (Fig.10.6). Byzantine monks lived in the honeycombed cells since 300 CE onward, escaping Roman persecution. They decorated their rock-cut churches, monasteries with religious frescoes evincing the fervor of early Christianity. Byzantine art and architecture find its best expression in the Goreme Open air museum monastic complex. It preserves ten such churches at the top of the hill, among other chapels and dwellings dating between 10th to 12th centuries. The church interiors feature lovely frescoes and paintings narrating popular biblical themes. However, photography of interiors is prohibited.

As an aside, there are many cave hotels in Cappadocia, catering to tourists seeking novelty and once-in-a-life-time experience. There are also impressive shops in spacious and cozy caves with all creature comforts.

The Underground Ancient City of Kaymakli—A Cozy Home for Fleeing Byzantine Christians

The Kaymakli City used to buzz with frenzied activities in its hay days. A tour of the hidden city makes one wonder how the ancient settlers scored such a feat. Turkish sources state that the earliest history may date back to 3000 BCE.[62] The Phrygians, i.e., Cappadocian Greeks (Indo-European people), first developed the city during the period 780–1180 CE to escape from Muslim Arab raids during the Arab-Byzantine war.[63] Originally, it used to shelter people from impending attacks of invaders, later it morphed into a thriving 85m-deep city through expansion (Fig. 10.6). To date, only 4 out of 8 underground floors of the city have been unearthed, where different floors had distinct uses: 1st floor—a barn, a stable, and a tandoor to bake bread; 2nd floor – a church, a cemetery, and living quarters; 3rd floor—kitchens, food warehouses, rooms, and wine-making from grapes, and also a copper smithing workshop; 4th floor—jars and supply warehouses, etc. The design permitted many thousands of people to live there with amenities like ventilation shafts, communication systems through pipes, and interconnected tunnels. The ventilation stack was at the center of the city, so that the entire space within received fresh air. The author saw firsthand the efficacy of the ventilation system while inside. The Kaymakli city had connection to all the houses in the region by secret passages, and also to the other notable underground city, Derinkuyu, by a 9-km-long passage. The inhabitants could tap water wells from inside. There are between 150–200 such other-worldly invisible cities in the Cappadocia region.[62]

The walls and roofs in the Kaymakli city enable geologists to examine the unaltered fresh texture of the ignimbrite (Fig. 10.6—image 7), a lithified rock, much softer than granite, basalt, and andesite; it's a porous mass of volcanic particles floating in a finer groundmass. Ignimbrite (tuff) is an excellent building material because it's lighter, durable, easy to carve, and has good insulation property. Ancient civilizations in Rome, Peru, and Easter Island, i.e., Rapa Nui (Moai statues), all used ignimbrite as building material.

Changing the Subject to the New Avatars of the Tethys—The Mediterranean, Caspian, and the Black Sea

In the old world, the area between Eurasia and Africa/Arabia plates had not yet looked like today. The Alps had been growing in height since 35–20 Ma ago, as Eurasia and the African plates were butting heads. Isolated bodies of seawater were lingering in the irregular intervening space between the plates, as the Tethys Sea was disappearing. The Mediterranean, Black Sea and Caspian Sea are isolated remnants of the vanishing Tethys.

The Messinian Salinity Crisis—When the Mediterranean Just Evaporated

The Mediterranean Sea suffered a unique global event at the End-Miocene. It went bone-dry in a very short time. Geologists call this rare event "Messinian Salinity Crisis," after the topmost Miocene stage (7.24–5.33 Ma). In 1970, an undersea drilling operation encountered

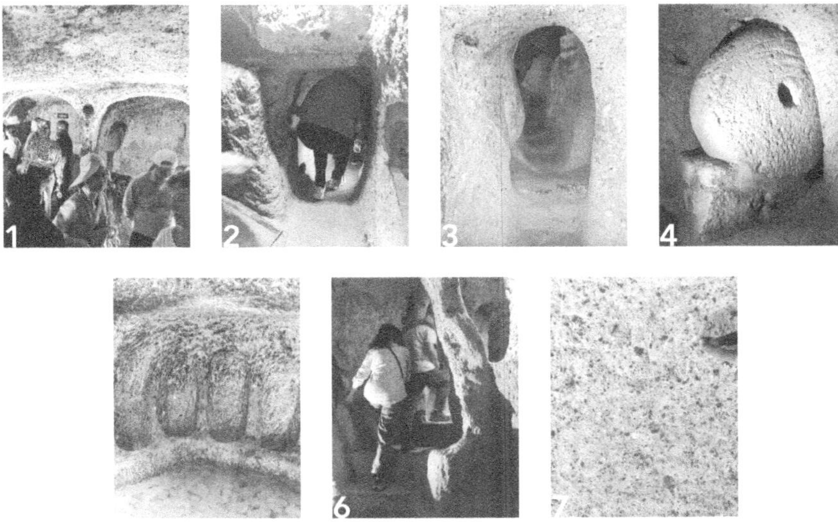

10-6 View of the Kaymakli city **1** Tourists at the entrance area **2** A person crouching in a low tunnel **3** A clear view of a narrow passage **4** A huge (~500 kilos) disk-like sliding stone that used to block the city entrance from invaders **5** View of a neat granary cellar **6** Visitors in a wider passage inside **7** Texture of ignimbrite or tuff showing dark lapilli and lithic fragments, set in off-white, fine-grained groundmass.

an unexpected 3-km-thick salt deposit, barely 100–200 m beneath the seafloor. How did the salt get in there? Evidently, once upon a time the sea had evaporated.

What caused so much evaporation? A global cooling event at the end-Miocene caused 50–70 m sea level drop.[9] The free flow of seawater from the Atlantic used to balance Mediterranean's normal salinity. The narrow strait of Gibraltar was the lifeline to make that happen. The drastic sea level drop cut off the Mediterranean from the Atlantic Ocean.[10] The location of the sea in a hot environment, just north of the Sahara Desert, did not help either. The drying up produced the 3-km salt column in just 0.7 Ma. Another aggravating factor—the northward advancing African plate pushed, raising the Mediterranean seafloor. The desiccation started first in shallow marginal basins; next, the theater of evaporation expanded, centered on the deepest part. Finally, the climate and fluctuating sea level made the Mediterranean a brackish (partly salty) lake.[11] Luckily, later the rising Atlantic Ocean refilled the Mediterranean again.

The Acropolis of Parthenon Fame is a Marble Memento of the Tethys

Few of us realize that the immortal attraction of Parthenon in Athens, Greece, is the product of a 'nappe' of Paleogene orogeny. In a nappe or overthrust, tectonic compression moves strata at a low angle above a younger stratum. The overthrust placed the older Acropolis Limestone above the autochthonous Athens Schist (it's a slightly metamorphosed sandy-marl facies, not a real schist).[8a] Prolonged erosion later of the main thrust sheet would have produced the '*klippe*' of the isolated Acropolis hill. A *klippe* is the tiny (geologically speaking) remnant of a nappe. Several similar *klippes* dot the greater Athens area. The famous marble of Athens, Greece, came through the violent end of the Tethys Sea. The marble of the majestic Parthenon comes from the Mt. Pentelicus, 16 kms northeast of Athens (Fig. 10.7).

Two global events did the Tethys in. The Eocene India-Eurasia collision blocked the connection between the Atlantic and Pacific oceans. Later in the Miocene, Africa/Arabia and Eurasia collision shaped the Mediterranean Sea, a western remnant of the Tethys Ocean. Vestiges of Tethys rocks are scattered widely along the Alpine-Himalayan Range and vicinity.

'The East African Rift System'— A Continent is Splitting Apart in Plain View

The East African Rift System (EARS) is a unique geological spectacle unravelling right in front of us. What is the driving force behind the most seismically active rift system today? Why is it located near the Horn of Africa? Active since the early Neogene, the rift fissure boasts impressive statistics—it is 6,400 km long, from Jordan, through Ethiopia, Kenya, Uganda, Tanzania, to Mozambique. It is a long chain of discontinuous valleys, bounded by faults (Fig. 10.8). Mantle plume-related volcanism is happening at the East African Rift

10-7 **1** An overthrust placed the 100-Ma-old Acropolis Limestone—a fine-textured marble of Tethys origin above the younger (72-Ma-old), autochthonous Athens Schist. **2** Close-up of the Athens schist **3** Thrust breccia **4** Acropolis limestone.

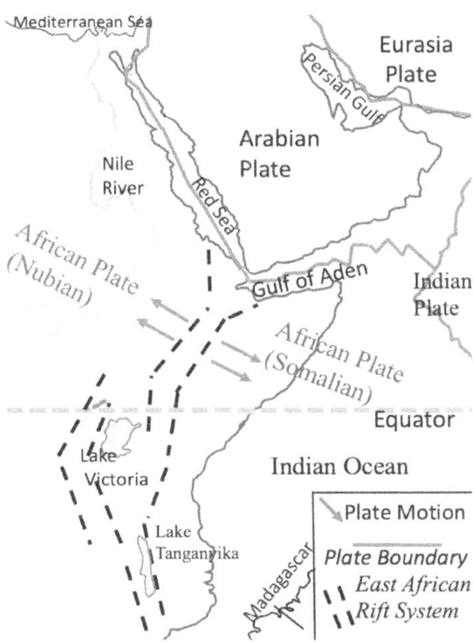

10-8 Sketch map of the East African Rift system, and the Eurasian, Arabian, Indian, and African tectonic plates. It runs from Jordan's Aqaba inlet in the Red Sea to Mozambique for 6500 kms.

System due to extreme crustal thinning. An Oligocene flood basalt eruption earlier coincided with the opening of the Red Sea and the Gulf of Aden. Along the African rift, the two sides are moving apart 6–7 cm/year, making the crust prone to rupture. The lava, high seismicity, and earthquake are hallmarks of the rift. Towering volcanoes like Mount Kilimanjaro, Mount Kenya, etc., dot the area. Presently, three active volcanoes are spewing lava, though 50 eruptions occurred in Ethiopia alone.[13]

Earthquakes often occur near the Afar region border faults in Ethiopia.[12] The last strong earthquake occurred in 2005, measuring 6.8 on the Richter scale. The epicenter was right in the middle of Lake Tanganyika between the Democratic Republic of Congo and Tanzania.[14]

The large lakes—Tanganyika, Victoria, and Malawi, are each nestled in individual rift basins, as they all formed on thinned crusts along axial fractures. The elongated Lake Tanganyika holds many global distinctions. It is the world's longest lake, holding 16% of freshwater reserves. It is the second oldest freshwater lake, the second largest—660-km-long, 50-km-wide, and the second deepest at 1,470-m. It laps the shores of Tanzania, Burundi, Democratic Republic of Congo, and Zambia.[15] The bottom 1,200 m of the lake is biologically 'dead' because of oxygen-depletion and high level of hydrogen sulfide.

The trend of Lake Tanganyika and bordering hill ranges follow the general rift-valley orientation. It originally formed as three separate smaller basins between 12–2 Ma ago. The lake becomes a single large one, only during highwater stage.

The Myth-rich Andes Story—The Longest Mountain Range, and the 'Andesite'

The Andean Mountain belt running north-south along the west flank of South America, is an imposing monument—the backbone of entire South America.[16] The grandeur of the Andes is awe-inspiring, as the author experienced while working in several Andean countries as a geologist. Early pre-Andean uplift involved local stable Precambrian rocks during the Cretaceous. The mountain owes its origin to plate interaction between the east-moving, denser, less buoyant Pacific Nazca plate diving under the west-moving lighter South American plate. Eventually, the down-going oceanic Nazca slab melts, as it reaches the critical temperature-pressure zone of deep mantle. The mountain grows, aided by the andesite lava flows rising through deep volcanic conduits and vertical fractures. Andesite is the volcanic equivalent of its coarser variety diorite (Fig. 3.3). Both are 'intermediate' rock types, i.e., subequal felsic (feldspar-quartz), and mafic (pyroxene-hornblende) minerals.

The ~4,000-m-high Andes is the world's longest (~8,000 km) mountain chain, between latitudes 10⁰N–57⁰S. It is the narrowest at its southern tip (150 km) in Tierra del Fuego ("Land of Fire") in Argentina, and the widest in the Bolivian Central Andes (600 km). Farther north, it narrows down again in Peru and Ecuador. Approaching Colombia, it splays out into three distinct cordilleras—Western, Central and Eastern, only the last continuing into Venezuela.

The Andes varies markedly from south to north in physiography, rock variety, etc. The 900-km-long southern Andes below 46⁰S, includes the Patagonia Ice Fields, Argentina. The Antarctic Peninsula may be the southward continuation of the Andes across the Drake Passage. The Central Andes is about 2,500 km long, in Argentina, Chile, Bolivia, Peru, between 46⁰–23⁰S latitude. The North Andes is 4,500 km long, from Ecuador to Venezuela up to 10⁰N.

The Highest Andean Peak—The Aconcagua, Argentina

The Aconcagua, close to the Chilean border, is about ~8,000 m high. It is over 200 km away from the deep-sea Peru-Chile Trench in the Pacific Ocean. The height difference between the Aconcagua Peak and Peru-Chile Trench is 15,000 m. Aconcagua, a conical stratovolcano of lava and volcanic debris, had grown during the 25-Ma-old Andean orogeny, at a rate of about 8-mm/year. Paleozoic red beds, tillites, Jurassic-Cretaceous marine sediments form its base. The tectonic push from the west to east along multiple thrust faults raised the Andean peak.

The Atacama Desert by the Andes—World's Oldest Preserved Mummies

The Atacama Desert is a geologically and economically important area of the Andes. It falls mostly within the Chilean territory, while Argentina, Peru, and Bolivia also share slivers of it. Earlier, Chile had a monopoly of world-wide nitrate trade, as Atacama holds the world's largest sodium nitrate reserve.

The Atacama hosts the famous 'Chinchorro mummies,' 282 in total, discovered in 1917 at a burial site near the Chile-Peru border.[17] The mummies are older (~5070 BCE) than the Egyptian ones. The 3-Ma-old desert faces the Pacific Ocean and continues for 1,000 km north to south. During the global aridity peak, desertification started about 15–10 Ma ago, and fluctuating semiarid conditions set in thereafter. The hyperaridity descended on the desert in the Pliocene.[18]

One may ask why the Atacama Desert is located where it is. For one thing, its latitude is farther south than other tropical/subtropical deserts like Sahara, Kalahari, etc. Additionally, it is located directly on the Pacific coast, next to the tall Andes Mountain. Some argue that the westerly location of the desert is the problem, i.e., the wrong side of the Andes, with respect to the prevailing wind pattern. Moisture-laden southeasterly trade winds hit the Andes, cools down and rains on its eastern flank. The Atacama Desert falls in the westerly rain shadow zone. Secondly, due to high pressure, descending moisture warms up and does not cause any rain in the desert area. Also, the cold Antarctic-hugging Pacific Humboldt current makes the onshore winds too cold to pick up moisture.[19]

However, others opine that the prevailing hyperaridity dating back 3 Ma was likely due to a pervasive global cooling event, creating deserts like Atacama, Sahara, and others. This idea does not hold water either. Computer simulations of past climate suggests that the 7-Ma-old Sahara appeared when the remnants of the Tethys vanished due to plate movements.[20]

Up to 900-m-high Atacama is partly mountainous, and 50 times drier than California's Death Valley.[21] Temperature variation is marked, 40 °C daytime and 5 °C at night. Following the lead of the

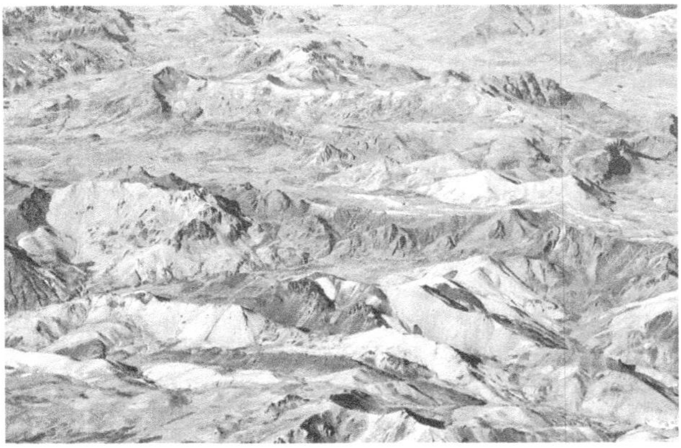

10-9 A bird's-eye view of the Peruvian cordillera near Cuzco, Peru. The different shades in the scenery are as noteworthy as the absence of vegetation or snow.

Viking Mars Missions, researchers performed soil tests in 2003 in Atacama's Mars-like Yungay region for signs of life, if any. They came back empty handed; it was devoid of life.

The Bolivian Altiplano or 'High Plateau'—A Gift of the Andes

The economically significant Altiplano contains one of the largest lithium reserves in the world. It is an extensive 3,650 m high plain, 965 km long. It lies between the Western and Eastern Cordilleras near the Bolivia-Peru border and pinches out toward Chile and Argentina. The Altiplano basin covers an area of 154,175 km^2. Unique only to the Bolivian Andes chain, it was probably a foreland 'sag' that developed in front of the rising 'proto-Andes.' Inversion and uplift of the Miocene foreland assemblage would have produced the Altiplano.

North of Bolivia, the Peruvian Andes varies in width from nearly 500 km in the south to about 200 km in the north. In southern Peru, the same Bolivian cordilleras continue. One unique aspect of the Peruvian cordilleras is that the parallel ranges converge at a midway point in Peru. It forms a knot, at a place named 'Nudo de Pasco,' or Pasco Knot.[25] A similar 'Vilcanota Knot' appears at Peru's south end, near the Lake Titicaca.

As one flies from Lima to Cuzco, the flight path passes above the Peruvian cordilleras. Near Cuzco, the Andes Mountain appears totally barren as the snow cap receded recently, and shows variegated colors, in the absence of vegetation and ice cover (Fig. 10.9). Surface denudation of the peaks is visible. The colors may be due to oxidation of iron oxide minerals (hematite, goethite, limonite), pyrite (FeS_2) and green chlorite and olivine.

The Peruvian Nazca Desert—Amazing Geoglyphs

Continuing from the Chilean Atacama, the Nazca Desert is the site of an old civilization highlighted by numerous geoglyphs on the desert floor. The area is 400 km south of the Peruvian capital Lima, where the Nazca people left behind a rich, priceless legacy of an old Andean culture. Quaternary alluvial sediments overlie the Neogene bedrocks in the desert area. The Nazca glyphs of 200 BCE–600 CE vintage, are concentrated in a 40 km^2 area. The lengthy, perfect geometric designs reflect the artistry of the people, showing diverse motifs of bird, mammal, crocodile, and human, including decapitated heads. How could they accomplish it on such a large scale (120–300-m-long), that it is best viewed from the sky above? Archeologists noticed that they scraped off a 10–30-cm-thick, iron oxide-coated top layer on the desert floor to expose the underlying yellow-grey subsoil. The design incisions are 0.3–2 m wide.[22, 23]

Conjectures abound on the significance of the artistry—astronomical, religious, irrigation/water supply, and social issues, etc. Whatever the purpose, the glyphs are a unique, timeless treasure displaying our ancestor's imagination and creativity.

Machu Picchu—The Andean Gem Immortalized in Stone

The famous Andean city Cuzco, and the iconic Inca cultural gem Machu Picchu, are situated on the Eastern Cordillera of Peru. Machu Picchu is one of the seven wonders of the world—a World Heritage Site since 1983 (Fig.10.10). The Incas had built Machu Picchu (530 m x 200 m) between 1450–1460 CE, with mostly 250-Ma-old local Andean granite (Fig. 10.11). They used mortar-less, dry-stone (ashlar) technique to build the sanctuary. The chiseled stones of the imposing citadel are so perfect that even a sheet of paper would not pass between the blocks. A great example of artistry is 'The Tem-

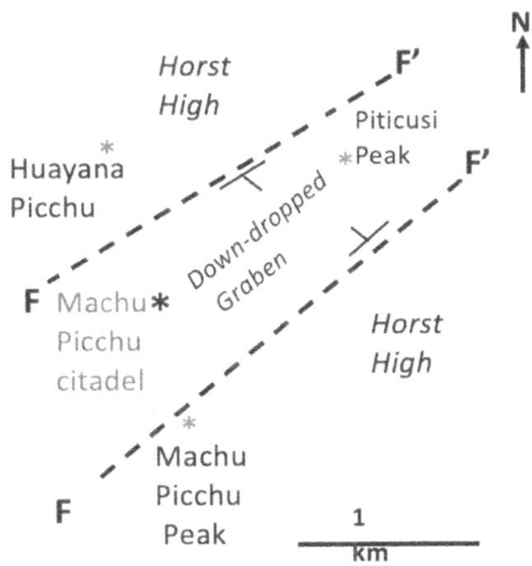

10-10 Majestic Machu Picchu citadel lies in the graben valley. Flanking horsts (highs) are Huayna Picchu and Machu Picchu peaks. Schematic faults after ref.[29]

ple of Three Windows.' The original quarry site sits nearby, though mostly ignored by tourists. Machu Picchu means the 'Old Peak' in Quechua language.

The Incas abandoned the sacred mountain-top sanctuary soon after the Spanish conquistador Pizzaro captured Cuzco in 1533 CE. Surprisingly, the Inca kingdom fell without a fight.[26]

Machu Picchu was unknown even to the colonial Spanish conquistadores. In 1911, Prof. Hiram Bingham of Yale University, USA, accidentally stumbled upon the Inca gem, while looking for the lost city of Vilcabamba. Mist and mystery shrouded it for so long, unknown to the world, and even to Peru. In 1916, Prof. Bingham and his group took away samples—shipped 74 boxes of Inca artifacts for study and systematic cataloguing in the Yale laboratory. It was supposed to be a temporary stay there. Yale finally returned all the artifacts to Peru only in 2011, exactly 100 years after its discovery. Years of Peruvian insistence finally bore fruit.[27]

10-11 Steep stairs in the Machu Picchu citadel, Peru, made out of the 250-Ma-old granite from local source. The roofless dwellings (foreground) are also made of the same granite blocks.

Machu Picchu edifice stands on a grey-white Permo-Triassic granite batholith (a large igneous rock body) that intruded a Permian-Jurassic rift system. The Andean Cordilleras did not exist at the time. Eocene tectonic compression of the ancestral Andes led to inversion, i.e., uplift, of the granite and older rocks (Fig. 10.11). Two peaks, on either side, frame the Machu Picchu citadel in the rift-graben valley—'Huayna Picchu' to the north and 'Machu Picchu Cerro' to the south.[29]

The Northern Andes—Heading Toward the Equator

The Andes in southern Ecuador is the narrowest, but the three cordilleras reappear farther north. The Pacific Nazca plate subduction has created some very active volcanoes, like the highest active volcano Cotopaxi (5,897 m) in the Cordillera Central. The inactive Chimborazo is even higher. The active Pichincha peak (4,784 m) overlooks the Capital—the city of Quito.

They Celebrate the Equator in Ecuador

The Spanish name 'Ecuador' means Equator, that passes through the country. The imaginary line of zero-degree latitude divides our spherical Earth into two equal halves. The hamlet of the 'Mid-point City of the World' ('Ciudad Mitad del Mundo'), 26 km from Qui-

10-12 A small stretch of the 40,000-km-long Trans-Andean Inca Road, the 'Qhapaq Ñan' in Ecuador.

to, commemorates the imaginary line. A 30-m-tall monument of local andesite stands here. A yellow line marking the mid-point at the monument's base highlights the two hemispheres.[30] Everyone enjoys the novelty—planting one foot each on the north and south hemisphere at the Mid-point City. It reminds us of the arbitrary N-S prime meridian (0^0 longitude—Greenwich, London, UK) separating Earth into the eastern and western hemispheres.

Ecuador also showcases many Incan fortifications. Inca Empire extended from Argentina up to Colombia. 'Ingapirca' ruins in Ecuador ('Machu Picchu of the North') represent the most famous citadel. They are unique, a blend of two distinct architectural styles of Inca and local Cañari group. The Cañari had started to build the Ingapirca edifices with mortar without polishing the blocks. Later the Incas finally finished the construction, including the famous, ceremonial, elliptical Sun Temple.[31]

The Pre-Columbian Trans-Andean Road—How Did the Incas Achieve It?

The Inca kingdom built an exceptional cultural milestone to connect the vast empire—the Trans-Andean Road. The 40,000-km-long, "Qhapaq Ñan" or royal highway, a UNESCO World Heritage Site, goes along the crest of the Andes (Fig. 10.12). It connected the Incan capital Cuzco to southerly Mendoza, Argentina, and northerly Quito, Ecuador. It even reached Colombia near Cali.[32] The Spanish conquistadors took advantage of the road regularly and renamed it "Camino Real" (Royal Road). A similar Incan pre-Colombian marvel is the 'Coastal Road' they had built. It started in Santiago de Chile hugging the coastline, headed to Lima, then reaching Tumbes near the Ecuador border. Bridges, stairways, stations dotted the roadway.[33]

The Colombian Andes—Where the Best Emeralds Lie Hidden

North of Ecuador, the Andes Range splays out into three Colombian cordilleras. The north flowing Cauca and Magdalena River valleys separate the three cordilleras. The 2,000-m-high, Western Cordillera does not extend far north. Cities like Medellin and Cali sit on the east flank of the Western Cordillera. The Central Cordillera is longer, more massive, and higher. About 15 active volcanoes adorn its crestline, including the snow-capped Nevado de Huila (5,750 m) and Nevado del Ruiz. The 1,200-km-long, Eastern Cordillera bifurcates forming a long intermontane valley about 2,700–2,800-m high, south of Bogota, the Capital city. The active Nevado del Cocuy Volcano (5,493 m) marks the highest point of the Cordillera.

The Colombian cordilleras formed successively, from west to east, and each displays distinctive suite of rocks. The western one formed in the Mesozoic because of the colliding Pacific and South American plates. The allochthonous mafic volcanic rocks and deep-sea metasediments occur there. The Romeral Fault System marks the suture zone between the oceanic plate of the Pacific, and the South American continent.[34]

The Central Cordillera had reached high relief in the Paleogene, due to interactions of the Pacific Farallon, Caribbean, and South America plates.[35] The site of the youngest Eastern Cordillera used to be a low-lying basin receiving sediments from the west and the east.[36] After the Miocene tectonic inversion, the sediment-filled saucer-like basin flipped into the towering Eastern Cordillera. The 8–16-km-thick strata became folded, faulted, and uplifted. The Mio-Pliocene Andean orogeny proved to be the final uplift phase for all the cordilleras.[34] The prized Colombian emeralds occur in an Early Cretaceous black shale and limestone between the Central and Eastern cordilleras (Fig. 10.13).

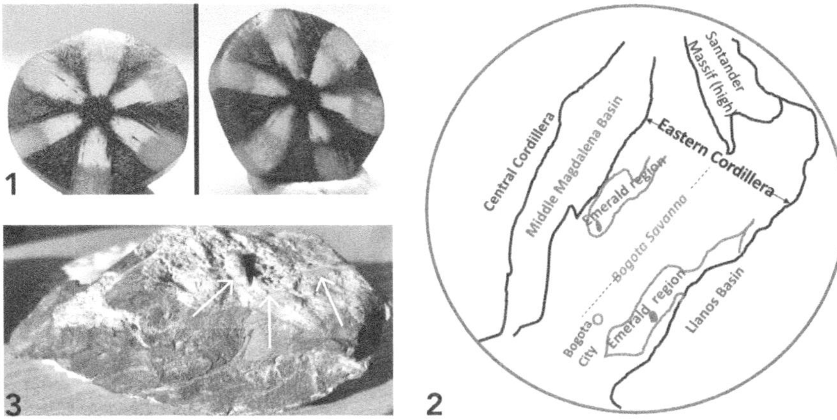

10-13 **1** Two views of the famous Trapiche emerald from Muzo mine, Colombia, with prominent asterism due to mineral inclusions. **2** A sketch map of emerald occurrence in the Eastern Cordillera. The two emerald regions are ~110 km apart, in the intermontane 2,640-m-high Bogota savanna. **3** Deep hydrothermal fluids created the slender, natural emerald crystals (white arrows), embedded in the Early Cretaceous pyritic black shale and limestone.

The Venezuelan Andes—Where the Caribbean Waters Beckon Nearby

The Eastern Cordillera bifurcates north of Cucuta, Colombia, into the Venezuelan Sierra de Perijá and Mérida Cordillera, both wrapping around the Lake Maracaibo basin. Vast river plains of the 'Los Llanos' lie to the east of the Andes in Colombia and Venezuela.

The Mérida Cordillera is about 600 km long and 100 km wide. It consists of two parallel ranges—Sierra de Culata and the easterly Sierra Nevada. The Sierras of granite gneiss and other rocks nestle the Chama River valley in between. The Mérida city is perched on the high, flat, fertile terrace of the Chama valley. The only Venezuelan Andean glacier, Humboldt, descends from near the tallest Bolivar Peak on the Sierra Nevada Range (Fig. 10.14). A cable car line ferries tourists from the 1,630-m-high, Mérida City to the 4,978-m-high Bolivar Peak, 12-km away. Interaction of tectonic plates of Pacific Nazca, Caribbean, and South America gave rise to the Venezuelan Andes. The final Plio-Pleistocene orogenic push created the Andean orographic high separating the Maracaibo basin from the southerly Bari-

10-14 The 4,978-m-high Bolivar Peak, Mérida Andes. The disappearing Humboldt glacier is sourced by a nearby peak. The outcrop is a metamorphic assemblage of Precambrian schist, gneiss, etc., with near vertical schistosity.

nas-Apure basin. Before the uplift, the whole of western Venezuela used to be a single large basin from Maracaibo to Barinas-Apure with the ancestral Andean basin in the middle. Tectonic forces inverted (uplifted) the ancestral Andes basin producing the Mérida Cordillera, now a lofty barrier between Maracaibo and Barinas-Apure basins. The orogeny exposed diverse rocks of the deep-seated Precambrian, and younger ages. The northern Andes merges into the east-west trending Venezuelan Coast Range on the north margin of South America, by the Caribbean Sea.

Mountain Belts Create Foreland Depressions and River Basins

What do great mountain chains—Himalayas, Alps, Andes, and the Rockies have in common other than stature? All of them gave rise to some of the world's biggest rivers in the adjacent foreland basins. A foreland basin is a topographic low, developing parallelly in front of an uplifted mountain belt. A mountain and adjoining low-lying frontal area form a favorable milieu to spawn rivers and water ways. Take for example, the Andes Range, running north-south on the western margin of South America (Fig. 10.15). The Amazon, Orinoco and numerous tributaries in Brazil, Peru, Ecuador, Venezuela, and Colombia, and rivers in Argentina, originate in the Andes. Some Asian examples are the Ganges, Brahmaputra, Indus rivers and their Himalayan tributaries in the foreland basin in the Indian subcontinent. In Europe, many rivers like the Iller, Lech, Traun, etc., formed in a foreland setting, north of the collision zone of Eurasia and African plates. The Alpine rivers flow north to join the east-flowing Danube River before it ends in the Black Sea.

Closer to North America, "The Big Muddy" Missouri River with a drainage area of one-sixth of the country, owes its origin to the foreland basin of the Rocky Mountains. It is the longest river in the USA, meeting the Mississippi River at St. Louis. The main U.S. tributaries are—Yellowstone, Little Missouri, Cheyenne, Niobrara, Platte, etc. The Canadian tributaries are Milk and Poplar rivers. The Eocene Laramide orogeny gave rise to the Rocky Mountains, causing the retreat of the Western Interior Seaway. Soon the ancestral Missouri River and tributaries were draining the slopes of the Rocky Mountains. The current course is believed to be only about 115,000 years old.[37]

Going Deep South—Amazing Amazon River: 'The River Sea'

The Amazon River, a giant waterway in a foreland basin setting at the foothills of the Andes Mountain, and beyond, is unique in many respects. Its course spreads over eight countries—Brazil, Peru, Ecuador, Bolivia, Colombia, Venezuela, Guyana, and French Guyana (Fig. 10.15). The name, given by Spanish soldier Francisco de Orellana in 1541, comes from the Amazon race of female warriors in Greek mythology, supposedly inhabitants of modern Ukraine. The Queen Penthesilea was a mythical Amazon woman, who took part in the famous Trojan War, in Homer's classic, the *Iliad*.[38]

The Amazon is the world's largest river by discharge volume (about 6,591 km^3), and basin size (7 million km^2), and the second longest river after the Nile in Egypt; 20% of all global freshwater discharge flows through the Amazon. It has bagged some impressive statistics—its flow is seven times larger than the next six world rivers combined, and 10 times that of the Mississippi River. Its drainage basin is marginally less than the area of the lower 48 states of the USA (8.1 million km^2).[39] Rainforests occupy two-thirds of the Amazon basin, a priceless treasure for the living world. No wonder that when Portuguese sailors discovered the Amazon, they could not believe what they were witnessing and yelled "O Rio Mar" or "Oh, The River Sea."

The Amazon Drainage Dates Back Millions of Years

The 11-Ma-old Amazon River, with a 10-km-thick sediment column, is a direct consequence of the rising Miocene Andes (Fig. 10.15). The modern Amazon took shape about 2.4 Ma ago. Samples from a 4.5-km-deep borehole yielded precise age of river sediments.[40] The voluminous Amazon discharge has diluted the Atlantic Ocean salinity even 100 km offshore.

One theory posits that an ancestral drainage in the area used to flow westward about 10 Ma ago, and not towards the Atlantic. The modern Amazon River did not exist at the time.[41] Computer modeling of past Neogene climate variation corroborated the westerly flow of rainfall drainage to very large inland lakes. The lakes would have drained northward into the Caribbean Sea across Venezuela. At the same time, the rising Andes captured more moisture, yielded increased rainfall, produced more erosion of the mountain during the last 10 Ma or so. It contributed more sediments, filling-up the vast lake region on its east flank. The lake region gradually morphed into wetlands, and thereafter, into a dry raised topography. The erosion of the Andes and the silting up of the trough did the lakes in. In course of time, the basin gradient reversed, as the Andean eastern foothills became higher than the Atlantic basin margin.[42] Finally, the Amazon River was born. It took its present shape in the Pleistocene, flowing into the Atlantic ever since.

In 2018, others proposed a new paleogeographic model.[43] A physiographic high—the 'Purus Arch,' existed in the middle of the Amazon basin between 20–10 Ma ago. The high area separated east flowing ancestral Amazon, from a westerly flowing Solimoes river. The Solimoes and other rivers drained into a lake region, and then northward toward the Caribbean.

The rising Proto-Andes, between 10–4.5 Ma, created the transcontinental Amazon River, flowing east into an ocean embayment first, and later flowing steadily into the Atlantic Ocean. In summary, a mighty river drains into a foreland basin at the foothills of a rising mountain belt.

10-15 Physiographic map of the Amazon River basin mostly in Brazil but spreading over 7 surrounding countries.

The Discovery of a Giant Aquifer in the Subsurface of the Amazon River

In 2011, researchers discovered a vast body of underground water, 4 km below the surface—it is 300 km wide and running west to east just like the Amazon. The team was analyzing old subsurface temperature data from oil company boreholes; they noted to their amazement, that the temperature at a depth of 4-km below the Amazon River was much cooler than expected. This is because Amazon water has been continuously percolating downward through permeable sedimentary layers. It makes subsurface water cooler than normal for that depth. The Amazon River moves a mean volume of 200,000 m^3/second (velocity 10^{-7}m/sec), while the underground waterbody at a depth of 4000 m, flows at a crawl's space near the coast; it moves a volume of 3,287 m^3/second (velocity 10^{-9}m/sec, i.e., 218 m/year). The discharge of the giant slow-moving aquifer of the Amazon basin is to blame for the low salinity in the offshore Atlantic Ocean.[44]

The Orinoco River—The Proto-Andes Had Blocked Its Northerly Course Earlier

The Orinoco River, Venezuela, originates near the Venezuela-Brazil border among Precambrian rocks including the Roraima Table Mountains. The Colombia-Venezuela boundary passes along a north-south stretch of the upper Orinoco River. It then veers northeasterly forming a large delta in the Caribbean Sea, near the island nation—Trinidad and Tobago. Hard to imagine, earlier the river used to flow directly northward through the Falcon State into the ocean in the Mid-Miocene. A cache of vertebrate fossils in Falcon provides testimony of the proto-Orinoco River. The rising Venezuelan Proto-Andes blocked the northerly Orinoco route 12 Ma ago, when the river established its present northeasterly course.[45]

The Unique Casiquiare River—A Link Between the Mighty Amazon and Orinoco Rivers

The 320-km-long, Venezuelan Casiquiare River links two major hydrographic basins—Orinoco and Amazon (via Rio Negro), especially during the rainy season. The Rio Negro waters periodically drain into the Orinoco River or vice versa, depending on the higher flood stage of either basin. The Casiquiare flows through dense tropical rainforests where it is often difficult to see the banks because of inundated overbank floodplains. My personal journey from the Orinoco to Rio Negro via the Casiquiare river is an indelible memory of the mid 1970's. A sector of the Orinoco headwaters may permanently switch to the Rio Negro-Amazon catchment area, through river capture process.

In Asia, the Gangetic Plain—Site of the Holy Ganges River

In Asia, the Himalayan Frontal Thrust separate the Gangetic foreland basin from the Himalayan foothills in India. The basin is an offspring of the Miocene Himalayan orogeny. It cradles the sacred 'Ganga' River of the Hindus, originating in the Gangotri Glacier. The river cuts through the Siwalik Ranges, skirt by the two holy sites of Rishikesh, and further downstream Haridwar. The steep dipping Late Miocene Siwalik hills represent ancient Gangetic plain sediments, uplifted by the colliding Indian plate. The 150–500-km-wide, and 300,000 km^2 Gangetic plain, occurs south of the deformed Siwalik ranges.[46] As the Himalayas rise, the foreland deepens, while its axis migrates south. The basin receives sediments mainly from the Himalayas.

The fertile Gangetic plains remained the heartland of an ancient civilization through historic times. Emperors during the historic era had their seat of government on the densely populated Gangetic plain. Later, new waves of foreigners invaded the country and ruled India from there. Most notably, Muslims from the northern steppes of Asia, and the British from across the sea, established their capitals on the Gangetic plain. Other European colonists also had their beachhead in Calcutta and elsewhere for some time. Many important cities like Delhi, Allahabad, Varanasi, Patna, and Calcutta dot the Indo-Gangetic plains.

The Ganges River winds through 2,500 km of alluvial and delta plains till it reaches the Bay of Bengal. Important Himalayan tributaries are Yamuna, Gomti, Gandaki, Koshi, Teesta, etc. The Meghna and Brahmaputra rivers also join the Ganges in Bangladesh. Calcutta (Kolkata) sits at the north tip of the Ganges Delta that extends 355 km along the coast. The total drainage area is more than one million km^2. The ~5-km-thick sediments mainly consist of layers of clay, silt, sand, marly limestone, and numerous layers of peat and lignite.[47] The 'Sundarbans' hosts the world's largest tidal mangrove swamp/forest, home to the famous Royal Bengal Tiger.

The Largest Submarine Fan in the Bay of Bengal—A Himalayan Gift

The giant Ganges-Brahmaputra River system has created the world's largest submarine fan system, the 'Bengal Fan'—3,000-km-long and about 1,000-km-wide. The fan splays out from the mouth of a deep submarine canyon, like a radiating Chinese hand fan. The Bay of Bengal coastal sediments feed the fan to expand in all directions. The rivers carry a billion ton (10^{12} kg) of Himalayan sediments every year to its final deep-sea resting place.[48]

The annual monsoon rain in India, Bangladesh, and neighboring countries is the clue to the source of sediments in the rivers. The word 'monsoon' may come from the Arabic 'mausim' or the Dutch word 'monsun.' Monsoon also visits other countries in Africa, Asia, Australia, etc. The southwest wind brings moisture from the Indian Ocean to India that finally reaches the unsurmountable wall of the Himalayas. It soars and cools, coming down as snow and torrential rain, deeply eroding rocks and producing abundant loose sediments. Ultimately, the sediment load finds its way into the Ganges, Brahmaputra, and their tributaries. Some get trapped in the alluvial plain, some stored in the delta. The rest, which is really a lot, mostly fine sediments, end up in the deep-sea fan, via a submarine canyon.

The 17-km-thick Bengal Fan is the thickest sediment body in the world. It has developed on the 2,580-m-deep Bay of Bengal Ocean floor. The oldest sediments are Early Miocene in age—a time when the Himalayas were rising rapidly.[49] The fan lies west of the N-S trending Ninety-East Ridge. Density currents funnel down the delta-edge sediments through the 'Swatch of No Ground' canyon to the abyssal ocean. The canyon has a flat, 5–7-km-wide valley floor, and a depth of 1,200 m beyond the shelf edge.[51] Sinuous submarine channels traverse the upper reaches of the Bengal Fan. Seismic images revealed a meandering, 10-km-wide, and 250-m-deep channel.[52] It is an ecological paradise for cetaceans like whales and dolphins.

Nile River, Africa—The Longest River with a Deep Geological Connection

The 6,500-km-long Nile River, commonly thought to be Late Neogene in age, flows through multiple countries (Fig. 10.16). The White Nile branch originates in the southerly Lake Victoria, and the Blue Nile comes from the southeastern Ethiopian volcanic highlands (Fig. 10.8). The two branches meet in Sudan before flowing through Egypt in the last leg of its journey to the Mediterranean. The stable south to north course of the Nile up to the Mediterranean Sea has long baffled scientists. Big rivers normally change course through geologic time. The Nile is rather unusual as it remains mostly entrenched in

10-16 The placid Nile River near Aswan dam. The hillock forms a typical 'Tor' structure (jointed, stacked, and separate blocks) due to weathering of Aswan granite.

its valley for millions of years. Recent research reveals that the Nile course stability is about 30-Ma-old, and it is due to a very deep-rooted cause.[52a] A deep mantle convection cell triggered uplift of the upstream Nile region including the Ethiopia-Yemen dome where the Blue Nile originates. A corresponding subsidence or sag of northern Egypt and the eastern Mediterranean Sea maintains the topographic gradient from south to north. The mantle collusion may be the key to the persistent topographic slope through time, stabilizing the path of the Nile River.

STONY TESTIMONY

LAVA, GRANITE, SALT, AND ASSORTED ROCKS

The Rocks from Fire

Vigorous plate tectonic movements caused volcanism at rifts, subduction boundaries, and collision zones. Neogene-Quaternary lavas occur in Iceland (along mid-oceanic spreading ridges), the Carpathians of Europe, Middle Atlas in Morocco, East African Rift System through several African countries, the Cascade Mountains, and the Basin and Range Province, USA. The Ring of Fire Belt has remained a focal point of volcanism around the Pacific (Fig. 9.15). Other areas include the Kunlun in the Tibetan Himalayas, Andaman Trench in the Bay of Bengal, the Caucasus-Arabia segment of the Alpine-Himalayan Belt, the Southern Rockies, the Caribbean Lesser Antilles, etc. Hot spot lava outpourings marked the Galapagos, Samoa, and the Hawaiian chain of islands.

The Western U.S. Had a Neogene Lava Bath—The Columbia River Basalt

The Mid-Miocene 210,000 km^2 lava field was the trigger for the abnormally high temperature during the 17–14.5-Ma-old, 'Middle Miocene Climatic Optimum.' As one of the youngest continental flood basalts in the world, it formed tall, 5-sided columnar, fence-post palisades, and marginal subaqueous pillow lavas in the states of Washington, Oregon, and Idaho.

Granite Intrusions

Neogene granite intrusions occur in some places, including the 18–3.4-Ma-old granite in the Everest,[54] the Patagonian Andes batholith, Argentina, and the 12-Ma-old, Panama arc-related pluton. Other examples are the Miocene granite in the North Apennines, Italy, and Turkey.[53] The youngest intrusive is the 10–0.8-Ma-old, Kurobegawa Granite, Japan.[55]

Assorted Neogene Sedimentary Rocks

The Neogene sedimentary rocks—sandstone, mudstone, limestone, and evaporite of terrestrial, shallow marine and deep-water origin, are widespread all over the world. We will mention only a few examples here. Glacial sediments like tillite, moraines, etc., appeared as the climate became colder, and both the polar regions had permanent ice sheets. In Asia, the Indo-Gangetic plain stores river sediments from the eroding Himalayas. The Early Miocene Ganges sediments are shallow marine-deltaic, whereas younger sediments are mainly fluvial. As an aside, arsenic contamination from the Himalayan 'arsenolite'

(As_2O_3), is a serious ground water problem in several Indian states.[56]

In North America, the Atlantic margin as well as the Gulf Coast, up to the Yucatan peninsula in Mexico, contain thick sequences of sandstone, shale, limestone, etc. Toward Florida, rocks are mostly limestone, whereas sandstone and shale dominate near the Mississippi river and delta area. The Gulf of Mexico holds >10-km of sediments. Neogene sedimentary rocks host many important onshore and offshore oil reservoirs in the United States, Mexico and South America including Venezuela. The Middle Miocene Cerro Pelado and Socorro Formations, Venezuela, show excellent development of coastal and shallow marine sandstone and mud rock with abundant fauna (Fig. 10.17). The North Sea basin, Europe, shows a mostly Miocene basin-fill sequence.[57] In Poza Rica, Mexico, the Miocene coastal fluvial and shallow marine rocks—sandstone, conglomerate, mudrock, impure limestone, etc., of the Early Miocene Tuxpan Formation, are exposed on the Poza Rica—Gutierrez Highway section (Fig. 10.18). The images 2 and 3 show rocks of near-shore shallow marine origin. How do we know the marine origin of the strata? Abundant fossils of sea creatures like bivalve, gastropod, echinoid, etc., confirm the origin of the rocks. We note several lenticular, shallow scour depressions filled with dark porous sandstones, especially in the upper part of image 2. The unusual black color is due to growth of moss and lichen in the porous, fossiliferous sandstone. The ubiquitous small, isolated whitish carbonate concretions, and white horizontal calcareous sandstone layers, especially stand out below the darker strata.

10-17 The shallow marine cross-stratified sandstone sequence overlies a weathered and oxidized mottled mud rock (probably a paleosol). The Cerro Pelado Fm., Falcon, Venezuela, produces hydrocarbon in some fields.

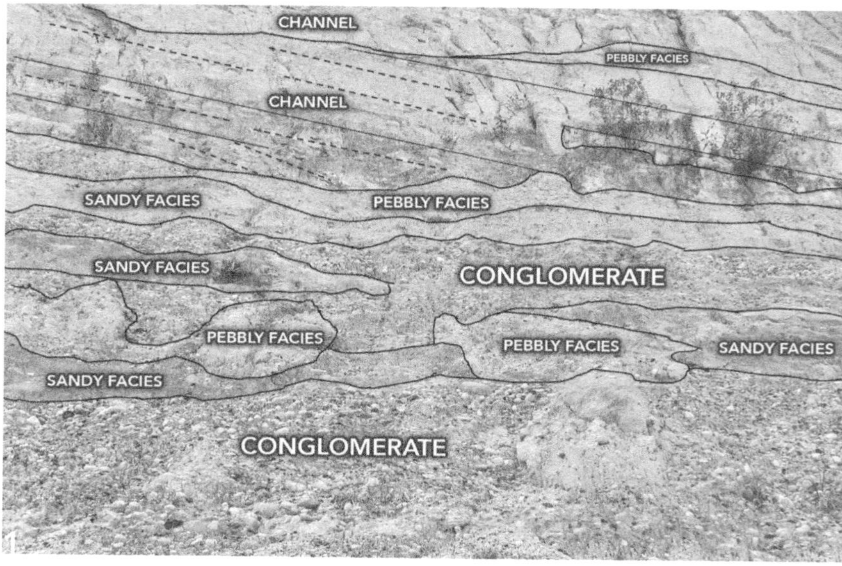

10-18 Views of fluvial and shallow marine Miocene strata exposed on the Poza Rica—Gutierrez Zamora Road section, Veracruz state, Mexico **1** It shows interfingering fluvial scour and fill, lenticular sandstone and conglomerate strata (~6-m-thick), Early Miocene, Tuxpan Fm. The main channel in the upper half shows sandy cross-beds (highlighted by black straight lines) of probable tidal origin. Note the laterally variable, fluvial, lensoid conglomerates in the lower half. **2** A marine sequence of interbedded sandstones of pale and dark color in the lower half. In the upper half, the dark color of the numerous lenticular black units is due to lichen/moss growth in porous sandstone. **3** A lensoid, channel-form, fossiliferous, and porous black sandstone (8-m-wide and 1-m-thick) sitting on a 15-cm-thick, whitish grey fossiliferous, calcareous sandstone.

NEOGENE SALTS

A Primer on All Things Salty

Some interesting statistics exist on salt and everything salty. One kilogram of seawater will give 35 gm of salt. The oceans contain 1.4 sextillion kilograms of salt (1.4 with 20 zeros after it). The salt mass in ocean water is almost the same as the mass of the Moon.[58] Salts are mostly sodium chloride, minor potassium chloride. Gypsum and anhydrite are sulfate salts. Some complex salts combine K, Mg, and Ca with chlorine and/or sulfur in different combinations. Chemist Usuglio was the first to evaporate seawater in 1884, to determine the sequence of mineral formation.[59] The basic rule is that the least soluble mineral precipitates first. At 50% of the original seawater concentration, calcium carbonate ($CaCO_3$) precipitates. Gypsum ($CaSO_4 \cdot H_2O$), halite, and K/Mg chloride/sulfate salts form later at about 20%, 10% and 5% seawater concentration respectively. Anhydrite ($CaSO_4$) appears later. Sylvite, a potassium chloride, and other complex K/Mg salts precipitate at the end.

Miocene Mediterranean Sea Salt—A Culinary Treasure

We recall an especially thick evaporite sequence occurs mostly hidden under the shallow Mediterranean Sea; some of the same salt is being actively exploited in several Spanish and Italian surface mines today. The Mediterranean Sea salt, high in magnesium, is prized by both chefs and consumers because of its many beneficial elements. However, all Mediterranean salt is not the same.[60] Salts from Sicily (Italy) and Spanish Lorca, reveal chemical variation within the ancient basin. The Sicilian salt formed in a basin-center position, while the latter formed in a basin-edge location.

Formation of the 1,520-m-Thick Louann Salt in Gulf of Mexico

The Jurassic Louann Salt is the source of all salt structures in the Gulf of Mexico. What is the origin of salt? The intriguing 160-Ma-old geological story began when the matrimony of North and South America ended in a protracted divorce. Initially, a large shallow 'rift' formed, as the two sides were slowly moving apart. It grew larger with time to accommodate more saline infillings. The prolonged evaporation for over 3.5 Ma precipitated the 1,520-m-thick layered 'Louann Salt' blanketing the sea floor. In Mexico, they call it 'Campeche or Isthmian Salt.'

Deep Salt—Mother of Gulf of Mexico Diapirs and Domes

The post-Louann salt history is more complex and somewhat puzzling to fathom. Numerous subsurface salt structures pierce the sedimentary strata, playing a crucial role in creating important oil and gas reservoirs. Mobile salt gave rise to the structures at various times and depths of the Gulf's sedimentary package. Some Louann salt is now dispersed onshore and offshore, forming more than 600 salt structures—domes, diapir, column, pillar, pillow, etc.[66]

The unique properties of salt dictate how the structures originate. Salt reacts to two principal forces—movement due to gravity, and tectonics. In the former, overburden loading, i.e., sediment layers above the salt, exerts vertical pressure on salt strata. Simultaneously, a density difference arises. Compaction makes overburden layers denser, whereas salt density remains constant. Salt is buoyant, homogenous, impermeable, and is not compactible. It deformed plastically under the massive overburden of the Cretaceous-Recent strata. Tectonic forces squeezed the ductile salt through less mobile overburden layers along faults or fracture planes. Salt flowed opportunistically giving rise to the dome, diapir, etc. The basin is full of Neogene-Quaternary salt diapirs from shallow subsurface to 1,600 m under the seafloor.[67]

Salt structures are often enormous—salt domes may be up to 8-km-across and ascend thousands of meters, reaching up to 200 m below the ocean floor. They may attain great heights as they shoot up from the Jurassic Louann-Campeche Salt Bed.[68] Some salt diapir size is mind-boggling—72-km-long, 8-km-wide, 9,230-m-high, i.e., taller than the Everest.[69]

10-19 Salt domes in the Zagros Mountains of Iran. The tectonically uplifted Jashak dome at the center is 12 kms long and 4 kms wide. The collision of Eurasia and Arabian platforms brought underground salt formations up to the surface, forming mountains.

Other famous salt basins occur in the North Sea, offshore Brazil, etc. The Zagros Mountains of Iran have three rare emergent Eocene salt domes, and many other multi-hued, spectacular salt structures (Fig. 10.19). Two of the salt structures have formed 'salt glaciers.'[68] Salt glaciers formed when the rising salt spilled on the ancient ocean floor.

Salt and Shale—Keys to the Gulf of Mexico Hydrocarbon Bonanza

Oil and gas exploration in the Gulf of Mexico largely depends on salt deformation structures.[71] Most salt domes and diapirs are far removed or detached from their original underlying Louann source bed because salt is enormously mobile. Oil accumulates in sandy reservoirs on the flanks, and under the bulbous head of mushroom-like domes and diapirs (Fig. 10.20). The mud layers above form a protective 'seal' for trapped oil in the reservoir.

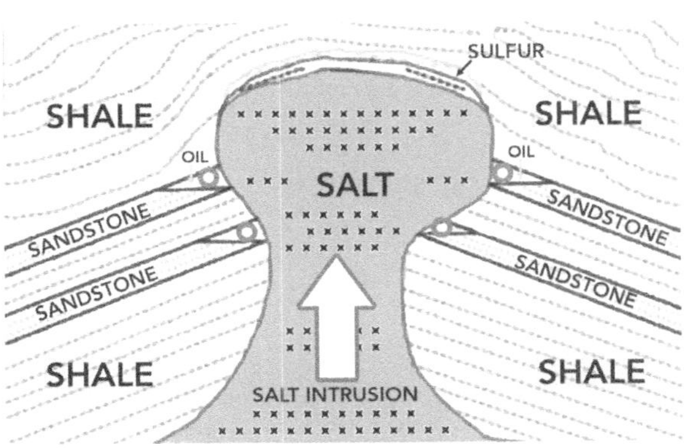

10-20 Model of a salt diapir and how sandstones form oil and gas reservoirs near salt structures in the Gulf of Mexico, and elsewhere. As a salt column rises from the source below, adjacent sandstone and shale strata become inclined. The oil column ('O') abuts against the impermeable salt wall.

Most producing oil fields occur in the Neogene sands around salt structures. In the Salinas del Istmo basin, Gulf of Campeche, México, many salt diapirs and domes are present facilitating formation of hydrocarbon reservoirs.[67] (Fig. 10.21) What is the sand source? Marine currents transport sands from far away, and deposit them in submarine fans and channels. In the new millennium, a 10,685-m-deep, oil well discovered a giant Paleogene reservoir in a sandy submarine fan. The reservoir produced 300,000 barrels of oil equivalent/day in 2009.[70] Similar important structures are shale/mud diapirs and domes (Fig. 10.22). The mud layers often behave like salt, due to subsurface pressure and liquefaction. The southern Gulf Coast shows many surface mounds where mud diapirs are exposed. The author had worked on both onshore and offshore salt and mud diapirs in Mexico. Neogene hydrocarbon reservoirs in the Macuspana Basin, Mexico, have formed on the flanks of mud diapirs and domes. The source mud of the domes is older Paleogene shale.[72]

The 'Lucas Gusher' and the Texas Oil Boom—Thanks to the Spindletop Salt Dome

In 1901, explorers drilled the first successful exploratory onshore well near Beaumont, Texas. They struck oil at a shallow depth—350 m, around the Spindletop salt dome. The 'Lucas' gusher (founder's name), was a 30–50-m-high mega-fountain of oil and gas lasting for nine days. As the gusher spouted, 'the startled roughnecks' ran for their lives when several tons of drill pipes came flying out of the hole.[65] The borehole produced 100,000 barrels/day. The Spindletop discovery was a game-changer, starting the Gulf of Mexico oil bonanza, and a century-old economic boom in Texas and Louisiana.

MIRACLE OF LIFE

The Neogene Mammals Everywhere

Life in the Neogene, especially later, was like today with a few exceptions. For one thing, the mercury was on a steady downward

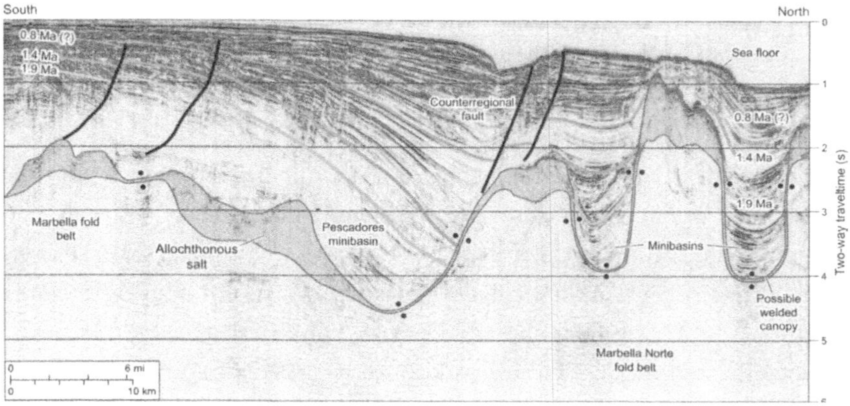

10-21 2-D regional seismic profile shows allochthonous salt (dark grey-fill) in the offshore Salina del Istmo basin, Gulf of Campeche, México. Deep ocean to the right (North) and the land is to left (South). Depth of Pescadores mini-basin above salt is ~7 km. Two mini-basins have formed in the sag between salt structures on the right. Note the uneven nature of the seafloor deepening from the south to north direction.[67] *(from Gomez-Cabrera and Jackson, 2009, AAPG Mem. 90. AAPG Copyright © 2009. Reprinted by permission of the AAPG whose permission is required for further use).*

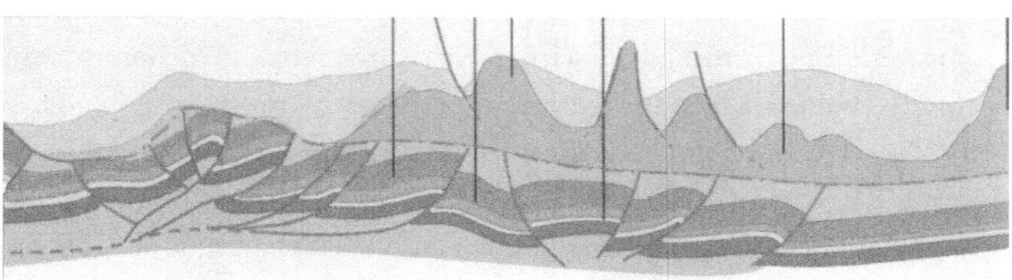

10-22 View of shale diapirs (grey domes in the upper half of the image, some marked by normal faults) in the Macuspana basin, Tabasco, Mexico. The folded layers below the dash-and-dot line along the middle represent rocks of different ages. Compression had given rise to multiple thrust faults shown by dark curved lines in the lower half. The black vertical lines represent different well locations. *(from Chavez et al., 2009, AAPG Memoir 90. AAPG Copyright © 2009. Reprinted by permission of the AAPG whose permission is required for further use).*[72]

spiral from the Oligocene onward, barring a short abnormal warm Mid-Miocene interlude. By Pliocene, the two polar ice caps were in place, with radiating glaciers. The global cooling trend peaked in the Pleistocene.

The Isthmus of Panama established connection between the two Americas, opened the door for two-way migration of animals and plants. One important outcome of Neogene tectonics, and sea level fall is that several land bridges had sprung up, connecting isolated landmasses. During the latest Miocene, the dried-up Mediterranean Sea facilitated an accessible land route for animals between Europe and Africa. Much later in the Quaternary, sea level fall created the Bering land bridge, enabling live traffic flow between Siberia/Asia and North America.

The Neogene-Holocene "Age of Mammals," saw a burst of mammalian expansion everywhere. The earlier exit of dinosaurs proved to be a boon for mammals who could diversify merrily and became the largest Neogene creatures. Carnivores like the sabre-tooth tiger, lion, and giant herbivores—wooly mammoth, rhinoceros, and giant deer, etc., spread far and wide.

Grazing animals became dominant in the Neogene because the cooler climate fostered extensive grasslands. Horses became very common. The ruminants (with two stomachs) and other grazing animals evolved adequate teeth and jaws for living on grasslands. It was a survival need as grasslands do not provide as much nutrients as forests. The carnivorous predators like dogs, cats—big or small, also became adept in chasing their grazing preys with speed and agility.

Birds flourished in the Neogene though sadly, some South American giant birds became extinct. The ancestors of modern flightless birds like emu, weighing up to 23 kg, did not survive beyond the Neogene.[74] Crocodiles, snakes, turtles, also found their niche and diversified.

The Great American Biotic Interchange—The Marvel of the Panama Isthmus

The Panama land bridge made free flow of fauna and flora possible between the two Americas, enriching the biodiversity of both continents (Fig. 10.23). A. R. Wallace, the 'father of biogeography' mentioned the event as the 'Great American Biotic Interchange' (GABI) in 1876.[75] While armadillo, porcupine and marsupial opossum found a cozy niche in the north, forebears of cat, dog, horse, bear, llama, and many other animals ended up in the vast plains and valleys in the south.[76] Prior to the Isthmus-bridge, the Pacific and Atlantic marine fauna used to mingle freely through the 'Great American Seaway.' The Isthmus proved to be unique—a land bridge between continents, and a permanent barrier between two oceans. In the post-isthmus era, marine fauna became distinct in each ocean. Even before the Panama Isthmus, fauna reacted to impending isolation as the submarine island arc was rising. As early as 6 Ma ago, a particular shrimp population ('snapping shrimp') started to evolve distinctly on either side, because of deeper water in the Pacific, and shallower in the Caribbean.[6] Benthic, i.e., bottom-dwelling foraminifers, also reflected the environmental changes and isolation of the two oceans, after 6.7 Ma.[77]

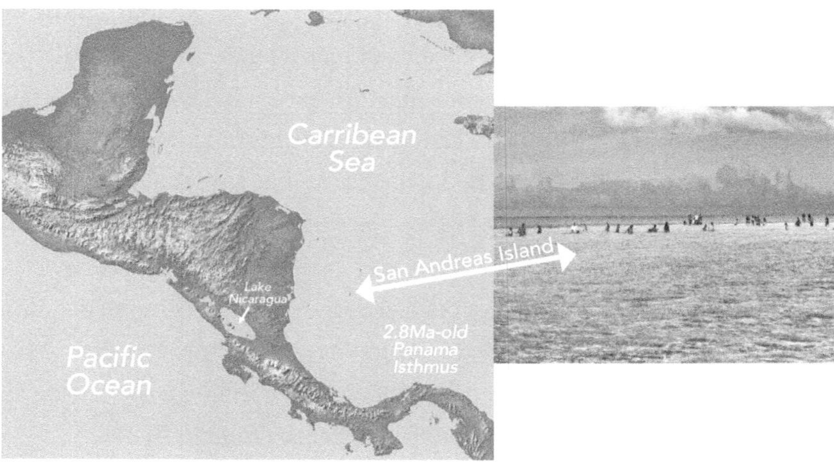

10-23 left Topographical image of Central America showing the Panama Isthmus, Lake Nicaragua, and the smaller Managua lake above. The two dots within Lake Nicaragua are the volcanic Ometepe Islands. The tiny Island of San Andres (Colombia) shows up as a mere dot in the Caribbean on the map (arrow). **right** The crystal-clear waters around San Andres show the incipient emergence of a creamy whitish marine bar in very shallow water.

Homesick Sharks Stayed on in the Nicaraguan Lake

Two lakes—Nicaragua and Managua, are remnants of a pre-isthmus ocean (Fig. 10.23). Lake Nicaragua's Spanish name 'Mar Dulce' (freshwater sea) reflects its gigantic size. Many habitual salt-water creatures—shark, tarpon, swordfish, and sawfish are now comfortably frolicking around in the new freshwater environment of the lake. Tagged bull sharks routinely travel from Lake Nicaragua to the ocean through the San Juan River rapids, unfailingly returning a short time later.[78] The lakes, connected by the unique Tipitapa River, represent an ancient ocean embayment, that became isolated from the open ocean by continuous lava flows and tectonic activities.

Marine Animals in the Neogene—
The Whale, Dugong and Megalodon Shark

Whales, already an exclusively marine animal, thrived in the Neogene. Hippopotamus appeared in the lower Miocene and continue to this day. DNA data shows cetaceans like whales are genetically related to artiodactyls—hippos and pigs. Yes, all whales, dolphins, porpoises have descended from a terrestrial mammal like *Pakicetus*.[73] Dugong or sea cow debuted in the lower Miocene. It is the only marine herbivore mammal, and the closest relative of elephants (*Proboscidea*). Another aquatic group, the *Pinnipedia*—sea lions, seals, and walruses, appeared in the Late Oligocene—Early Miocene. The megalodon shark, a dominant aquatic creature, first appeared in the earliest Miocene, becoming extinct in the Pliocene.[61] It was the biggest shark ever –18-m-long, with razor-sharp 18-cm-long teeth (Fig. 10.24).

Other marine creatures like pelecypod (bivalve), gastropod (snail), coral, and others thrived in the Neogene (Fig. 10.25). Many famous Miocene reefs occur in the Mediterranean region. The Early Miocene San Luis Reef and its deep-water debris flow deposits of the Pedregoso Fm. in the Falcon State, Venezuela, is also noteworthy (Fig. 10.25, image 3).

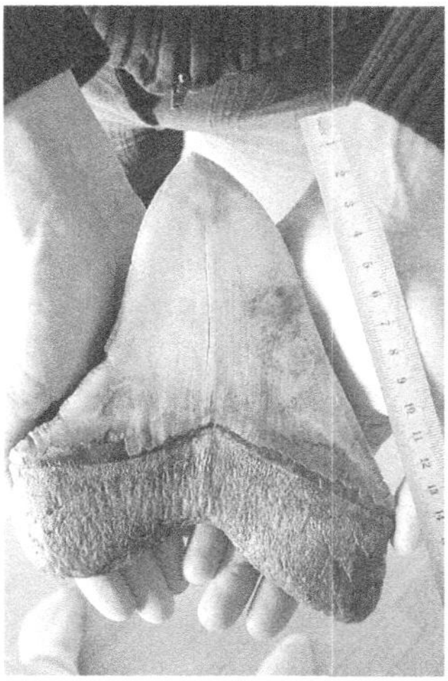

10-24 Fossilized tooth of the Megalodon shark of Miocene age from the Atacama desert, Chile. The tooth is 18 cm diagonally with a base of 13 cm.

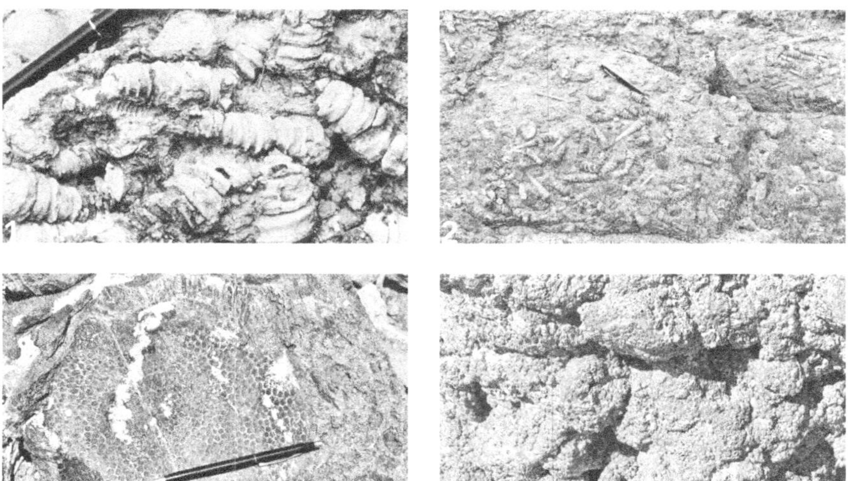

10-25 Some Miocene faunae from the Falcon region, Venezuela (1, 2, 3, 5), and Algarve, Portugal (4): **1, 2** Coquina beds of gastropods (*Turritella*) from the Late Miocene La Vela Fm. **3** Coral fragment from San Luis Reef, slumped into the deep-water Pedregoso Fm. as debris flow **4** Bryozoa colony on Algarve beach **5** Molds of pecten-like bivalves and gastropods, including *Turritella*.

The Plant Kingdom—A Glimpse of Today

The cooler Neogene climate fostered deciduous trees that shed leaves annually like the modern maple. Vast grasslands supported herbivores, and indirectly their predators. The new marine algae—the tall kelp grass, appeared in the cool Neogene waters and provided essential cover and food source for some of the sea animals.

Primate Evolution and Our Early Ancestor's Trail—A Long Time Waiting

The debut of our distant forebears is the most transcendental event shaping the Neogene/Quaternary world. The path was far from straightforward, with many twists and turns, and frequent aberrant dead ends.

Primate evolution story is rather fragmentary and complex, starting in the Paleocene (58 Ma) when prosimians—tarsier and lemur, showed up. Later *catarrhine* and *platyrrhine* primates diverged in the Oligocene (Fig.10.26). They have distinctive nostrils—the simian, flat-nosed *platyrrhine* of the new world (South/Central America), and *catarrhine* of the old world. The latter, with narrow downward facing nostrils, again branched into two broad superfamilies—old-world monkeys, and the *Hominoidea*,[24] encompassing all apes—great or less-

10-26 left Gray langur ('Hanuman')—old-world monkey, Rajasthan State, India. **right** Cappuccino—new-world monkey, Liberia beach, Costa Rica.

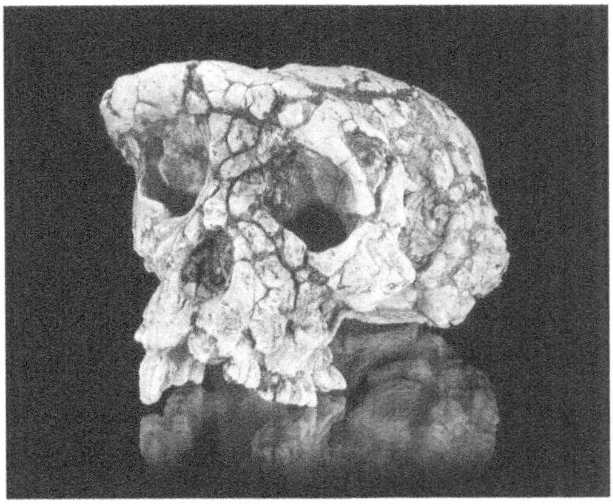

10-27 Skull cast of *Sahelanthropus tchadensis*—One of the hominin ancestors of the *Homo genus* in Chad, Africa.

er.[50] All were tree dwellers then—prosimians, monkeys, and apes. The *Hominidae* great ape family, and *Hylobatidae* lesser ape (gibbon) branched off in the earliest Miocene. The *Ponginae* great apes (orangutan) diverged from the *Hominidae* great ape family in the early Middle Miocene. The African *Proconsul* and Indian *Ramapithecus* were Miocene apes of unknown affinity. Some argue the 13-Ma-old, *Nyanzapithecus*, discovered in Nadupet, Kenya, may be our traceable original hominid ape ancestor.[50a] Fast forward to *Sahelanthropus tchadensis* ('*Toumai*') from Chad (Fig. 10.27), and *Orrorin tugenensis* from Kenya, both ~7–6-Ma-old, showed some primitive traits of hominin apes. Later, volcanic ash layers in Ethiopia yielded the 4.4-Ma-old, hominin female skeleton of the *Ardipithecus ramidus*.[28]

The Cousins of the *Homo genus*—
The *Pan*, and the More Distant *Gorillini genus*

The precise date of divergence of hominin and hominid apes is still without consensus. During the Late Miocene (~8–10 Ma ago) the great ape family branched into 2 hominid subfamilies: 1—The stem ancestor of *Australopithecus*, *Homo*, *Pan genus*, and 2—The great ape *Gorillini genus*. Later (~7–6 Ma ago), the ancestral hominid ape again split into 2 groups—the primitive hominin (i.e., human lineage including Australopithecus), and the hominid *Pan genus*.[79] The latter evolved into 2 *species*—*Pan troglodytes* (chimpanzee) and *Pan paniscus* (bonobo). One may ask what differentiates the *Pan genus* from the hominin apes. The former is not bipedal, walking on knuckles, and has much larger canine teeth like the apes orangutan and gorilla. The discoverers[82] of *S. tchadensis* in Chad argued that the *Pan genus* was older than the transitional *Sahelanthropus* ape that were bipedal hominins and had smaller canine teeth.

In summary, hominids are non-human great apes, i.e., gorilla, chimpanzee/bonobo, orangutang, and all ancestral great apes. The transitional bipedal hominins were *S. tchadensis*, and *Orrorin tugenensis*, whereas the *Ardipithecus genus* may be the first primitive hominin. However, there is a lack of unanimity on the hominid-hominin transition topic.

Australopithecus was the first unmistakable bipedal hominin ape that might have finally transitioned into the *Homo genus*. So began the >2.5-Ma-long solitary journey of the *Homo genus*, climaxing into the crown jewel—*Homo sapiens*, only about 200,000–300,000 years ago.

Genome study reveals that we share 98.6% of DNA with chimpanzee and bonobo, and 98% with gorilla. We are closer to the *pan genus*. We did not descend from gorilla or pan; they are sister genus.[32]

We will review human evolution in Chapter 12, because *H. sapiens* arrived late in the Quaternary.

CHAPTER HIGHLIGHTS

- The cooling trend peaked at End-Neogene with snow finally crowning the Arctic Pole.
- Modern continents in place. All mountain belts—Himalayas, Alps, Andes, etc., were still rising.
- The debut of the Panama Isthmus, connecting the two Americas.
- Hot spot volcanoes (Galapagos, Hawaii, Samoa), Pacific Ring of Fire lava outpourings, mid-ocean ridges, and Caribbean Lesser Antilles.
- Presence of a 5-Ma-old, 3-km-thick salt layer in the once dried-up Mediterranean Sea.
- Rising salt structures from the ~1.5-km-thick Jurassic Louann Salt in the Gulf of Mexico create numerous hydrocarbon reservoir traps.
- The East African Rift System splitting apart the African continent in plain view.
- The Andes story with glacier, volcano, desert, altiplano, and emerald gemstone, etc.
- Neogene rivers—Nile, Amazon, Orinoco, Ganges, etc., occupying foreland basins.
- Debut of many modern faunas and plants.
- Expansion of mammals, birds, and reptiles; cooler climate favoring grazers and predators; rise and extinction of the Megalodon shark.
- 8–10 Ma ago, divergence of our ape ancestor from the *Gorillini genus*.
- ~7–6 Ma ago, the ancestor of *Australopithecus* and *Homo genus*, again separating from the Hominid *Pan genus* (Chimpanzee and Bonobo).

WHAT'S NEXT?

Chapter 11 unravels Quaternary paleogeography and its fascinating rocks. In the glacial climate, both poles donned icy crowns. Short interglacial periods like the present offer earthlings some relief.

top A gigantic (>1.5-m-across) calcareous concretion of Miocene age in seaside Zorritos, southern Peru. Local people retrieved the sphere with a crane from a creek bed near the Pacific coast, painted it like a football, and put it on a pedestal for all to admire **bottom** Liesegang bands and rings due to weathering and iron oxide cementation of alternate concentric layers of sedimentary rocks – Oligo-Miocene Pecaya shale, Falcón State, Venezuela.

11

Quaternary Period: Cycles of Ice and Fire

Pleistocene Epoch (2.58 Ma – 11.7 ka) and Holocene (11.7 ka – Present): Glacial-Interglacial Cycles

The Pleistocene and Holocene epochs showcase many interesting geologic, geomorphic, and climatic events, a few of which especially stand out for their rarity and significance. Myriad natural Quaternary geological monuments are in pristine condition, despite intense weathering and erosion. Most Quaternary animals and flora are present today.

Numerous glacial-interglacial cycles marked the last 2.58-Ma period. Normally, a glacial cycle lasted about 90 ka, while an interglacial lasted much shorter, around 10 ka. About 10 such cycles visited Earth during the last million years. We are presently enjoying a balmy interglacial spell. The warm interludes occur when more intense summer solar radiation hits the Northern Hemisphere.

PALEOGEOGRAPHIC HIGHLIGHTS

Nature Active as Ever

When the Cenozoic Earth finally sprinted to the Quaternary Period near the visible tentative finish line, most things were in place. The mountains, river valleys, oceans, and most Neogene fauna and flora were present, though some lost forever. Plate tectonics have been active, and major mountains are still rising because of it. Tectonic plates continue to move apart at Mid-oceanic rifts, and in turn, get consumed at subduction zones. One thinks of the Andes where the Pacific Nazca plate is sliding under the western flank of South America.

The Pacific Ring of Fire keeps spewing fire and fury. Likewise, the Pacific and the North American plates are active, sliding-past each other along the San Andreas Fault. Many crust-mantle fissures along the mid-oceanic ridges are still exhaling tons of hot magma from deep inside the womb of Earth. Numerous mantle plume hot spots keep belching lava too, generating a new chain of volcanic islands—one island at a time, like the Hawaiian, Samoan and Reunion Islands. About 1,242 active volcanoes adorn the Pleistocene landscape.[1]

Mountain ranges are not sitting still either. Tibet is rising 10 mm/year. The Himalayas rose 1cm/year, or 10 km in the last 1 Ma. Such mind-boggling vertical growth is invisible because the hapless Eurasian plate receiving the gut-punch is stretching rather than deforming vertically.[2] Constant erosion and weathering are also taking its toll on mountain heights. The Andes uplift, in comparison, seems more muted, up to 3-mm/year during Plio-Pleistocene.[3]

The Quaternary Icehouse Reminds Us of the Distant 300-Ma-old Permian One

Climate change is the most remarkable aspect of the Quaternary. Oscillating glacial advance and retreat left its indelible footprints on landforms. Glaciation was pervasive in Antarctica, Asia, Europe, the Americas, and Australia. The Wisconsin Glacial Stage (75–11 ka) rep-

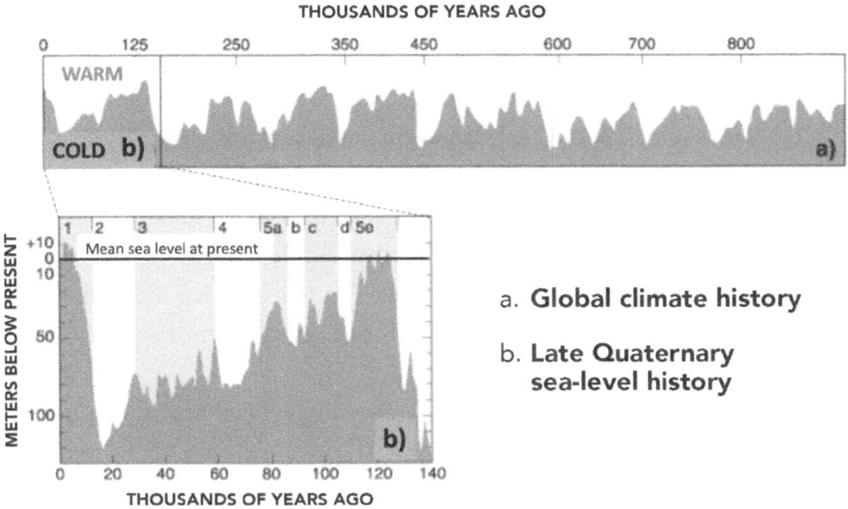

11-1 Top climate curve (a) shows cold and warm cycles during the last 900 ka. Bottom (b) shows two lowest sea levels (~140-ka- and ~20-ka-old) correspond to glacial peaks.[3a]

resents the last global icy event.

The Illinoian (191–130 ka) and pre-Illinoian (2.5 to 0.2 Ma) stages preceded the Wisconsin Stage. These two glacial stage names are rather misleading. Recent studies clearly point to more than 20 glacial advances during the Pleistocene, based on $^{18}O/^{16}O$ ratio in planktonic micro-foraminifers. These plankton incorporate heavier oxygen isotopes (^{18}O) in their shells when the water is ice cold. The Laurentide, and the westerly Cordillera ice sheets, covered Canada and the northern USA, when glaciers reached far enough south to cover the area of New York City. The landmass of Iceland, Greenland and northern Canada formed an icy wonderland.

The Wisconsin terminal moraines largely covered or eroded earlier tills. These are now scattered all over the eastern seaboard of North America—from Long Island, New York, to the Oak Ridges Moraine, Ontario, and farther north. The Great Lakes of North America, and Niagara Falls are the handiwork of the glacial event. Likewise, many large lakes dotted the Basin and Range Province of western USA. The whole area from the Laurentide ice sheet to the Arctic, and even northerly Siberian and Scandinavian ice sheets, was one large frozen landmass, the British Isles included. The 10–13-ka-old glacial moraines in the Venezuelan Andes are partly equivalent to the Wisconsin event. Far south, the Patagonian glaciation of Chile and Argentina are also coeval. The thick Himalayan valley glaciers retreated many 10's of kms after the Last Glacial Maximum (Fig. 11.1).

The global sea level change may partly depend on glaciers. Melting or forming of large ice sheets on land will add or subtract ocean water volume, affecting sea level change. That is why ice melting at the poles is alarming. Thermal expansion or contraction of seawater may also cause sea level rise or fall.

Tectonic activity is the third cause of sea level variation. When two landmasses collide, the intervening sea vanishes, triggering a global sea level fall. We recall the India-Eurasia collision and demise of the Tethys Sea.

The 125–11.7-ka-old Last Glacial Period caused a periodic yet overall sea level fall worldwide. Earlier, about 130–125 ka ago, sea level was 4–5 m higher than today. The ~20-ka-old Last Glacial Maximum (LGM) caused sea level fall by 120–125 m as it locked up so much water.[4] The U.S. Geological Survey estimates that ice would have covered 25% of Earth at peak LGM. Now it covers only 10% of Earth, and that too is vanishing fast due to likely anthropogenic excesses.

In nature, constant change is the name of the 'game,' though geological ones may take a prolonged period to play out. Just like the LGM gripped Earth, it gradually loosened its hold. Sea level began to rise again as Earth started warming up. It rose through discrete temporal spurts, not in one go—19 ka, 14.6–13.5 ka, 11.5–11 ka and 8.2–7.6 ka. We are still enjoying the balmy interglacial that started 11.7-ka ago (Fig. 11.1b).

The Last Glacial Maximum—Desert Expansion, Look at the Sahara and the Taklamakan

The Pleistocene-Holocene ice age blanketed large landmasses with a 3–4-km-thick ice cover. Also, the Last Glacial Maximum caused palpable expansion of the Sahara Desert and sand fields. A study of 4 deserts, and 4 sand fields in China confirmed that the ~20-ka-old, Last

Glacial Maximum was indeed a catalyst in spreading extreme dry conditions. Deserts like the Taklamakan, expanded between 10–39%, while the sand fields grew by 20–38% in 3 cases and 270% in another. Conversely, the wet and warm, 9–5-ka-old, Holocene Optimum interval, made desert areas shrink by 5–20%, and sand fields vegetated.[5]

The Sea-Like Proglacial Lakes Agassiz and Ojibway—Their Discharge Diluted Oceans

A proglacial lake forms by meltwater trapped by ice sheets or moraines. Rapid sea level rise during the Holocene 8.2–7.6-ka interval, is linked to similar glacial melting. The vast Agassiz and Ojibway lakes originated from melt water. The former covered much of Manitoba, Ontario, Saskatchewan (Canada), and Minnesota and North Dakota (U.S.), while Ojibway drowned large parts of Quebec and Ontario (Fig. 11.2). The 12-ka-old, Lake Agassiz, at the toe of the Laurentide ice sheet, was 258 m deep, and inundated nearly 1,000,000 km^2 area. Discharge of lake flood waters emptied an enormous 105 km^3 of ice-cold freshwater, and rapidly diluted the northern oceans 8.4-ka ago.[4] It raised global sea level by about 1–3 m, altering Atlantic salinity and circulation, triggering the 8.2-ka-old global cold event. Hudson Bay

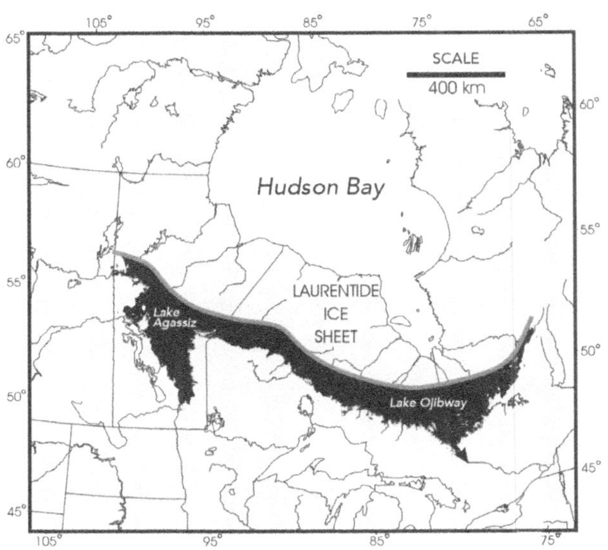

11-2 Creation of the Pleistocene Agassiz and Ojibway proglacial lakes (~2,000-km-wide), from the melting of the Pleistocene Laurentide icesheet in North America.

seafloor drill-core data confirms the catastrophic lake discharge, and its profound impact on the Holocene climate and organic life.[6]

Raised Beaches: Earth Rebounds After Glaciers Melt Away

The 3-km-thick Laurentide ice sheet made the crust buckle. However, the land gradually rose back ('glacio-isostatic rebound') when the weight of glacier disappeared. Curiously, the crustal rebound did not keep pace with the rising sea level. Shallow seas inundated large parts of North America about 8 ka ago.[7a] The prehistoric Champlain and Tyrell Seas flooded inland valleys of the St. Lawrence Seaway, Canada, and Hudson Valley lowlands, USA. Since then, the seas had receded from the inland sites located far away from the modern coastline. The long-gone Champlain Sea left behind a trove of whale, seal, and other marine fossils in raised, high-and-dry, ancient beach deposits. It's amazing that the seas had encroached nearly 300 km inland from the present shorelines.

The raised beaches and marine terraces in the Canary Islands, Spain, are also noteworthy, varying up to 70 m above the present sea level.[8] Sea level fall, and tectonic uplift in post-glacial time produced the 12 terraces. The current isostatic uplift rate is high—10 mm/year seen in Scandinavia. Impressive marine terraces also outcrop in New Zealand and the Caribbean, especially in the Dutch Antilles Islands. In Barbados, a raised, coral-rich marine terrace formed in the 120–115 ka interval,[9] when the sea level dropped 37 m.

Sea level rise may cause beach erosion and active cliff retreat. During the Holocene Sea level rise, waves actively eroded the softer layers of sea cliffs, making the overlying harder ledge precariously project out. Finally, the overhang collapsed causing cliff retreat. Many beaches are bordered by high sea cliffs in the UK., California (U.S.), and elsewhere. The Holderness coast of England is the most eroding (2-m/year) shoreline of Europe. The adjoining cliffs of soft chalk and glacial till have retreated 4 km since the Roman times, and the North Sea had swallowed 30 villages. Repeated landslides occur along the cliffs.[25] The sea cliff in Dover, England has been retreating now up to 3 cm/year. Absence of a beach next to cliffs hasten cliff re-

11-3 One of the many Lagoa golden beaches in the Algarve region, Portugal, framed by 25–30-m-high, vertical sea cliffs of limestone and shale, is constantly shedding large blocks of rock to the beach facing the North Atlantic Ocean.

treat. Similarly, active erosion of the Miocene Algarve soft cliff rocks and sea caves is changing the coastline today. The sea cliffs of impure marly limestone of Algarve (Portugal) are notable for their steepness, continuity, and regular rock falls to the beach below (Fig.11.3).

Land Bridges Across the Sea—Discovery of the New World

One of the windfalls of the last ice age was the discovery of hither-to-unknown landmasses by our ancestors. Examples of new land routes due to sea level fall are the Bering Strait between the Americas and Asia, Doggerland in Europe, 'Ram Setu' or Adam's Bridge between India and Sri Lanka, Sinai Bridge between Africa and Asia, and the Asia-Australia bridge. One common thread among all the 'bridges'—these are gifts of the Last Glacial Period and specifically, the Last Glacial Maximum.

Beringia—Welcomed the First Visitors to the New World

The Bering Land Bridge or Beringia is now a narrow 90-km-wide strait, a tongue of the Pacific Ocean. It stretches from the eastern tip of Asian Siberia to the western tip of Alaska, North America. A sea level drop of more than 120 m about 20 ka ago exposed the floor of the strait. It provided a unique opportunity for humans and animals alike to cross the strait. Plant species also spread both ways.[10]

The event heralded the arrival of native Siberian population to North America, who eventually populated the whole stretch from Alaska to Tierra del Fuego. It played a momentous role in the early settlement of prehistoric peoples in the Americas. The discovery of 61 perfectly preserved fossilized footprints of adolescents and women in New Mexico in the White Sands National Park, USA, throws light on the time of human arrival. Grass seeds near the footprints yielded a radiocarbon age between 23–21 ka. The timeline suggests that some humans had crossed the strait before the Last Glacial Maximum event. Impassable glaciers covered the whole of Canada and northern half of the USA at the time.[11] The strait was about 50 m deep, like now. Fossils of 18-ka-old land plants on the submerged Bering bridge confirm the temporary emergence of the strait.[10]

The Doggerland Settlement—A Prehistoric Culture Swallowed by the Rising North Sea

The 26,000 km^2 landmass, connecting Europe and the British Isles, is now 15–36 m under the North Sea. It was an inviting place where prehistoric humans thrived during the early interglacial period, about 10 ka ago. It showed an attractive natural landscape of hills and dales, rivers, lakes, marshes, coastlines, and beaches. Underwater photos of ancient trees, forests and human dwellings are proof of the organized layout of the place. Some parts lie only about 5 m below the water surface today, near the Netherland coastline. A catastrophic flood event would have submerged Doggerland's earthly treasures about 8.5 ka ago. Apparently 'Storegga,' the super-giant landslide of the 290-km-long Norwegian shelf to the north, had slid violently into the ocean. The catastrophe created an unprecedented tsunami leading to the aquatic burial of Doggerland.[13] It remained as a small, isolated island until about 5 ka ago, when it entered its watery grave as the sea level rose. Fishing trawlers have found some interesting evidence of the vanished culture on the sea floor—remains of mammoth, lion, rhinoceros, and hunting artefacts. Fishermen often find ancient bones and tools on the sea floor. A rare 1931 discovery was a 4–10-ka-old, ornate, barbed, antler-point for harpooning fish within a lump of peat.

The Mythical Ram 'Setu'—India-Sri Lanka Bridge Leading to the Sacred Adam's Peak

The Ram 'Setu' (Adam's Bridge') is another intriguing example of the global tentacles of the Quaternary glacial age and how it enriched the storehouse of local myths and folklore. The Indian Poet Valmiki wrote the Hindu epic '*Ramayana*' back in 5th century BCE. The central character, Prince Rama from Ajodhya, North India, had a bridge built across the Palk Strait to reach Sri Lanka. He did it with the help of his monkey brigade, led by his devotee Hanuman, a primate langur. Rama supposedly marched into Sri Lanka, rescued his kidnapped wife Sita from the demon king Ravana and killed him. Many Hindus vouch for the veracity of the revered Rama story.[15]

The British East India Company cartographers in 1804 coined the name Adam's Bridge, and the Adam's Peak in Sri Lanka, perhaps to please the believers of Abrahamic religions. Adam's Peak is a unique holy place, revered by adherents of 4 major religions—Buddhism, Hinduism, Christianity, and Islam. Adam's Peak, also known as 'Sri Pada,' is a hill, sacred for Buddhists and Hindus.[12] Apparently, Buddha visited the place in 500 BCE. To Buddhists, the sacred, conical, 2,243-m peak, showcases Buddha's holy footprint, measuring 1.8 m x 0.8 m. Hindus think it is Lord Shiva's footprint. To Christian and Islamic believers, Adam's Peak is the place where Adam came down on Earth from the Garden of Eden. Supposedly, Adam observed penance for a long thousand years, standing all the time on one foot to atone for his serious celestial moral lapses. Adam's long suffering—standing on one foot, left its indelible mark etched on stone—a mega-size footprint. Legend has it that Adam was 10 m tall. Landing on Earth, Adam came down from the peak, crossed the Adam's Bridge from Sri Lanka to India, ostensibly in search of his beloved Eve.

What Does Science Say About the 'Ram Setu' (Adam's Bridge)?

The 48-km stretch of the 'bridge' lies between Pamban Island off the coast of Tamil Nadu, India, and Mannar Island off the coast of Sri Lanka. The Geological Survey of India (GSI) reported from radiocarbon data that both Rameswaram and Talaimannar existed, during the 7-18 ka period. The older date falls within the Last Glacial Maximum, while the date of 7 ka falls within the Holocene warming period. GSI's coral thermoluminescence dating indicated the Adam's Bridge to be only 500–600 years old.[14] The Indian Space Research Organization (ISRO) noted 103 small patch reefs, aligned in a linear pattern dotting the stretch.[15] NASA satellite MODIS captured an image of the 'bridge' of limestone shoals with an abundance of algal greens in the area.[16] NASA stated that satellite images do not prove that the linear shoal is man-made or otherwise (Fig. 11.4).[17] Water is very shallow, rarely more than 1 m deep, making navigation impossible now. Crossing the strait on foot in the historic past was possible as late as the 15th century CE. A local temple's records show that the strait was emergent and dry before a cyclone destroyed it in 1480 CE.

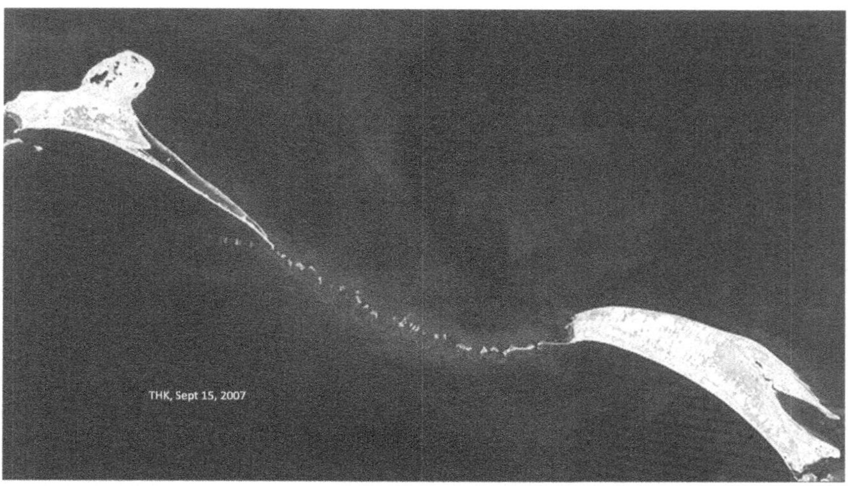

11-4 Landsat 5 image of the Palk Strait and a string of sand shoals between Pamban Island, India, and the easterly Mannar Island of Sri Lanka, itself an island.

The origin of Adam's Bridge or Ram Setu is shrouded in mystery. One difficulty is that there is little scientific field data on these shoals and reefs. For example, what is the nature of the patch reefs, cited by ISRO? Do they represent in-place biological growth of organisms like corals, or are they sandbanks or shoals of reef debris and sands? Future work in the area may provide some definitive answers.

The bridge may be a double tombolo as some suggested earlier. A tombolo or tied island forms when an island is connected to the mainland by a narrow linear strip of sediments. It is a typical barrier island formed during the rising Holocene Sea level. The author knows of a similar tombolo from the Paraguana Peninsula, Venezuela. A 20-km-long neck of a narrow sedimentary strip 2-km-wide, connects the peninsula to the mainland Venezuela.

The Sinai Land Bridge—Baboons Found a New Home in Arabia

It is a bridge separating the Mediterranean to the north and Red Sea to the south. Sinai Peninsula is the only part of Asia that is also part of Egypt, Africa. The 125-km-long strip of the Isthmus of Suez connects Asia and Africa. Baboon dispersal to Arabia happened via the Sinai bridge. Others argue baboons migrated from Africa to Arabia in the geologic past, via a temporary land crossing that appeared in the Red Sea.[19] Cenozoic sea level drop caused the closure of the Bab el Mandeb Strait at the south end of the Red Sea. The Red Sea evaporated to a dry salty ocean floor, isolated from the Gulf of Aden.[20]

The Asia-Australia Bridge—Sailors Conquered the Last Frontier

Asian maritime explorers probably had reached Australia 65–40 ka ago. A difficult question is how they got there, judging by the wide separation between the two continents now. Pleistocene glaciation apparently had played a crucial role in making this happen. Land bridges facilitated human migration to Australia, an unexplored island continent. Australia's coastline extended up to the Timor Sea at the time. Still, the 90-km stretch of ocean was a formidable barrier for the mariners.

Ice sheets extended far into the sea during the Last Glacial Period (71–59 ka ago), creating two massive landmasses—*Sunda* consisting of all southeast Asia including Indonesia, and *Sahul* comprised of Australia, New Guinea, and Tasmania. The early voyagers would have pulled off this mammoth task of island-hopping to Sahul by two viable routes. One possible route was via the closely spaced Sulawesi islands, then landing in New Guinea, already a part of Greater Australia. Later, they would have dispersed along the coast to other regions. The second route might have been via Timor and landing in northern Australia. The latter route sounds realistic, as northwest Australia preserves many old archeological sites of early newcomers.

One difficulty today is that the post-glacial thaw has radically changed ancient coastlines. The rising Holocene Sea level has largely submerged and reconfigured past landmasses. It drowned large parts of Sunda, spawning thousands of southeast Asian islands, especially in modern Indonesia. Sahul also assumed a separate identity as New Guinea, Australia, and Tasmania; the last cut the cord only 10 ka ago.[24]

The Danakil Depression, Ethiopia—An Unearthly Place

The drifting of Africa and Asia created the 400 km^2 Danakil sag at the north end of the Afar Depression. The Afar area at Hadar, Ethiopia, was the discovery site of 'Lucy,' the famous 3.2-Ma-old *Australopithecus afarensis* fossil. Extremophile microbes survive in the inhospitable Danakil waters (Fig. 11.5). Local conditions eerily mimic Martian climate, hinting at the possibility of similar biosignatures in Mars.[21]

Once a part of the Red Sea, Danakil is now dry, though lying 100–150 m below sea level. Rifting of the African, Arabian, and Somali plates created Danakil as an incipient ocean basin. Some unique features are the multi-hued sulfurous super-acidic hot springs; water is corrosive like an acid, with ultra-low pH of 0.2. pH means 'Power of Hydrogen,' measured by free hydrogen and hydroxyl (OH) ion content of a liquid. Other interesting oddities of the Danakil are the Erta Ale volcano (a rare lava lake crater), Dallol cinder cone crater (45 m below sea level), the unusual Karum and Bakili salt lakes, salt domes, and extensive salt pans. Miners extract salt, a precious commodity,

11-5 The surreal landscape of Danakil depression (125 m below sea level) and its multi-hued mineral deposits in the inhospitable terrain of the Afar Triangle, Northern Ethiopia.

from the salt pans around the lakes. The area is very hot, average 35–40°C year-round, occasionally up to 50°C.[22]

The Danakil highlights the incessant plate tectonic activity and rifting. Geologists are in luck to have a ring-side seat, as it offers a rare glimpse of an active rift. Danakil became isolated from the Red Sea, when copious freshly erupted lava sealed off the narrow neck of an inland sea. The area morphed into a lake that soon evaporated in the roasting heat of the Afar Triangle. NASA posits that geologically it is a transitory phase. The Red Sea will again flood the Danakil depression in future, as the area is actively subsiding.[23]

STONY TESTIMONY

Quaternary Sediments of Ocean, Lake, River and Glacier

The main Pleistocene rock types are glacial deposits, volcanic lava, and abundant carbonate sediments. Sand, silt, and clay also abound in various sedimentary environments—eolian, fluvial, deltaic, lacustrine, shelfal, and deep-sea. Evaporites occur as a secondary mineral.

One very special natural wealth some countries enjoy, especially Canada, United States, Chile, Argentina, China, and countries in Europe, is the large number of lakes of glacial origin. Ice sheets and glaciers had so much erosive power that they excavated many deep basins out of hard rocks. Later, glacial melt waters filled them forming lakes, some as large as an inland sea.

The Niagara Escarpment is the handiwork of deep erosion of soft strata underlying a resistant hard Paleozoic rock. It consists of mainly Silurian limestone, dolostone, sandstone, shale, etc. (Fig. 11.6). The 1000-km-long escarpment, av. 228-m-high, is a majestic, arcuate physiographic feature. It starts near the western shore of Lake Michigan, passing north through the Manitoulin Island (Canada); it comes down and continues along the south shore of Lake Ontario traversing Ontario, Canada and upstate New York, USA. The topmost resistant Silurian Lockport dolostone layer protects the softer erodible rocks below. Pleistocene glacial melt waters along numerous

11-6 Interstratified limestone, mud rock and sandstone sequence (>30 m) of the top part of the Niagara Escarpment exposed above the tunnel at Niagara Parks Power Station, Canada.

spillways from the top dolostone have given rise to spectacular gorges and waterfalls all along the length of the escarpment, including the famous Niagara Falls.

The Great Lakes of North America hold the largest freshwater reserve in the world. Lake Superior, with an area of 82,103 km^2, is the largest lake in the world by surface area. Prolonged Quaternary glacial digging shaped the Great Lake depressions, younger than 14 ka. The lakes got totally filled with glacial melt water by 4-ka BP. Canada has the largest number of lakes in the world, covering nearly 7.6% of land surface. Glacial downcutting and erosion can produce plenty of sediments. Okanagan Lake, western Canada, preserves 800 m of sediments representing a single phase of the Wisconsin glacial event.[26] Lakes aside, highly porous glacial moraines store enormous amount of subsurface water.

Glaciation is episodic, controlled by Milankovitch astronomical cycles. Long glacial spells rhythmically alternate with much shorter, warmer interglacial breaks, like the one we are enjoying presently.

Some Diagnostic Glacial Rocks and Landforms

The signature product of a glacier is till that becomes tillite when lithified. Tills/tillites are disorganized assemblage of debris of all sizes—boulder, cobble, sand, silt, etc. A diagnostic feature of glaciolacustrine origin is 'varve.' When a glacier retreats, fine particles settle as mm-cm scale 'varve' in proglacial lakes. A varve couplet typically consists of color-contrasted layers—a coarser, pale summer silt layer, and a dark winter clay lamina. Varves are ubiquitous in glacial lakes—both modern and ancient. Earlier, the author had studied Gondwana glacial varve in India.

Esker and drumlin are diagnostic glacial bedforms. Eskers are subglacial channels at the base of moving glaciers and ice sheets. When ice disappears, the long, tunnel-form, eskers remain preserved as steep sided winding ridges, filled with glacial debris.

The drumlins (Fig. 11.7), about 1–2 km long, 300–600-m-wide and 50-m-high, occur in swarms or as drumlin fields. These large, parallel, and elongated fluted structures, resemble inverted spoons in profile, or may show a 'basket of eggs' morphology. Nearly 4,000 drumlins dot the famous Canadian Peterborough Drumlin Field, Ontario.[27] Western New York State, and many places in Europe like Germany also show impressive drumlin development.

The U-shaped Tunnel Valleys exist deep within glaciers carrying melt water mixed with rock debris. These are very large features—max.100-km-long, 4-km-wide, and 0.4-km-deep. When any glacier retreats, it leaves behind long, deep, parallel to subparallel valleys, that may later morph into deep river valleys or more characteristically, as finger lakes. The seven large finger lakes of upstate New York

11-7 The partly submerged, inverted spoon-shaped, Georges Island is a drumlin of glacial origin in the middle of Halifax harbor. It is a cultural heritage site in Canada with a lighthouse and some old installations from historic times. Other Nova Scotian drumlins of rich local history are the Citadel Hill, McNab's Island and Fort Needham.

are great examples of Tunnel Valleys (Fig. 11.8). They are arranged in a subtle arcuate form converging to a northerly source area. The largest and deepest Lake Seneca, measuring 61-km-long, 3-km-wide, and 188-m-deep, is touted as the lake trout capital of the world. Lake depth may be a telling example of the erosive power of Pleistocene glaciers. The fertile Niagara region, Canada, and New York Lake region happen to be important areas for vineyards and wineries.[28]

Another important feature of glacial erosion is 'moraine'—made of disorganized debris, from rock flour to boulders as big as a multistoried house. The moraines are of several different types, depending on the location. In 'lateral moraines' glaciers push tills toward the valley walls, preserved as long ridges. A 'medial moraine' forms when lateral moraines of two glaciers meet down-valley, forming a dark middle ridge within white ice. A glacier may leave behind other types of moraines: irregular patches of 'ground moraine,' a 'terminal moraine' at journey's end, or a 'recessional moraine' with subparallel back-stepping ridges.

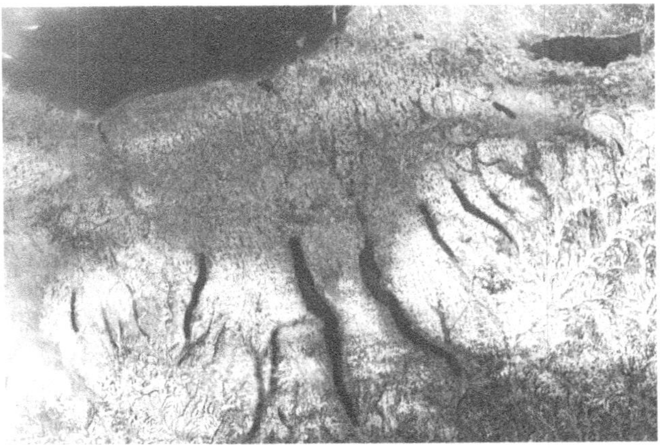

11-8 Satellite view of the Finger Lakes, New York, USA., south of Lake Ontario. The large, elongated Cayuga Lake in middle right, is 133 m deep.

The Oak Ridges Moraine in Ontario, Canada, is a world class example (Fig. 11.9). The economically important, ecologically sensitive, geological landform is of Late Wisconsin glacial vintage, only 12 ka old. It is up to 150 m high, 2–30 km wide and stretches over 160 km from the Niagara escarpment in the west to the Trenton River system in the east. It is not a continuous ridge, it developed in 4 discrete glacial stages. It runs parallel to the north shore of Lake Ontario, touching some of the northernmost areas of the Greater Toronto Area, Canada. The Ontario authorities regulate urban development in the Oak Ridges Moraine because of its water resources, agricul-

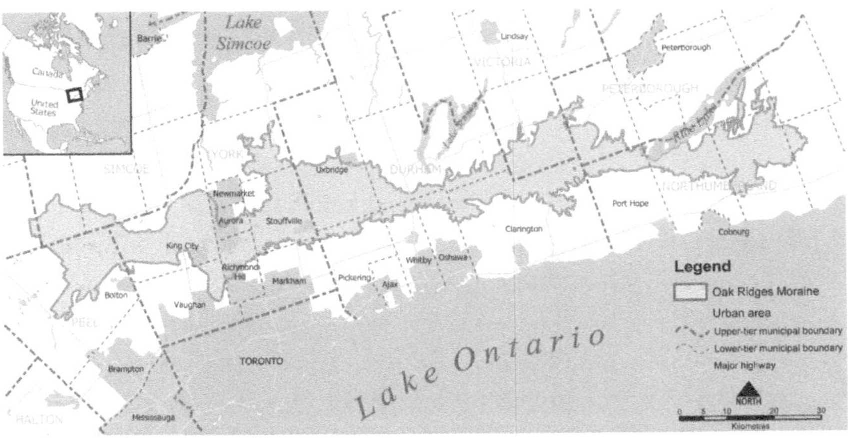

11-9 Oak Ridges Moraine, north of Lake Ontario and the Greater Toronto Area, Canada.

tural importance, biodiversity, and recreational facilities. It feeds 65 river systems both to the north and south, the majority ending up in Lake Ontario.[29]

Other glacial structures are striated pavements, erratic boulders, etc. Striated pavements with linear grooves and lines arise from continuous grinding and rasping action of entrained clasts and boulders, parallel to the ice flow direction. Glacial valleys tend to be U-shaped.

'Loess' deposits (German—'loose') consist of silt and clay of glacial origin. Winds may transport the fines away from the glacier's toe, deposit them in areas where it dies down. Loess can be a few cm to 100 m thick. It can also be of purely eolian origin, unrelated to glaciers. The yellowish loess, rich in minerals and nutrients, forms fertile soils in many places in the world. Silt and fine sand may form the bulk of it, while glacial rock flour and clay may make up 20% of loess. Its soft texture contributes to high crop productivity. Some world-famous loess deposits occur by the Yellow River, China, the Great plains of North America, parts of Russia and Kazakhstan. The North American loess silt owes its origin to the last two glacial advances. The rich Loess Hill deposits—Iowa and Missouri, USA, came from the Missouri River valley.[30] Terraced cultivation maximizes productivity and prevents erosion of fine textured, soft loess soil.

The Scarborough Bluffs—A Reminder of the Chalk Cliffs of Yorkshire, UK

In 1793, Elizabeth Simcoe, wife of the then Governor of Canada, had a first glimpse of the towering cliff, while on a boat journey on Lake Ontario. Its height impressed her, she likened it to the majestic Chalk Cliffs of Scarborough, England. The name stuck both for the cliff, and the future Scarborough City that came up later in the Greater Toronto area nearby. The origin of the eye-catching bluffs revolves around the compelling history of the vanished Lake Iroquois. It was a much larger avatar of the present Lake Ontario, both having the same overall shape. The Scarborough Bluffs around Toronto represent a 15-km-long remnant shoreline of the Iroquois lake-delta. The ancient proglacial lake level was a good 30 m above today's Lake Ontario because the St. Lawrence River mouth became dammed by ice during the last glacial stage. Once melting glaciers retreated about 13.4 ka ago, Iroquois drained southward through the Mohawk River into the Hudson Bay. The marked water withdrawal exposed the 90–100-m-high, vertical Scarborough Bluffs.[31]

The scarp face of the bluffs is unstable, receding steadily over the years. Seasonal freezing and thawing of scarp material contribute to further instability and spalling. Landslides of large chunks of sediments including huge glacial boulders happen frequently, including one in August 2020.[32] Lake Ontario is constantly lapping the base of the escarpment and wearing away the soft material making the slope unstable. Also, urban constructions on top of the bluffs increase the risk of landslides. The 13-ka-old sediments are loose, not yet lithified.

The Rocky Mountain Foothills Erratic Train, Alberta, Canada

The 'Erratic boulders' are large blocks, carried by glaciers far from the source area. An 'erratic' is always out of place. The 900-km-long 'Foothills Erratic's Train' is a textbook example. The 'Foothills Erratic's Train' along a N-S foothills strip of the Prairies, consists of a linear concentration of thousands of quartzite boulders originating from a peak in the Jasper National Park in the Rockies. Apparently, a colossal rockslide dumped the 'erratics' onto the Cordilleran ice sheet, that then transported them out onto the plains. The merged Cordilleran—Laurentide ice sheets carried the erratics train on a long journey, even carrying the famous 'Big Rock' to Okotoks, Alberta.[33]

Erratic Boulder—'The Big Rock' Hitched a Cozy 400-km-long Ride on Glaciers

The 'Okotoks Erratic' in the Alberta countryside, Canada, is the world's largest erratic, now in two pieces (Fig.11.10). The 41 m x 18 m x 9 m 'Big Rock' is a 16,500 tons quartzite block, the size of a 3-storied building. The merged Laurentide and Cordillera ice sheets dropped it off 12–30 ka ago. It had a 400-km-long, comfortable cool ride to its destination—Okotoks.

11-10 The 'Big Rock'—Okotoks Erratic, Alberta, now broken into two blocks, near Okotoks town in the Calgary area, Canada. It traveled 400 kms on a glacier to reach Okotoks, Alberta.

Once a glacier retreats, transported 'erratic' boulders stay on as silent, yet eloquent testimony of the passage of the vanished glacier. The Central Park of New York hosts an 'erratic' schist block from far away. The Laurentide Ice Sheet placed the gift there many thousand years ago.

Perito Moreno Glacier, Argentina—A Rare Glacier Bridge Over a Lake

The 'Los Glaciares' National Park is a special, geo-touristic attraction (Fig. 11.11). Here the Perito Moreno Glacier finishes its journey in the large land-locked Lago Argentino. The 30-km-long, 5-km-wide and 170-m-thick, Andean glacier originates in the 2,100-m-high, Southern Patagonia Ice field. The unique feature of the glacier is that it periodically debouches into the Lago Argentino. The glacier literally divides the lake in two, as its snout traverses the whole width of the lake, becoming 'grounded' on the opposite shore. It creates a veritable icy bridge, cutting off circulation between the sides. The lake's Brazo Rico arm continues to receive melt water from a southerly glacier, eventually raising its water level 30 m above the rest of Lago Ar-

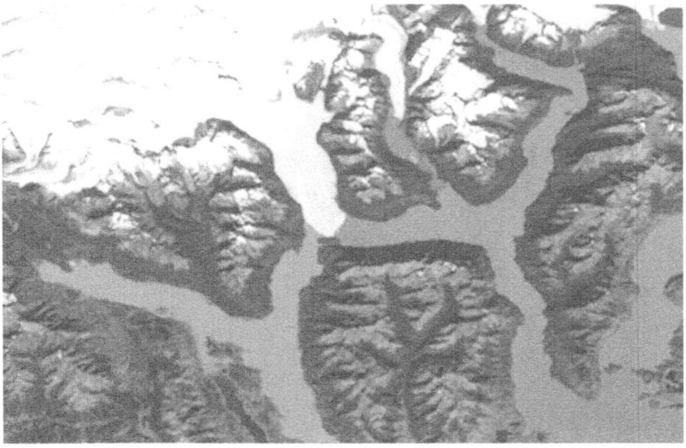

11-11 The Perito Moreno Glacier snout divides the lake. **left** The Brazo Rico arm. **right** The Lago Argentino lake.

gentino. The situation becomes precarious when pressure on the ice-bridge builds up to a critical point. Finally, the glacier-dam ruptures and slowly collapses. The recurring phenomenon attracts tourists for the unusual experience every few years.[34]

Niagara Falls, North America—A Gift of Glaciers and Lakes

The Niagara River flows from Lake Erie to Lake Ontario, which in turn flows out to the Atlantic Ocean via the Saint Lawrence Riverway.[36] The river traverses 58 km between Lake Erie and Ontario,

11-12 View of Niagara River from the plane window. The river between the USA and Canada, quietly ends its 58-km-long journey from Lake Erie into Lake Ontario near Niagara-On-The-Lake, Canada (left side).

with a height drop of 99 m (Fig. 11.12). The Tonawanda branch creates the American Falls dropping 21–34 m, while the main Niagara channel near Chippawa, Canada, descends 57 m from the Horseshoe Falls. Water tumbles down at 2,400 m^3/sec from a crestline of 670 m, creating a 35-m-deep plunge pool. All waters are of ancient lake origin, less than 1% renews annually.

Nature's wonderland—the Niagara Peninsula, has an interestingly chilly history of how the falls materialized. The world-famous tourist hotspot is not geologically old. During the Wisconsin Glacial period, the 2–3-km-thick, mountainous Laurentide Glacier stayed put on the larger Niagara region. The ice melted away about 12 ka ago, and the retreat revealed huge basin-size depressions that glaciers had excavated on the rocky bottom. Melt water gradually filled the basins and the Great Lakes took shape—Superior, Huron, Michigan, Erie, and Ontario. The lakes hold 18% of the world's freshwater reserve. From Lake Superior to Lake Ontario a low natural slope exists, facilitating flow from Superior to Ontario via other lakes.

Glaciers occupied the lake sites at least 6 times during the last 0.78 Ma.[37] The back and forth of ice sheets produced lake level fluctuations leading to channel cutting and isostatic rebound of the land. The glaciers preferred the path of least resistance and followed existing routes of pre-glacial river valleys. The pre-existing physiography and lithological weak spots controlled the shape of the lakes, as glaciers gouged out the softer shaly bedrock. The Lake Superior is sitting on a soft Precambrian bedrock on the Midcontinent Rift, USA. Lakes Michigan and Huron are perched on top of the Paleozoic Antrim Shale and Salina Group respectively. The Lakes Erie and Ontario occupied the two distinct valleys of the Late Ordovician Utica and Queenston shales. The thickness of ice that carved out the lakes, controlled the depth of each lake—Superior 410 m, Michigan 285 m, Huron 223 m, Ontario 86 m and Erie 65 m. As an aside, Lake Erie happens to be the thunderstorm capital of Canada.

The Niagara River is the link, carrying the gift of pristine green waters from Lake Erie to Lake Ontario (Fig. 11.12). About 10.5 ka ago, geological causes like ice movement, isostatic rebound, etc., had temporarily rerouted the melt water along a northern route. As a result,

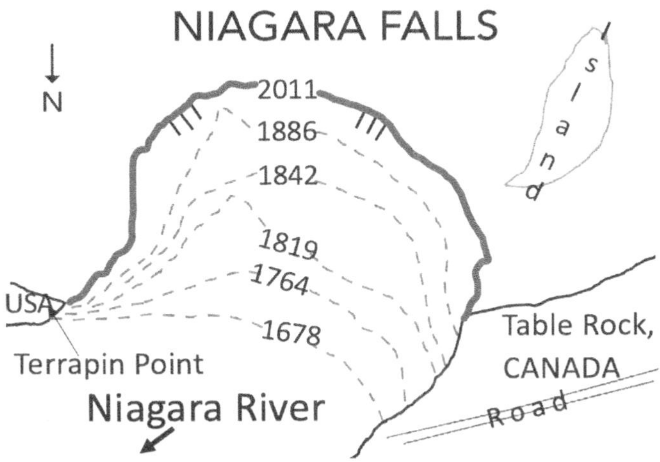

11-13 Headward erosion of the Niagara Horseshoe Falls in recorded history from 1678 to 2011(~present day). The Falls moved backward along the Niagara River about 11 kms in the last 12,000 years.

the southern route starved and struggled to flow.[38] Lake Erie became half its present size, while the Niagara River was carrying only 10% of its present load. The malnourished state lasted till 5 ka ago, when melt water again started flowing through the Niagara River.

The headward erosion of the waterfalls is directly linked to river discharge—the more the discharge, the faster is the headward erosion. The river gorge moved backwards very slowly, <1 km in the first 5.5 ka due to prolonged lean discharge. It moved backward 8 km in the last 5 ka during high discharge. The Falls moved back 11 km from the Queenston escarpment (Fig.11.13).

Why does the crest line of the waterfall move backward? The river gorge cuts through the Silurian and Ordovician sedimentary formations. The resistant top Silurian Lockport dolostone layer normally protects softer underlying Rochester Formation shale from erosion. But when the top layer itself becomes vulnerable due to fractures, or excessive erosion of the softer layers below, it may finally collapse. Large blocks spall off into the plunge pool. As the top layer succumbs, the crestline moves backward causing headward erosion of the waterfall.

The emerald-green color of the water is due to an abundance of dissolved minerals, and rock flour produced by the friction of rock particles, and 'cavitation.' As water under high pressure rushes down, it forms bubbles. The bursting bubbles send shock waves on contact with rocky floor and pulverize it producing rock flour.

Climate change is an unknown variable that may affect Niagara River in the distant future. Some models predict that ongoing isostatic rebound may severely compromise the existence of the American Falls in the next two thousand years. The Horseshoe Falls will probably erode back 6 km in the next 15 ka. Exposed hard limestone and soft shale layers will create a series of rapids replacing the now familiar gorgeous waterfall. Fast forward 50 ka—alas, probably no tumbling falls anywhere, only a placid river gliding by.[38]

So Many Glacial Events in Earth History—The Causes?

We may wonder why glacial events happened on Earth: The Huronian and Cryogenian of the Precambrian, the Late Ordovician (Andean-Saharan), the Late Paleozoic (Karoo), the Paleogene (Antarctic), and the ~2.58-Ma-old Pleistocene when the Arctic also became icy (Fig. 11.14). Several factors likely could have triggered glaciation at different times:

1. Volcanic ash, dust, and sulfur dioxide are well-known culprits, as they can block solar radiation on Earth.
2. Waxing and waning of animal life can cause CO_2 and temperature variation. Modern atmospheric CO_2 uptick is partly due to excessive number of farm animals.
3. An Ice age occurs when summer temperature in the northern hemisphere languishes indefinitely in the freezing zone. The warmer interglacial periods happened during greater summer solar radiation in the Northern Hemisphere. Curiously, a stepwise fall in temperature ushers in a glacial period, but its end happens abruptly.[41]

4. Mountain building may cause a temperature drop because they attract ambient moisture, form snow, and cool the climate.
5. Interstellar clouds between Earth and the Sun may curtail sunlight.
6. The Milankovitch astronomical cycles (Fig. 11.15) explain the amount of incident sunlight on Earth's surface and the onset of the icehouse stage. Several factors influence sunlight variation—wobble of Earth on its axis, variation of Earth's distance from the Sun due to variation of its elliptical orbit, and Earth's orbiting tilt angle (22.1°–24.5°). A lower tilt angle causes less incident light, and cooler temperature. Especially, changes in Earth's orbit path and tilt angle may determine the onset of ice ages.[40]

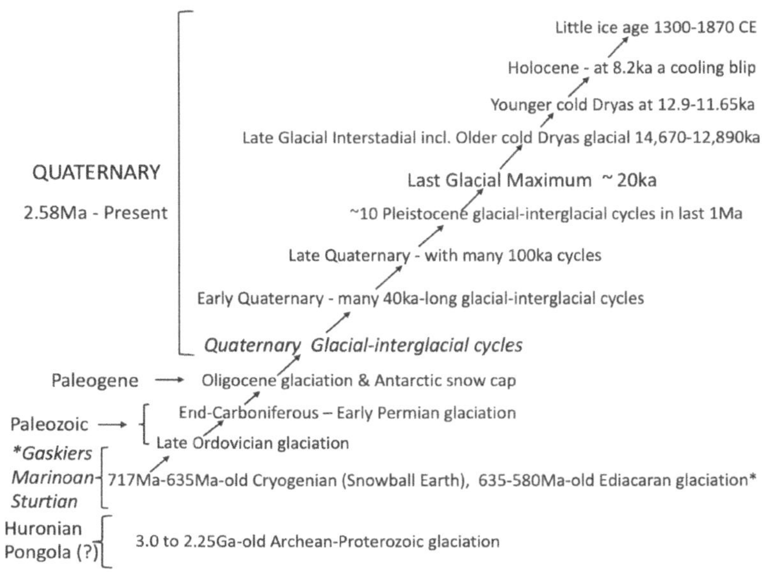

11-14 Glacial climate cycles [40, 40a]—Archean to the Present. 5 Precambrian events were followed by the Paleozoic, Paleogene, and the Quaternary ice ages.

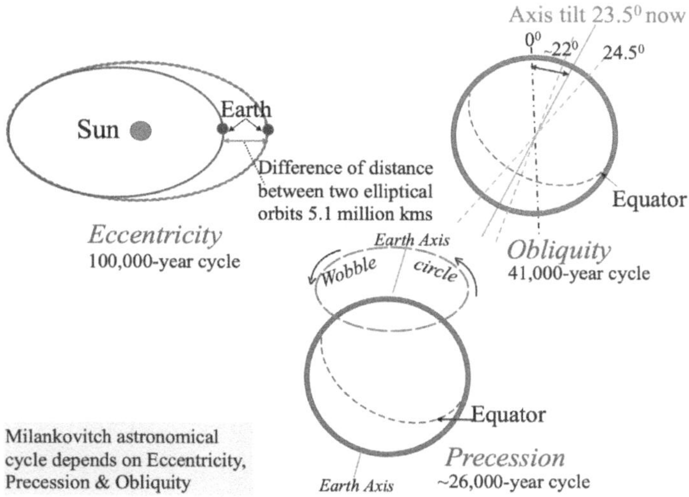

11-15 The Milankovitch cycle[42] explains the amount of sunlight falling on Earth, based on axial tilt angle, Earth's wobble around its axis and distance from the Sun.

Rocks of Lava and Ash Origin: The Pacific Ring of Fire

The Ring of Fire eruptions make lava rock a principal constituent of the Quaternary strata in the Philippines, Japan, Indonesia, Aleutians, Central America, Mexico, etc. Other places include the Andes, U.S. west coast, British Columbia, and Yukon Territory of Canada. The Yellowstone super-volcano and the Big Island of Hawaii, USA, are not part of the Ring of Fire. Only the Big Island of Hawaii is presently erupting in the Hawaiian chain; earlier, the now inactive Molokai, Maui, and Lanai, etc., had erupted during the Quaternary Period. We recall a mantle hot spot is the likely source of the erupting lava.

The Hawaiian Eruption, 2018

The Kilauea volcanic eruption was a glowing example of an event that unfolded before our eyes. It wrought some major physiographic dislocation upon the island. Not only did it change the local coastline, but also affected the ecosystem of the area. The 'Puna eruption' buried part of the Hawaii Route 137 on its way to the ocean. It was the most destructive eruption since the 1980 Mount St. Helens on mainland USA. New lava covered 35 km^2 of land and carved out 3.54 km^2 new land in the Pacific Ocean.[45] The outpourings made considerable impact on the island, as the news media and the U.S. Geological Survey covered it for months before it waned.[46, 47] The Halemaumau summit crater began to drain along cracks in the large lower rift zone in May followed by a 6.9 magnitude temblor. Soon, the magma found an easier route along fissures at the foothills when the lava level inside went down a jaw-dropping >300 m below the crater floor.

The emptied magma chamber produced >500 m of summit collapse in the end. A little later, a lava river was flowing through multiple fissures to the ocean, destroying homes, fields, and properties along the way. The lava emitted an enormous volume of SO_2/each day. A startling thing happened in June—the lava river encountered the island's largest waterbody on its way—the 6-m-deep Green Lake. Red-hot lava vaporized the whole lake in just 90 minutes, and totally filled it with lava in 5 hours. The eruption created 'vog' or volcanic fog, a suffocating mixture of CO_2, SO_2, and water. On quenching rapidly in coastal waters, the glowing lava also created toxic 'laze,' a mixture of hydrochloric acid, glass particle and steam.[49] Hawaiian eruptions sometimes produce a volcanic dome fountain, rarely up to 500-m-high (Fig.11.16).[43, 44]

11-16 Red-hot, 20-m-tall lava dome above the black lava field—1969–71 Mauna Ulu eruption of Kilauea volcano, Hawaii.

The Yellowstone Supervolcano—If It Ever Erupts Again

The Yellowstone Supervolcano, a mega-caldera measuring 3,960 km^2 in Wyoming, USA, is part of the Rockies. Three giant past eruptions marked the supervolcano—2.1 Ma, 1.2 Ma and 0.63 Ma ago, the last one created an enormous caldera and the Lava Creek Tuff.[50] One smaller eruption happened 174 ka ago. As the supervolcano belched magma the last time, the hollowed-out chamber collapsed to form the caldera. The whole park area is part of the caldera. The Yellowstone erupts two types of magma—rhyolite and common basalt showing palisades of 6-sided columns (Fig. 11.17). Spectacular geysers including the 'Old Faithful,' myriad hot springs, and fumaroles are living proof of the superhot cauldron lying below the surface (Fig. 11.17). Why is the Yellowstone supervolcano given such a distinction, while the Mount St. Helens of 1980 fame is just a volcano? A supervolcano has a magnitude 8 on the 'volcanic explosivity index', i.e., magma

11-17 left A roadside outcrop of basalt with distinctive columnar joints in Yellowstone Park. Note the difference in joint pattern in the main lava flow above, and the lowermost inner layer at the road level with a blockier and stout joint pattern. This may reflect compositional difference between the flows. *Photo: Sudhin Dey.* **right** Old Faithful Geyser tapping a heat source of a magma body at shallow depth.

eruption exceeding 1000 km^3 of lava outpourings.[51] The three eruptions were 6,000, 700 and 2,500 times larger than the volume from the Mt. Saint Helens in 1980.

Recent seismic tomographic imaging of the subsurface caldera revealed two stacked magma chambers separated by a sill of solid volcanic rock. The upper chamber with abundant mixed gas has greater potential of explosive eruption. It has an area of 3,600 km^2, with 5–15% rhyolite melt and lies only 5–17 km below the surface. The 20–50-km-deep lower chamber, is 4 times bigger, but only with 2% of basalt lava melt.[52] Chances of new eruption is low, but it will be an unprecedented calamity, if it ever does.

The Long Valley Caldera, USA—A Dormant, High-Risk Volcano

This supervolcano near Mammoth Mountain, California, east of the Yosemite National Park, last erupted 0.76 Ma back. Known as Bishop's Tuff in the region, it is a rhyolite tuff of vesicular lava and consolidated ash. The ash traveled far and wide—from the Pacific Ocean to Nebraska, more than 2,000 kms away.[53] The eruption lasted several days and pyroclastic material—pumice, crystal, glass, and ash, produced a layer, 1,500-m-thick. The 32-km-long caldera is hiding humongous, 600 km^3 of semi-molten magma, in a deep chamber.[54] The Long Valley, Mt. Shasta and Lassen are some of the high-threat volcanoes in the United States.

The Toba Supervolcano Eruption—Apocalypse in Sumatra

The Toba Tuffs of Sumatra (Indonesia) documents another remarkable Quaternary volcanic event. The still active Toba super-volcano erupted magma forming a 3,500 km^2 caldera visible today.[56] The caldera hosts the 100-km-long Toba Lake—the world's largest volcanic lake. It erupted a cumulative 3,400 km^3 of silicic magma in four volcanic phases, the last one only 74 ka ago. The biggest eruption of the Toba Tuff caused widespread dispersal of ash and gases to far-off places like the China Sea, etc. Ash in alluvium in India correlates with the eruption.[57]

The cataclysm would have lifted billowing smoke, fine ash, and sulfur gas 37 km high in the stratosphere. The massive dust and sulfur aerosol clouds had likely blocked sunlight a great deal. The ensuing 3–5°C temperature drop produced a 'volcanic winter,' hastening the onset of glacial conditions, already underway at the time.[58] It's hard to imagine such a scenario today.

The Trans-Mexican Volcanic Belt—Magic of a Subducting Tectonic Plate

The east-west trending Trans-Mexican volcanic belt dotted with volcanoes of breath-taking grandeur plays a significant role in Mexico (Fig. 11.18). The Pliocene-Quaternary volcanic arc[59] appeared because

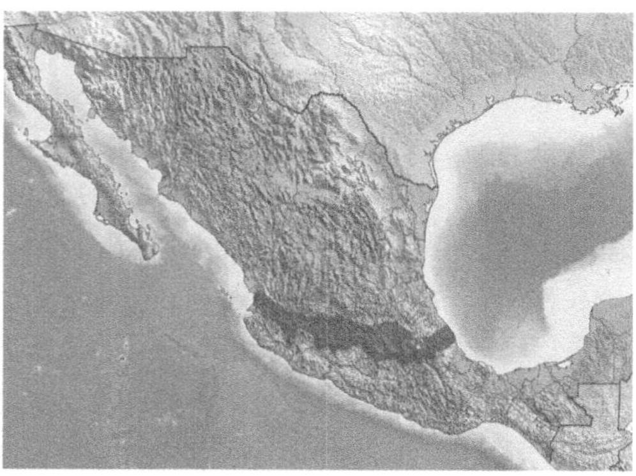

11-18 top The Trans-Mexican Volcanic Belt (TMVB) as a dark band, extends from the westerly Pacific coast to the Gulf of Mexico. Most active volcanoes like Orizaba, Popocatepetl, etc., occur on the southern flank of the TMVB. Mexico City is located in the middle of the eastern TMVB. **bottom** Tectonic features like the Middle America (i.e., Central America) Trench and subducting Pacific plates.

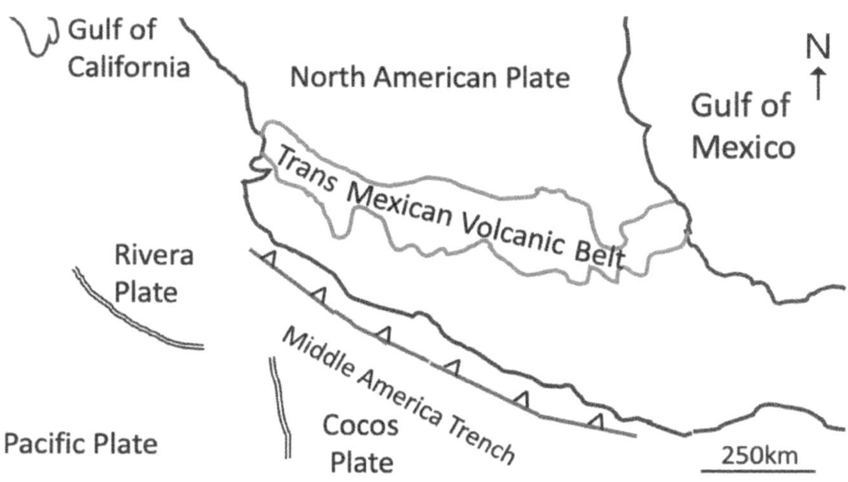

the Pacific Cocos plate, and its splintered Rivera plate, had been diving under the North American plate along the Middle America Trench (Fig. 9.1). The Cocos Plate has been undergoing partial melting at deep level and is rising as magma forming the chain. Frequent eruptions and earthquakes are hallmarks of the belt. The principal stratovolcanoes—El Pico de Orizaba, El Popocatepetl, El Nevado de Toluca etc.,[60] are located on the south side of the east-west belt, whereas the famous ore and precious stone mines/quarries occur on the north flank. The active volcano Popocatepetl started erupting andesite lava about 0.33 Ma ago.

The 1,000-km-long Trans-Mexican volcanic belt has 3 sectors: western, central, and eastern. Two historic and architecturally rich metropolis adorn the eastern sector—Mexico City of 22 million inhabitants, and Puebla of imposing colonial churches. The tallest mountains and stratovolcanoes of Mexico highlight the region. A borehole lava sample near Mexico City gave an age of 14 ka.

Lava flows had created the geologically important Mexico basin, that hosts the namesake metropolitan city, and the remnants of the two great Quaternary lakes—Chalco and Texcoco. The lakes played an important role in the Aztec, Toltec, and Mesoamerican cultures. The oldest lake sediments are up to 0.22 Ma.[61] Freshwater diatom from sediments below the City, revealed that deep lakes existed 100 ka ago, and marshy conditions prevailed during the last 10 ka.[62]

Varying the Subject, the Messy Quaternary Tar Seeps—'Nature Also Pollutes'[63]

Tar or oil seeps are known from time immemorial. The Bible—Genesis 14:10 says "Now the Valley of Siddim was full of tar pits, ..., some of the men fell into them and the rest fled to the hills"[64]; in Exodus 2:3, the plight of Moses' mother is described thus ".... she took for him a papyrus basket made of bulrushes and coated it with bitumen and pitch. Then she placed the child in it ...."[65] We all know about the Noah's Ark. But before Noah set sail from Egypt, ... "God said unto Noah .... Make yourself an Ark of gopher wood; make rooms in the ark, cover it inside and out with pitch"—Genesis 6:13–14.[66]

All ancient cultures knew about oil seeps, including the Babylonians, for the waterproofing of boats, ships, etc. The famous, 6-ka-old Hit site of the "Fountain of Pitch' by the Euphrates River, Iraq, shows evidence of its use as mortar. The Mesopotamians employed it as adhesive for setting jewels and mosaics, while the Egyptians knew of its myriad uses. In China, they exploited seeps to generate heat and light 2 ka back. Marco Polo had mentioned tar use for treating camels in the Caspian region. The Seneca tribe, New York, used to trade the magical 'Seneca Oil' from upstate New York for medicinal cures. Even the Neanderthals utilized it in the Paleolithic age.

Oil seeps were well known in the Dead Sea, Israel, and the Great Salt Lake, USA. Reported notable seeps include the La Brea Tar Pits in Los Angeles; the Santa Barbara Channel, a seaward extension of the Ventura Basin, California; the Gaspe Peninsula, Quebec, Canada; the Guanaco Asphalt Lake, Venezuela; and the famous Pitch Lake of Trinidad and Tobago. The Santa Barbara seep discharges 50–70 bbls/day, which flows as a stream of asphalt, sand, water, and gas into the ocean.[63] Seeps also occur in Barbados' Consett Bay, Gulf of Mexico, and the Caspian Sea off the Absheron Peninsula.[71]

Tar seeps are geologically significant as they may hint at potential underground oil resources. North America's first oil discovery happened in 1858, in a Canadian well next to oil seeps in the locality of Oil Springs, Ontario. The 1859 Drake well in Pennsylvania started the oil industry in the U.S. Oil seeps led to the discovery of the famous oil fields of Kirkuk, Iraq. The 1914 discovery of Mene Grande ('large seep') oil field by drilling the Zumaque-1 well on the eastern shore of Lake Maracaibo, ushered in the oil boom era in Venezuela (Fig. 11.19).

The Guanaco Asphalt Lake, Venezuela—An Apple of Discord

It is the world's second largest tar pit, also known as Bermudez Pitch Lake in eastern Venezuela, reported by the German explorer A. Von Humboldt in 1799. The '*Warao*' tribe waterproofed their canoes with seep since historic times. Pure asphalt continuously bubbles up to the surface. The 4.45 km^2 and 1.5–2-m-deep asphalt spring is covered by

11-19 The Mene Grande tar seep in the Zulia State, started the oil boom in Venezuela in 1914.

a surface crust. Vegetation and plants, including the 'moriche palms' (*Mauritia flexuosa*) protect it from natural weathering.

An American company used to mine the tar pits in the 1890's.[68] When the Venezuelan President at the time decreed a higher tar tax, the mining company balked. They supported his opponent in a local conflict, leading to the 'Asphalt War' of 1902–03. The dispute temporarily soured diplomatic relations between the two nations. Eventually, the Venezuelan president expropriated the company in 1908. However, mining operations stopped there in 1934.

The Famous Pitch Lake, Trinidad and Tobago

The La Brea Pitch Lake by the Gulf of Paria, about 1.1 km inland from the Caribbean shoreline, is the largest asphalt seep reservoir in the world.[69,73] The famous explorer Sir Walter Raleigh discovered it in 1595, and he took some asphalt home to caulk his battered ship with it. The seep, by the La Brea village, is awaiting approval as a UNESCO World Heritage Site (Fig. 11.20). Asphalt is an emulsion of bitumen, gas, water, sands, and minor clay.[70] The La Brea is not a static lake, it moves ever so slowly, as shown by flow lines on asphalt surface, and floating ancient tree trunks. It is a circular depression in the ground with an area of 410,000 m^2, and a depth of 76 m. It holds 9 million tons of pitch, sitting above a Late Neogene heavy oil reservoir.

11-20 An aerial (drone) view of the La Brea Pitch Lake, Siparia, Trinidad and Tobago.

The Origin of the Pitch Lake—From Geological to Mythological

How did the lake form? The first explanation is geological. Natural crude oil remains in pores of underground reservoirs, under the vertical pressure of overlying rocks. Deep-seated rocks, when tectonically uplifted close to surface, may release some of the overburden pressure by developing fractures at the top. Buoyant oil may leak out along the cracks/fractures or faults and emerges on the surface. The asphaltic oil comes from the Cretaceous Naparima Formation, Trinidad, time equivalent to the oil rich Querecual Formation of nearby Venezuela. Some large deep faults and fractures appeared as the Caribbean plate rammed against South America. The fracture network facilitated oil leak from the deep reservoir that slowly filled a mangrove lagoon on the surface. Patches of pitch on the Caribbean shore confirm that the lake had earlier extended up to the Gulf of Paria.

The second explanation on the seep's origin comes from indigenous folklore.[71] Many years back, Prince Kasaka of the Cumana tribe eloped with his ladylove Callifaria, a rival chief's daughter. Soon the angry Chief Callisuna took matters in his own hands, and promptly attacked the Cumana tribe. He recovered his daughter, bringing her home tied to a horse. The winged Arawak god got really upset seeing the sorry plight of princess Callifaria. He cursed and damned the entire village, had it sunk inside Earth and covered it with black 'Piche.' Perhaps, the assorted memorabilia like a fossilized giant sloth, a mastodon tooth, remains of a 4-ka-old tree, tribal artifacts, etc., lent credence to this tale.

Asphalt is mined from the center of the depression. New asphalt and gas continually bubble up from below to replenish the pool. Nature unfailingly refills the pitch lake at night, after 24 wagons of asphalt are extracted every day from this very special mine.[72] A 'renewable' source with a dirty twist. Pitch lake reserve now stands at less than 10 million tonnes after so many years of exploitation. An oil sand outcrop is mined at a nearby quarry.

The asphalt lake was instrumental in the discovery of several oil reservoirs in the early 1900's. Many tourists visit here because of its fame, including its supposed therapeutic benefits. Locals regularly enjoy the rejuvenating sulfurous bitumen therapy in lake waters.[72]

How Light Oil Becomes 'Pitch'?

When the subsurface oil leaks out, does it come out as tar or oil? Initially, it comes out as normal oil, but the lightest fraction vaporizes quickly. Bacteria loves to eat light oil and biodegrades it at a subsurface temperature of 80°C or less, producing an irreversibly altered product—tarry pitch or 'heavy oil.'[76] At least 7 oil-eating micro-organisms exist, including the voracious '*Alcanivorax*.' Same temperature condition may not whet all bacteria's appetite to the same degree. Both the Persian Gulf and Alaskan Chukchi Sea samples prove it.[74, 75] Two of the largest hydrocarbon reservoirs in the world produce only heavy oil—the Venezuelan Orinoco Tar Belt, and the Canadian Athabasca Oil Sands. These vast reservoirs exemplify the crucial role bacteria play in biodegrading light oil.

Mud Volcanoes –A Long Connection with Ancient Cultures

They are often associated with gas seeps in many places like the Baku region of Azerbaijan in the Absheron Peninsula. Mud cones of Touragai, etc., can attain heights of over 400 m, towering above the Caspian Sea.[76] The notable Baku seepage 'Eternal Fires' used to be a multifaith holy site for Zoroastrians and others for over two millennia.

Marco Polo had mentioned a Baku fire that is still burning for the last 4 ka on a short hillside stretch. Many other existing natural flames in the area were snuffed out in preference to commercial extraction.

The Arakan coast of Burma showcases 50-m-high mud cones with gas seeps. Trinidad and Tobago coast boasts of a 60-m-high mud volcano spewing gas-charged mud, and occasionally throwing up large blocks of rock (42 m^3) of Cretaceous-Paleogene age.[78]

Rocks and Landforms Beyond Glaciers and Volcanoes

The Quaternary sedimentary deposits are widely present as sand and mud in alluvial plains, fan deltas, lakes, beaches, barrier islands, delta lobes, shallow marine and deep-sea deposits, etc. Abundant shallow marine carbonate banks, imposing reefs and deep-water carbonate deposits are common. Other special lithologies include calcareous-evaporitic cave deposits, travertine, and tufa deposits. Peat, a coal precursor, is an important Quaternary marsh deposit. One advantage of Quaternary study is that we can observe sediments in different natural environments like rivers, deltas, etc. The Quaternary landforms are loftier because natural weathering agents have not got enough time to alter it beyond recognition.

The Quaternary Sediments are Not Yet Sedimentary Rocks

The sediments are mostly unlithified, i.e., unconsolidated. Lithification is simply the process of loose sediments becoming rocks (Litho = stone in Greek). The overburden pressure and compaction largely aid lithification of sediments. Compaction makes fine clay behave like a plastic, sticky paste. The sediments become cohesive and turn into rock.

New mineral cements, precipitating from hot fluids passing through pore spaces underground, also aid in binding loose grains. The cementation process takes a long time in the subsurface. Recent non-calcareous sediments are normally devoid of cement. However, one exception—the 'beachrock,' shows how loose sands may become a rock in plain view. When waves impinge on a rocky shore, repeated

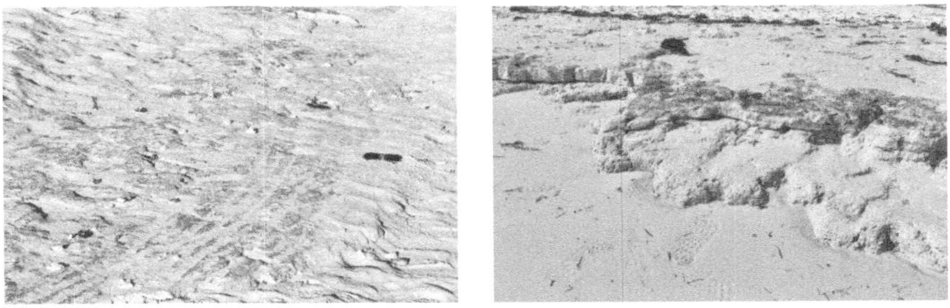

11-21 Two views of modern beachrock from the Caribbean shore. **left** The arcuate, 15-cm-thick beachrock stratum (sunglass at the edge of the bed). **right** Low angle seaward-dipping beachrocks, about 30-cm-thick, next to the wet loose beach sands. Freeport, Great Bahama Island.

24/7 sprays of seawater hit loose sands and percolate through pores; it quickly precipitates $CaCO_3$ in the pores and cements the loose grains. The beachrock in the Bimini Island, Great Bahama Bank, and elsewhere, is famous (Fig. 11.21).

MANY QUATERNARY GIFTS—LAKES, RIVERS, BEACHES, AND REEFS

Great Salt Lake—Remnant of Lake Bonneville, USA

The Pleistocene Lake Bonneville of western United States used to be large, like Lake Michigan, and much larger than the Dead Sea saline lake. It was a remnant of the Cretaceous Western Interior Seaway connecting the Atlantic and the Arctic Ocean. The Great Salt Lake, Utah, is a tiny part of the old Lake Bonneville.

The ancestral Lake Lisan in the Jordan Rift Valley, Asia, gave rise to the Dead Sea about 70 ka back in what is now Jordan, Israel, and the West Bank.[80] It is the deepest (300 m) hypersaline lake, marking the lowest point (>400 m below sea level) on Earth's surface.

Lake Titicaca—Many Distinctions, Endless Salt Flats, Playa Lakes

The famous Lake Titicaca, on the Andean Altiplano, at the Bolivia-Peru border, is the largest freshwater lake in South America. It is the highest navigable lake, on the second highest plateau after Tibet. Titicaca's position on a sloping Altiplano topographic sag, allows it to receive drainage from surrounding rivers. The Titicaca, along with lakes Poopo, Coipsa, and Uyuni, are part of the Altiplano intermontane basins, separated by transverse spurs.[81]

Lithium Salt in the Salar de Uyuni Salt Flats, Bolivian Altiplano

Salar de Uyuni salt deposit occurs in the Altiplano of the Bolivian Andes. It is nourished by Titicaca via rivers and the Poopo Lake. It is mainly sodium chloride, and other minor salts of potassium, magnesium, and lithium. Bolivia has 15% of the world's lithium reserve, mostly in the Salar. At 10,000 km^2, it is likely the largest salt pan in the world (Fig. 11.22). It covers an area of 10,580 km^2, at an altitude of 3,663 m, and holds 21 million metric tons of lithium brine.

The endless, flat 'salar' shines so brightly that Apollo 11 astronaut Neil Armstrong mistook it for a glacier from space. No wonder they call it the world's largest natural mirror. Salar de Uyuni is the dried-up avatar of a 30–42-ka-old, ancient inland lake. The blue green brine below the surficial crust holds the precious reserve of lithium, much in demand for varied industrial/electronic use.

Winds create parabolic halite dunes on top of the salt crust.[92] Miners once drilled 121 deep wells in the Salar to understand the sequence

11-22 Salt Harvest – The Salar de Uyuni salt flat on the Altiplano, Bolivia—a dried-up ancient Andean Lake.

of rock formation. They found 12 salt crusts alternating with 11 mud layers of normal lake origin. In the lower half, thick halite and thin mud layers alternate, whereas mud layers are thicker and halite beds thinner in the upper half. Halite layers mixed with detrital minerals formed in a playa lake environment. Bromine content increases from base to top.[93]

The local industry routinely drills wells to tap the brine which is then moved to evaporating open air pools. Lithium salts are concentrated through evaporation, later processed in plants.[64] Electrolysis of Li- and K-chlorides at 450°C produces metallic lithium.

Two ideas can explain the origin of the Salar de Uyuni—one is geological, and the other is of native indigenous folklore. For geologists, it is simply a salt pan, left after the desiccation of an ancient lake 30–42 ka ago. The main salts are halite (NaCl) and gypsum ($CaSO_4.H_2O$). The source of lithium is from Andean lavas.[39] The folklore tells that the goddess of a nearby volcano once angrily pulled her suckling baby from her breast, and in the process her milk, mixed with her tears, spilled out and formed the Salar.

Lake Xochimilco with Gondolas—Relict of the Pleistocene Texcoco Lake

It is the remnant of the ancient 50-km-long Texcoco Lake. The Early Holocene (12–6 ka) global warming caused snowmelt that used to feed the region to disappear. The shrinking Texcoco morphed into the present diminutive Xochimilco Lake.

Mexico City, an hour's drive—about 25 kms away from Xochimilco, now spreads over the ancient Texcoco lakebed. A network of waterways existed around 'Tenochtitlan,' the ancient Mexico City. The interconnecting 1.5m-deep canals are shallow, covering an area of 170 km^2. The lake, called 'Place of the flowers' in the 'Nahuatl' dialect, is a great tourist draw nowadays, at a stone's throw from the metropolis. A colorful festive vibe permeates the canal network; multicolored trajineras or Mexican gondolas, and chinampas (floating gardens) dot the scenery, while musicians and myriad vendors glide by.

The Chinampas is an old Aztec invention to grow fruit, flowers, and vegetables on an artificial terrain to feed the family. Skilled artisans weave reeds and use sticks to build large rafts with stakes around, which are securely anchored to the lake bottom. The lake mud and soil on top of the raft create a nursery bed ready for growing fruits and flowers.[82] It is now one of UNESCO's World Heritage Sites because of its deep cultural significance.

'Birth of the Universe'—The Flyer's Dance by the Xochimilco Lakeside

Another unique show 'Danza de Los Voladores' takes place nearby (Fig. 11.23). The 900–1200-CE-old Mesoamerican dance ritual, artistically depicts the birth of the universe. Five performers in colorful red and white attire and feathery headgear, climb to the top of a 30–40-m-high sturdy pole. The musician—who plays flute and drum, remains perched on top of the pole to play tunes the whole time. Other musicians on the ground accompany him throughout. The four flyers, each facing a cardinal direction, remain on a small wooden rotating platform high up. The daring musician '*caporal*' standing atop, represents the center of the universe. The four flyers tie themselves securely with strong ropes at the top. Then they jump from the platform and plunge into the air, start spinning, propelled by the rotating platform. All the time, the 'voladores' (flyers) are making dancing movements and gestures. It is a visual delight to see the graceful acro-

11-23 The Flyer's dance, from Papantla, near Poza Rica, Veracruz state, Mexico. They perform the ceremonial dance in many places, including Xochimilco, near Mexico City.

batics of the flyers against the blue sky, later gradually coming down to ground after many spins. The author had up-close experience of seeing it in Papantla, Veracruz State, the birthplace of the dance, performed by indigenous Totonac artists and elsewhere. Originally, the intent was to propitiate the fertility of Earth, worshipping the elements of the Sun, wind, water, and Earth.[83]

The dance, however, may not be unique to Mexico alone, as the author remembers similar popular flyers' dance in distant Calcutta, India, many years back.

The Restless Mississippi River—It's Delta Lobes Switched Position 7 Times in the last 8 ka

The Holocene Mississippi delta is river-dominated, where wave or tide action is negligible. Its shifting channels have given rise to frequent delta lobe switching (Fig. 11.24). Mississippi has switched its position at least 7 times during the last 8 ka. A river's life-long mission is to take its discharge and sediment load to the destination—the

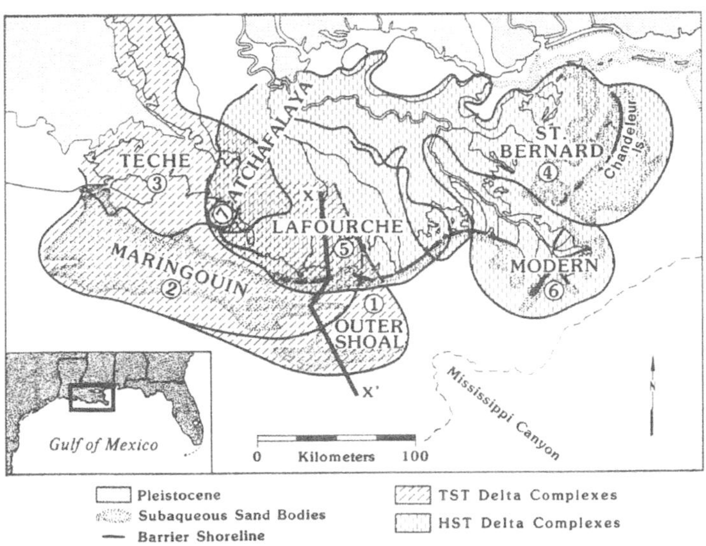

11-24 Mississippi delta lobes in the Gulf of Mexico, have switched position seven times during the last 8,000 years. Lobes 1 and 7 are the oldest and youngest respectively.

ocean or a lake. When a lobe becomes raised due to sediment pile-up, a river may find it difficult to flow because of low channel slope; it automatically seeks out a new path that facilitates its discharge. It breaches its levee, creates a new channel, and then a delta lobe forms farther down at a new location. Old delta lobes of the Mississippi River became slowly submerged due to subsidence, as sea level rose. Once under water, currents and waves modify and erode the lobe fringe and sculpt a barrier island—a poster child of sea level rise. The Modern Mississippi lobe—the #6 in Fig. 11.24, partly by design, is a bird-foot delta that reaches far out into the deep Gulf of Mexico.[84]

The Largest Alluvial Fan—Ever-Wandering Koshi River, A Ganges Tributary, India

The course of most large rivers has been stable during the Quaternary. One notable exception is in the Ganges River basin in India. A particular Himalayan tributary of Ganges—the Koshi River, has been restless, continuously moving laterally, due to sediment load and tectonic activity.

Tributaries may dump excess sediments on riverbeds and flood plains, often creating obstruction to their own flow downstream. The Koshi River from Nepal Himalayas has moved laterally 140 km in 250 years, always on the go for an easier route to its destination, the Ganges. Its 180-km-long alluvial fan is one of the largest. An alluvial fan appears when a river emerges from a narrow mountain gorge onto a low and flat, unconfined floodplain. In the past, a major neo-tectonic fault caused a westward tilt of the river's Himalayan catchment area and Koshi responded to it by shifting. Inclined Holocene terraces, and common earthquake tremors are signs of tectonic activities.[85] The Brahmaputra River, Bangladesh, also wandered ~100 km in the past 200 years.[86]

Quaternary River Terraces—Effects of Tectonic Activity

Many rivers build impressive high terraces along their steep valleys, with glacial outwash debris. As the mountains continue to rise tectonically, rivers carve their own alluvial deposits to form terraces with sediments of all sizes. Alluvial terraces are flat-topped, good for urban and agricultural developments. The author saw this firsthand in the Venezuelan Andes, where the Chama River, flowing in a valley between two mountain ranges, had built up a set of terraces. They are picturesque, flat, stepped terraces, extensively used for farming and dwellings. The 1,630-m-high, colonial city of Mérida sits on the topmost terrace about 100 m above the Chama River valley (Fig. 11.25). The Santo Domingo River on the south Andean flank has built 12 terraces in the last 200 ka, due to high frequency climate cycles.[79]

11-25 A high, cultivated Chama River Terrace, Mérida Andes, Venezuela. The river skirts around the steep terrace walls.

The 'Cotton Castle' of Pamukkale, Turkey—Snow White Stalactites, Hot Turquoise Pools

The singular place in the Lycus river valley boasts of the largest travertine deposits in the world. It sits just below the historic ruins of Hierapolis. Cleopatra, the Queen of Egypt, used to bathe in an exclusive Hierapolis pool, sourced by the same Pamukkale hot springs. The Hierapolis-Pamukkale site is a UNESCO World Cultural Heritage Centre, and the unofficial 8th wonder of the world. The geology, history, culture, and spirituality of the area is so rich that one can scarcely cite them all.[88, 89]

Tectonically, the Pamukkale hot springs are associated with the Denizli graben depression (basin) created between two faults. Mesozoic sediments including marine limestone filled up the Denizli basin. Later, uplift/inversion of the basin-fill and extensional faults/fractures marked the Pamukkale region. The tectonically active Pamukkale Fault lies to the north of the 'Cotton Castle' site within the Denizli basin. Undersaturated rainwater facilitated copious karstification of strata. Karstification means dissolution of carbonate rock producing plenty of pore space and cavities. At the time, continuous subduction of the African plate beneath the Aegean microplate (a fragment of the Eurasian plate) triggered intense Quaternary volcanic activity in the nearby Mt. Honaz. Rainwater percolating through the karstic rock underground became hot from deep magma heat. The hot water dissolved the limestone, carrying calcium and bicarbonate ions in solution. The water surfaced as 17 separate hot springs along fractures and faults. Degassing of dissolved carbon dioxide at ambient lower surface pressure and temperature precipitated calcium carbonate as travertine of the snow-white 'Cotton Castle.' The amazing array of cascading white terraces and turquoise blue rimstone pools has developed on sloping 200-m-high cliffs (Fig. 11.26). The hot springs formed 2.6 Ma ago, whereas the travertine deposits are ~0.6 Ma old. The Pamukkale travertines are much younger. Water temperature varies from 19–57°C (max. 100°C).[88] Cold water tufa is also present nearby.

The kings of Pergamon established the upper-level Hierapolis-Necropolis complex and its thermal spa in the 2nd century BCE. The recreational and healing property of the hot springs, and the fame of the ruins of the ancient city, temple, and bath, are the great attractions of the Pamukkale travertine plateau. Many massive earthquakes happened in the area in ancient times along the Pamukkale fault that destroyed the ancient city repeatedly. The city was rebuilt several times after strong earthquakes in 17 CE, 60 CE, and the 7th century, but the inhabitants finally abandoned it after the last earthquake in the 14th century.[90] The tectonic, seismic, hydrological, and geother-

11-26 1 The famous Cleopatra pool still in popular use in Hierapolis, Denizli province, Turkey **2** A terrace with natural cotton ball texture by the hot pool **3** Cascading white calcium carbonate stalactite and many hot turquoise rimstone pools **4** An intricate fan of travertine flowstone made by spring water. Images 2-4 from Pamukkale, Turkey.

mal milieu in Pamukkale created a unique setting for the spectacular travertine terraces.

Beaches Beckon—White Sands and Tropical Sun

Beaches are an essential part of our modern life because of their recreational and economic potential. They provide an excellent natural buffer zone between local communities and oceans during stormy inclement weather. Beaches are generally quartz sands though calcareous skeletal grains make up many inviting white sandy beaches, such as the Caribbean and other shores. In Hanauma bay, Honolulu, abundant patches of coral heads show up as dark areas amid turquoise blue waters contrasting with the yellow sandy beach (Fig. 11.27).

11-27 The Hanauma Bay, Honolulu, Hawaii. Abundant coral heads and bioherms show up as dark areas in the crystalline blue coastal waters.

What Natural Processes Give Rise to a Beach?

The raw material for sands is rock and bioclastic particles of all sizes; these break down into sand size through weathering, erosion, abrasion, and friction. Rocks may take thousands of years to disintegrate into sands, after they have been transported from source areas. Beaches occur usually along mainland and island shores but may also occur by some lake and river shores. The 14-km-long Wasaga Beach by the Georgian Bay of Lake Huron, Ontario, is the largest freshwater beach in the world. Marine bars and many islands also develop attractive beaches (Fig. 11.28). Incessant rolling action of waves and long transport make beach sands well sorted and rounded. The 250-km-long Praia do Cassino, Brazil, is the longest natural beach. In special cases, people can enjoy shorelines of exotic colors. The black magnetite and basalt sands, and beautiful green olivine sands in Hawaii are eye-catching. Also, some calcareous beaches of the Bermudas and Bahamas show beautiful pink sands due to the presence of *Homotrema rubrum*, an encrusting foraminifer.

Beach Ridges of Mexico, 100's of Kms Long—Result of Stepwise Sea Level Drop

The 'Beach Ridges' are testimony of Holocene sea level fall. A series of new strand plains emerged over thousands of years when sea level fell repeatedly. The long, linear, largely parallel, beach ridges progres-

11-28 Different types of beaches and marine bars near the Sado River estuary, Setúbal, Portugal.

sively become younger seaward. Two Mexican examples stand out for their size and continuity. The 200–225-km-long Pacific coast of Nayarit displays 280 close-spaced ridges 1–2-m-high, that successively advanced over 15 km seaward in the last 4.5 ka.[87]

The author had visited a noteworthy beach-ridge plain at the mouth of the Usumacinta-Grijalva River delta, Tabasco, Gulf of Mexico. The 300-km-long coast shows nearly 500 ridges that advanced 24 km in the last 6.5 ka due to sea withdrawal. Swash surges built the beach ridges with fine sands supplied by the delta.[77]

Spectacular Ooid Shoals in the Limpid Turquoise Bahamian Sea.

The subtidal shallow ooid carbonate shoals show many interesting features throughout the length and breadth of shoal—sands ridges, inactive ridges, sand waves, tidal channels, transverse shoulder bars, and troughs (Fig.11.29). The shoal appears to be deepening to the right, judging by the ink blue seawater. An integrated earlier study, based on remote sensing and other geo-located data (bathymetry, hydrology, etc.), added new information on the Lily Bank ooid shoals for use as analog for ancient oolite deposits.[91]

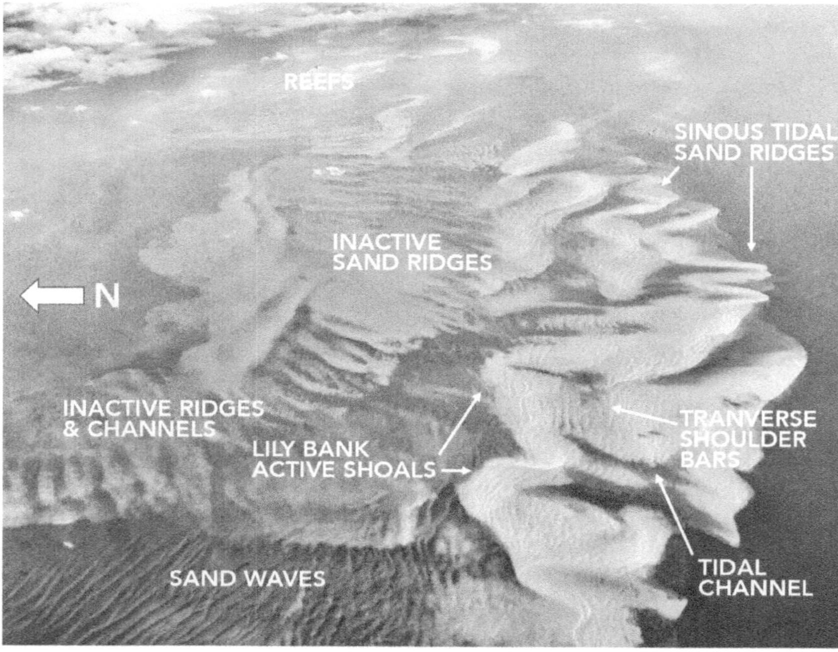

11-29 Aerial view of well-developed subtidal white ooid shoals of Lily Bank (part of Little Bahama Bank). They show different geomorphologic and sedimentologic features including sinuous tidal sand ridges several kilometers in wavelength, and several meters in height.[91] The Lily Bank's east-west extension is 10–12 kms.

Imposing Reefs—Colorful Marine Monuments

We discuss the amazing coral reef structures here because they are essentially rocks amid settling sediments (Fig. 11.30). Corals build reefs in tropical to subtropical, warm, sunny climate amid calcareous sediments in shallow seas. An individual polyp in a reef attaches itself to the substrate for stability. The reefs are also home to many plants and animals—sea weeds, algae, snails, scallops, fish, sharks, octopus, lobsters, etc. The largest non-mammalian vertebrate 'whale fish' (up to 16-m-long), living 1,000 m below the ocean surface, visit coral reefs during mass spawning time in the Great Barrier Reef (GBR), Australia.[67] Corals show a 18–20^{0}C isotherm distribution in oceans over wide geographic areas—the southern United States, northern South America, almost all-around Africa, southern India, southeast Asia, Europe, Australia, etc.

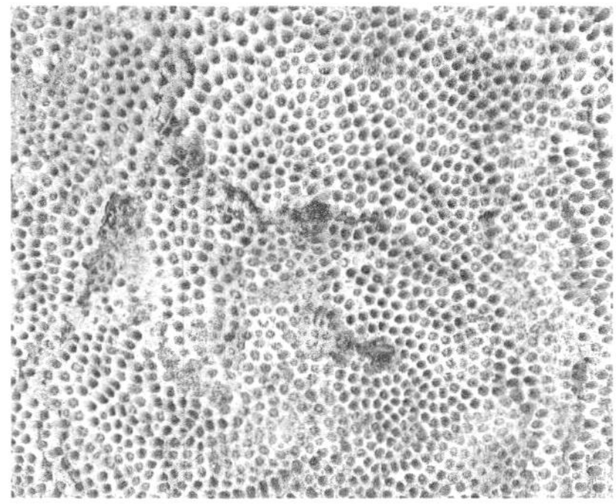

11-30 Pleistocene coral slabs, Liberia, Costa Rica. **left** vertical radiating view. **right** top view of *Porites* polyp (~1–2 mm-across) network.

How Do Tiny Corals Build a Reef?

The coral reef builders thrive in the 25–29°C temperature range. Any carbonate reef is the handiwork of millions of corals (Fig. 11.31). However, they do get a helping hand from tenant algae as the duo shows a symbiotic relationship.[55] A coral polyp, like a sea anemone, is a sac-like body, with a top central opening (mouth) surrounded by projecting tentacles, and a stomach inside the sac. It needs to attach itself to the seafloor for stability. It is a predator, stinging small preys with tentacles and then devouring them. However, it gets most of its food and nutrients from resident algae or *zooxanthellae* within the polyp. The coral polyp first makes life comfortable for the lowly algae,

11-31 Close up of calcium carbonate skeleton of 4 modern coral polyps (corallites) from the Caribbean shore, Cuba. A corallite is a tube with delicate vertical plates radiating from the center of each. Individual corallite tubes are joined together by horizontal plates.

giving them shelter and protection in its tissues, and something else vital for the coral's own survival. It gives them its own bodily waste of chemical compounds, essential raw materials for photosynthesis. Corals produce carbon dioxide and water through respiration, just like humans. In gratitude, the algae produce much needed oxygen and glucose through photosynthesis. The algae use CO_2 and H_2O molecules and light energy to produce molecules of clean oxygen (O_2) and glucose ($C_6H_{12}O_6$). They also produce amino acids and other compounds. The polyps readily use 90% of the algal gifts for growth and sustenance.[48]

We know how corals live and thrive. But how do the tiny creatures build the mighty barrier reefs? How do they secrete calcium carbonate minerals from seawater? Seawater has more bicarbonate (HCO^{3-}) than carbonate (CO^{3-}) ions. The polyps utilize the HCO^{3-} ions between its cells, they preferentially pump out the H^+ ions to enrich the proportion of CO^{3-} ions. When ambient Ca^{2+} ion bonds with the CO^{3-}, crystalline aragonite ($CaCO_3$) appears. The polyp sits inside an aragonite 'cup' it just created. Next, it lifts itself up to build another cup and the process goes on to erect the reef structure.[35] The polyp stays ensconced in the hard skeleton to withstand natural hazards—storms, currents, waves, boring (by fish, animals), and predators. The Mexican Yucatan Peninsula is a unique example of uninterrupted $CaCO_3$ precipitation over nearly 80 Ma from the Late Cretaceous to the Present.

The Great Barrier Reef, Australia—In Crisis Now?

Corals manage to build huge continent-size reefs. Two of the world's largest reefs are the 2300-km-long Great Barrier Reef, Australia, and the 900-km-long Mesoamerican Reef skirting around the Yucatan Peninsula in Mexico, Belize, Guatemala, and Honduras. Only the Belize Barrier Reef sector is 300 km long.

The Great Barrier Reef lies on the northeastern flank of Australia, a veritable nursery of marine life with 150 fish, and ~400 hard coral species calling it home (Fig. 11.32). The average coral polyp height is 1.5 cm. Stony coral polyp is 1–3 mm in diameter.

In 2020, a research vessel discovered a new isolated coral reef during marine exploration. The 20-Ma-old pinnacle reef, taller than the Empire State Building, New York, is isolated from the proper Great Barrier Reef itself.[18] It is 500 m high, with a 1.5-km-wide base, and its top reaches 40 m below sea level. Scientists noted a thriving colony of coral, shark and other species populating the reef crest. The

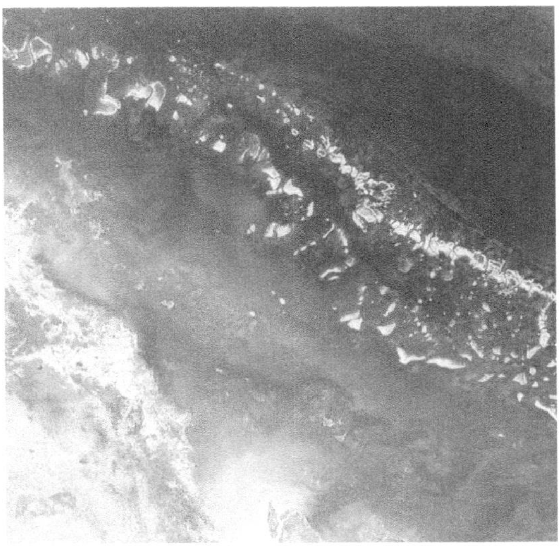

11-32 Satellite view of a part of the world's largest, 2300-km-long Great Barrier Reef, Australia.

news is especially welcoming, as a study revealed that half the corals of the Great Barrier Reef have died since 1995. The new reef joins 7 other detached reefs discovered in the last 100 years.

One alarming trend is the 'bleaching' blight, severely affecting the survival of reef corals. It reflects their largescale die-off in oceans, mainly in Australia. Data shows alarming mortality rate in different reef sectors: far North—26%, North—67%, Central—6%, and south—1%. Bleaching between 2015–2017 killed off a large part of the corals in the Great Barrier Reef .[7] Why bleaching happens, creating such havoc? The beautiful colors of corals we see are due to algae living in the polyps. The polyps may expel all tenant algae involuntarily, under stressed conditions. Climate-related temperature change, pollution, excessive sunlight, abnormal low tide, etc., may cause bleaching. A change of ~1°C temperature will provoke corals to reject the color-giving algae. With algae gone, coral reef loses its color, becomes white or 'bleached.' In advanced stages, corals may die off, because algae provide food and oxygen—vital ingredients for survival. The lesson—mitigate global warming to save the coral reefs.

Elsewhere an interesting find thrilled undersea researchers; they discovered the longest sea creature—a 45-m-long *siphonophore* in deep marine waters off Western Australia. A *siphonophore* is a multi-cellular colony that can collectively live, reproduce, float, and move through ocean waters.[18] These are hydrozoans like coral, jellyfish and sea anemone. The Portuguese man o' war is a siphonophore with a deadly sting.

- Several land bridges—the Beringia, the Asia-Australia bridge, and 3 others, product of sea level fall during the peak Pleistocene glaciation.
- Dilution of oceans and lowering of Holocene global temperature—a result of discharge of pro-glacial melt water of legendary North American lakes—Agassiz and Ojibway.
- Numerous, brief interglacial spells like the present one, common in the Quaternary.
- Glacial features and significance: drumlin, moraine, finger lake, tunnel valley, 'erratics.'
- Nature and origin of the glacial Great Lakes, Niagara Falls, Lake Iroquois, etc.
- Icehouse-related growth of deserts, sand fields, and raised beaches.
- Nature, scale, future likelihood, and impact of volcanic eruptions—Hawaii, Long Valley Caldera, Yellowstone (USA), the Pacific Ring of Fire, the Toba supervolcano in Sumatra, Indonesia.
- Origin and significance of tar seeps: Trinidad and Tobago, Venezuela, USA, etc.
- History, and origin of important geomorphic features—rivers like Mississippi, Amazon, etc.; lakes like Titicaca, etc.; beaches/beach ridges of Hawaii and Mexico; reefs like the Great Barrier Reef, Australia, etc.

WHAT'S NEXT?

Chapter 12 is all about the fascinating creatures that roamed the Quaternary landscape, and many still do. The most far-reaching event is undoubtedly the emergence of *Homo sapiens* on the scene and his deep impact on Earth.

The photogenic Benagil sea cave, Algarve, is on the south shore of Portugal. It is a magical milieu of green Atlantic water, golden yellow sands, and the blue sky peeping through a giant sinkhole ('the eye'), letting sunlight filter in and bathe the precious beach. Sea erosion, karstic weathering and wind has fashioned the unique sea cave-sinkhole in the layered 20-million-year-old stony edifice. Note the limestone-limy shale strata of the walls.

12

Miracle of Quarternary Life: *Homo sapiens* on the Scene

The Transition from Early Primates—It Took a While

The most consequential Quaternary event is the debut of the *Homo genus* and finally, the *Homo sapiens*. The *Homo genus*, evolving from the *Australopithecus*, took ~200–300 ka to produce its newest species, *Homo sapiens*, by nature's continuous trial and error and fine-tuned evolutionary process. The only species that asks questions about where we come from, and our place under the Sun. At the same time, no other species has so deeply impacted the environment, wildlife, flora, and the delicate ecosystem of the planet we call home.

When Did Our Direct Ancestors Emerge—The *Homo genus*?

We recall from Chapter 1 humankind appeared in the very last minutes of Carl Sagan's 4.57-Ga-long cosmic calendar year. The *Homo genus* arrived on Dec-31, only 6 minutes before the end of Sagan's cosmic year.[1] *Homo sapiens* entered only 2 minutes before midnight.

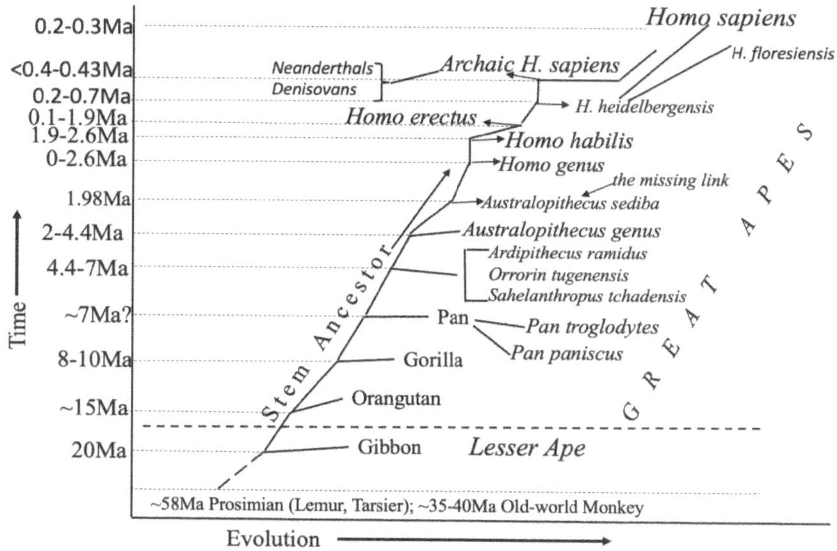

12-1 A simplified scheme of great ape evolution during the last ~20 Ma. The vertical axis (time, i.e., age) is not to scale. Humans appeared 0.2–0.3 Ma ago.

Our distant hominin ancestors would have been the *Sahelanthropus tchadensis*, *Orrorin tugenensis* (both ~7–6-Ma-old) and *Ardipithecus ramidus* (>4 Ma). Orrorin is 'original man' in Tugen language, Kenya. *S. tchadensis* and *O. tugenensis* showed both hominin and primitive hominid traits and they were bipedal and arboreal.[6] Darwin divined long ago that bipedalism made us who we are today. Indeed, our hominin lineage may go back to ~7 Ma. Our stem hominid ancestor had split from the *Gorillini genus* ~8–10 Ma ago. We also know our ape ancestor then split into two—the true hominins, i.e., *Australopithecus*/*Homo genus*, and the hominid *Pan genus* (chimpanzee, bonobo), between 8–5 Ma, more likely ~7–6 Ma ago.[2] The bipedal—arboreal *A. ramidus* was an early hominin *Australopithecus*. *Paranthropus*, an aberrant *Australopithecus*, met its evolutionary dead-end about 1 Ma ago.

We broadly discuss below the 20-Ma-long journey of the great apes, a theme not yet fully understood. Fig 12.1 is a summary of the complex subject of great ape evolution climaxing in the *Homo sapiens*. The figure also refers to primate evolution, hinted in Chapters 9 and

10. Many species of the *Homo genus* debuted in the Quaternary yet went into oblivion because they failed to adjust to the ever-changing Mio-Pliocene-Quaternary physical, and climatic milieu.

Sterkfontein, South Africa, may be the world's best-known hominin fossil site. It holds 1,000 fossil fragments, several hundred thousand animal fossils, 9,000 stone tools and 3 fragments of fossil wood. The site tells an intriguing, 3.5-Ma-old, Plio-Pleistocene, hominin development history from early hominins to the *Homo sapiens.*[3]

Australopithecus, the 'Southern Ape'—Likely Ancestor of the *Homo genus*

Australopithecus is most likely the direct ancestor of the *Homo genus*. The brain or memory capacity continued to increase from the *Australopithecus* to the *Homo genus*. Paleoanthropologists measure cranial capacity 1) by filling the skull cavity with a substance and measuring its volume, or 2) making an endocranial cast, and measure the volume of water it displaces. The *Australopithecus* showed a consistent value of 450 cm^3, while all *Homo* species—*H. habilis*, *H. erectus* and *H. sapiens*, etc., showed gradually higher brain capacity.[4] The brain size of *S. tchadensis*, *O. tugenensis*, and *Ardipithecus ramidus* was much less than a modern chimpanzee (400–450 cc), also slightly less than the *Australopithecus*' brain. The brain size of *H. erectus* (average 1000 cc) showed variation—850 cm^3 in the most primitive form, to 1,100cm^3 in the Javan specimen in southeast Asia. The *Peking Man*[5] brain size was like humans—1,300 cm^3. The *Homo genus* with larger brains became adept in facing physical hazards and climatic challenges. They prospered by making necessary tools and practical adjustments to their lifestyle.

Australopithecus is literally the 'southern ape,' referring to its origin in Africa. Hominin evolution of the *Australopithecus* and *Homo genus* is complex, not yet fully clear. However, the simplified evolutionary chain may be *Australopithecus anamensis* > *A. afarensis* / *A. garhi* > *A. africanus* > *A. sediba* (?) > *Homo habilis*, and so on. *A. anamensis* would have followed the distant *Ardipithecus ramidus*. D. Johansson and T. Gray discovered the famous 3.2-Ma-old "Lucy," one of the best documented fossil of *A. afarensis*, in Hadar, Ethiopia (Fig. 12.2).

The skeletons showed evidence of sexual dimorphism—males larger than females.[7] In Laetoli, northern Tanzania, a volcanic ash layer shows the 3.6-Ma-old, bipedal footprints of *A. afarensis*[8] (Fig. 12.3). They were also adept arboreal apes. The *Australopithecus africanus* is much talked about because of 'Mrs. Ples' and 'Taung Child' skeletons discovered in Sterkfontein. The 2.5-Ma-old 'Mrs. Ples' is the most complete skull ever found. However, paleoanthropologists had debated for a long time about the sexual identity of the skeleton. Finally, came the verdict in 2012—it was indeed 'Mrs. Ples.'[9] The 'Taung child,' discovered in 1947, had also created a decades-long dispute regarding its origin. Many thought it was an intermediate species between the *Homo* and *Australopithecus genus*. Finally, experts concurred—it's a ~3-Ma-old *Australopithecus africanus*. In 2006, a paleoanthropologist[10] reported telltale evidence of beak and talon marks in the fossil's eye area and skull, confirming that the 3-year-old 'Taung child' was the victim of an eagle-like predator.

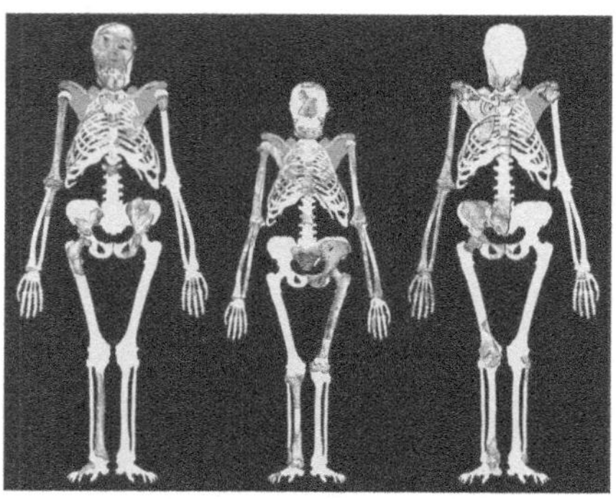

12-2 Skeletons of 'Lucy' *(A. afarensis)* flanked by two *Australopithecus sediba* fossils, Malapa site, South Africa.

12-3 The earliest 3.66-Ma-old, bipedal footprints (cast) in Laetoli of *Australopithecus afarensis*, a prehuman hominin, in Tanzania.

Australopithecus sediba— The Last Common Ancestor (LCA) or Missing Link?

The *Australopithecus sediba* may be the transitional fossil showing traits of both *Australopithecus*, and the anatomically distinct *Homo genus* (Fig. 12.2). Scientists discovered it by chance in 2008, when 9-year-old Matthew Berger accidentally stumbled upon it in the 'Cradle of Humankind World Heritage Site.'[11] He saw the intact skeleton of *A. sediba* near the Malapa fossil site, while walking his dog. The famous ~190,000 m^2 Malapa site, near Johannesburg, shows a maze of limestone caves with a trove of early hominin fossils. Scientists claimed that the *Australopithecus sediba* is the elusive 'missing link' primate.[13] The 1.98-Ma-old fossil shows traits of both *A. africanus*, and *Homo habilis*. It inherited several features from *A. africanus*—leg length, long upper limbs, narrow upper chest, small brain capacity, whereas its broad lower chest, pelvic structure, upright gait, skull details, etc., are like the *Homo genus*. It was bipedal, and occasionally arboreal for foraging and escaping from predators. All species from *S. tchadensis* to *A. africanus* could combine upright walking and tree-living, like the *Homo genus*. *Australopithecus sediba* may or may not be the missing link, but experts largely agree that the *Homo genus* evolved from the somewhat human-like *Australopithecine* apes.

Table 12.1 (multiple sources including those cited here) shows the age and evolutionary brain size increase from the *S. tchadensis* to the *Homo genus*.[17, 18, 20] All species of the *Homo genus*, except *H. sapiens*, became extinct during the last 2 Ma: *Homo habilis*, *H. erectus*, *H. heidelbergensis*, *H. naledi*, *H. ergaster*, *H. rudolfensis*, *H. florisiensis*, *H. neanderthalensis*. The Neanderthals used to roam around in Europe even <40-ka back, while *H. florisiensis* may not be extinct in the Indonesian Flores Island, claims an expert (2022).[24]

Homo habilis, the Tool-Maker—Our 'Handy Man' Ancestor

Anthropologist Mary Leakey and team discovered *Homo habilis*, the oldest *Homo* species, in the Olduvai Gorge, Tanzania, in 1960 (Fig. 12.4). The "Ledi-Geraru" jawbone found in 2013, may be a *Homo habi-*

Table 12-1: Hominin Brain Size

Hominin Species/ Age in Ma	**Brain Size/Capacity ($cc = cm^3$)**	**Remarks**
Homo sapiens, Present ~ 0.2	1350	Surviving Homo species
Peking Man or *H. erectus pekinensis* 0.5-0.75	1000-1300	Asian *Homo erectus*
Java man or *Homo erectus erectus* 1.6-0.6, latest 0.108	Up to 1300, avg. 1000	Asian *Homo erectus*
Homo neanderthalensis <0.04—0.43	1600	Partly *Homo sapiens* contemporary
Homo erectus 1.9—<0.5	~1000	Lived the longest
Homo habilis 2.6—1.9	<700	Earliest Homo toolmaker
Australopithecus sediba 1.98	420-450	Homo — *Australopithecus* missing link
Average Australopithecus > 2	450	Brain comparable to chimpanzee
Australopithecus africanus 2-3	~450	"Taung Child," "Mrs. Ples."
A. afarensis 3-3.8	~430	"Lucy"; unique Laetoli footprints
A. anamensis 3.9-4.2	~370	Bipedal & arboreal like *A. afarensis*
Ardipithecus ramidus 4.4	~350	Bipedal & arboreal
Orrorin tugenensis 6-7	>300	Our early ancestors, small canines Bipedal & arboreal
Sahelanthropus tchadensis 6-7	<350	

12-4 A forensic facial reconstruction/approximation of our handyman ancestor *Homo habilis*, the first species of the *Homo genus*, discovered in the Olduvai gorge, Tanzania, 1960.

lis.[30] The name comes from the Ledi, Geraru Ledi, and Geraru rivers in Afar, Ethiopia. The moniker '*habilis*' refers to the manual dexterity of *H. habilis*, the first handyman of sorts, and maker of numerous stone tools.[30] The species would have carried tools like flakes and pebbles to distant places for later use, a sign of human-like thinking. The larger brain would have needed more energy from an enriched diet including meat. *H. habilis*, a more social creature than *Australopithecus*, used to live in large groups of ~80 individuals. Primate group size is proportional to its brain size.[15] By modern standard, *H. habilis* was puny, 135 cm tall and lightweight, weighing ~32 kg. The transition from *Australopithecus* (*A. sediba*) to the *Homo genus* was fundamental, having far-reaching impact on the success of our species later. Several bodily changes equipped the *Homo* lineage for the challenges of the future world—feet with well-defined toe and heel, separated by an arch that allowed the foot to support body weight. Also, a modified pelvic structure supported the upright walking gait. More importantly, the substantial brain size increase influenced high intelligence, language skill, tool making, and problem-solving ability—traits that helped the species in challenging conditions.[6]

In summary, the *Homo habilis* had a larger, more complex brain, well suited to tool making. It had the long arms of the *Australopithecus genus*, but had less prominent eyebrows, or protruding jawline, also had smaller premolars—apt for omnivorous diets.[2] Fagan proposed an approximate timeline for Homo evolution and species appearance.[14] Opinions may vary about ages of species, but the timeline provides a useful frame of reference for the evolutionary ladder. The *Homo genus* and *Australopithecus* would have coexisted between ~2.0-2.6 Ma ago.

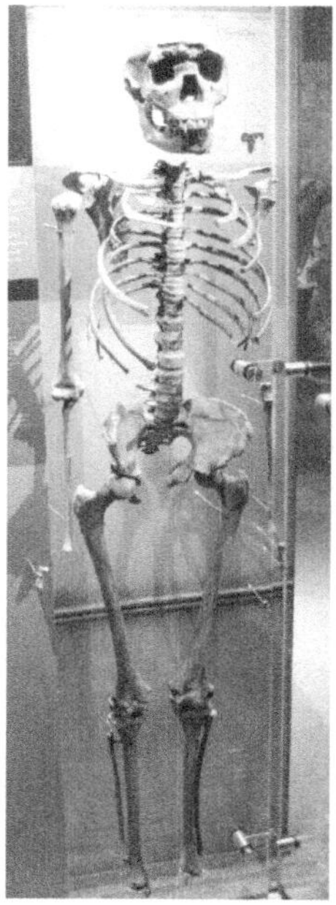

12-5 'Turkana Boy'—The most complete *Homo ergaster* (an African *H. erectus*) skeleton found by the Lake Turkana, Kenya.

Homo erectus—the Longest-Lived, Widely Traveled Ancestor

The *H. erectus* post-dated the *H. habilis*, and was the first ancestor having many similarities with humans—brain size ($3/4^{th}$ of humans), stature, bifacial tool-making ability, use of controlled fire, etc. In 1984, the Lake Turkana shoreline, Kenya, marked the discovery of the bipedal, 1.6-Ma-old 'Turkana Boy' (Fig.12.5). The skeleton had a primitive skull and jawbone, but otherwise, remarkably *H. sapiens*-like. It is the most complete early human-like fossil ever, with similar limb and torso proportions.

The youngest *H. erectus* fossils are only 117–108 ka old, found by the Solo River near Ngangdong, Central Java. A cache of 12 skull caps and 2 lower leg bones documents a catastrophic end of at least 12 individuals. The corpses had probably washed down the river during a flood event. The skeletons prove that *H. erectus* had lived the longest of all *Homo* species.[23]

Some think that *H. erectus* was the first to leave Africa ~ 1.5–1.6 Ma ago. They had reached new places, acclimatizing in forested Java, cold, temperate China, and Europe. Some argue climate change prompted *H. erectus* to leave Africa for 'greener' pastures, perhaps following preys. By this time, our ancestors were completely bipedal, able to traverse vast distances. They made elaborate tools and probably used controlled fire for domestic use.

The Transition from the *H. erectus* to *H. sapiens*

The 'complete replacement model' holds that pre-modern *Homo genus* in Africa gave rise to modern *H. sapiens* about 0.2 Ma ago. Another idea is that anatomically modern humans spread from Africa to Europe and Asia about 50–55 ka ago.[28] Everywhere (Europe, Africa, Asia), the longest-lived (1.9–0.1 Ma) *H. erectus* was evolving towards *H. sapiens*, showing attributes of both. Did they meet a dead-end? Many opine that the remnant African population was instrumental in the emergence of our species—first the *Archaic H. sapiens* and then *Homo sapiens.*[14] The transition from the *archaic H. sapiens* was probably ~160–200 ka ago or earlier, based on vital fossil clues in Africa and Asia.

The 'Archaic *Homo sapiens*' Failed to Climb the Final Rung

Chronologically, the transitional subgroup between the *Homo erectus* (or *Homo heidelbergensis*), and the *H. sapiens* species, appeared around 300 ka ago or before.[32] Two great examples of the *Archaic H. sapiens* are the *Neanderthals* in Europe, *Denisovans* in Siberia, and probably the *Solo man* in Java, Indonesia. All three modern human lineages are 200 ka or younger. The Khoisans in South Africa diverged 200 ka ago, while West African and Eurasian branches parted about 100 ka ago.[35] Our species coexisted for some time with *archaic H. sapiens*, like the *H. neanderthalensis* in Europe and Middle East, *Denisovans* in Asia, as well as the *H. erectus*. The mixed features prompted some to categorize *archaic H. sapiens* as *Homo helmei* in Africa. Some key fossil finds in the old world helped paleoanthropologists to document the transition from the archaic to the modern species—the 120-ka-old *Florisbad* cranium in Laetoli, Tanzania, the 160-ka-old, *Herto* skull, Ethiopia, and the 80–120-ka-old, *Skuhl* 5 skull in Skuhl Cave, Israel.[33]

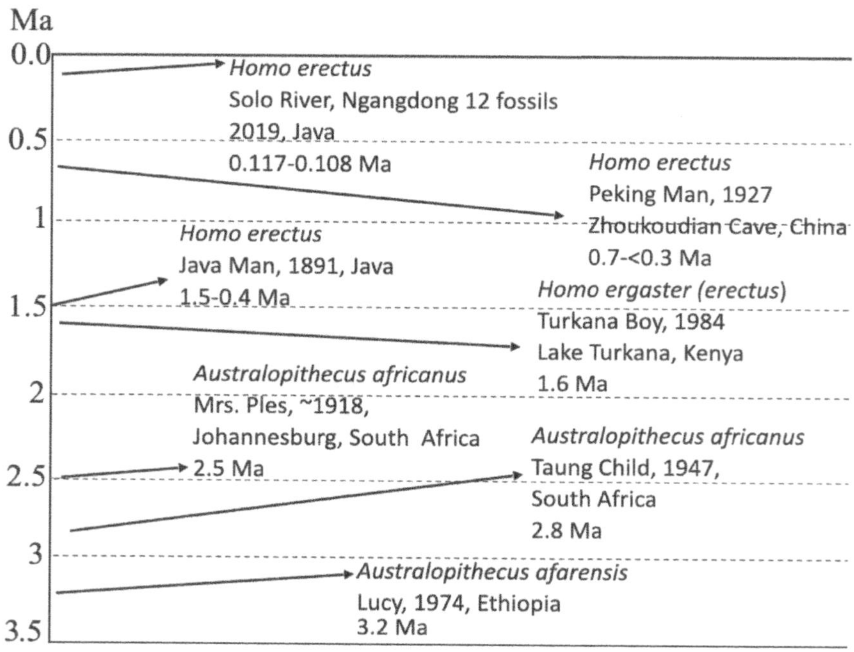

12-6 Famous Hominin fossils – Name, age, and details of discovery in Africa & Asia.

The 'Dragon Man' or *Homo longi* highlights the 146-ka-old 'Harbin cranium' of Dragon River fame in China. Its 1420 cc brain capacity, and overall similarity with *Homo sapiens*, convinced local experts that it is a new species, closer to *Homo sapiens*, than the *H. sapiens neanderthalensis*.[34] For others, the 'Dragon Man' is a *Denisovan*, not a *H. sapiens*. Indeed, the relationship among the *H. sapiens*, *Neanderthal*, and *Denisovan* is still not fully clear. Figure 12.6 shows some famous hominin fossils from Africa and Asia.

Neanderthals Had Larger Brains—What Happened to Them?

The *Neanderthals* were like modern humans in many ways. They probably originated about 400–500 ka ago from African ancestors. A second model argues that the *Neanderthals* of Europe, and the Asian *Denisovans* were autochthonous and coeval. Some people use the word

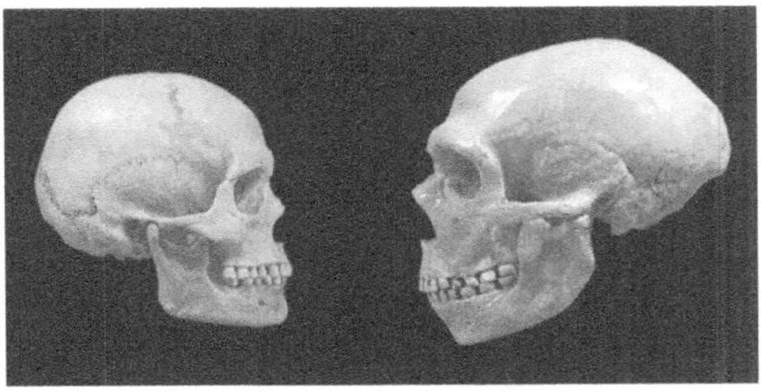

12-7 A comparison of skulls of a modern human (left) and Neanderthal (right) from the Cleveland Museum of Natural History. The latter's brain size is larger.

'*Neanderthal*' in an unjustified, pejorative sense—ugly, dim-witted, ape-like creatures. To begin with, *Neanderthal* brain size (~1600 cc) [20] was larger than *H. sapiens* (1350 cc). Because of our similarity in appearance and intelligence, experts prefer to call them *H. sapiens neanderthalensis* to differentiate them from *H. sapiens sapiens*, the modern human (Fig.12.7). They were humans of a different sort. The 2021 discovery of a 130-ka-old skull at Nesha Ramla, Israel, likely belonged to a *H. neanderthalensis.* They used to hunt small and large animals and maintain fire with fuel wood.[35]

The name '*Neanderthal*' came from a skeleton found in the Neander valley in Germany. They likely used language, and had many modern, simple skills—weaving, making clothes, uses of medicinal plants, a well-developed stone tool culture (Mousterian industry). Good hunters of mammoth, bison, and horse, they made well-designed tools for specific purposes.

Neanderthal burial sites in France suggest that they believed in after-life.[14] They were the artists of the 65-ka-old Spanish cave art. The elaborate 90-ka-old *Neanderthal* wooden digging tools came from a site in northern Spain.[36] The species became extinct rather recently about 40–30 ka back. Reasons may be one or a combination of factors—extermination, competition with *Homo sapiens*, a drastic climate change, a pandemic or disease. Italian scientists discovered nine male *Neanderthal* skeletons of probable hunters in the Guattari cave, amidst bones of elephant, rhinoceros, horse, great deer, etc. The cave is located only about 100m from the Tyrrhenian Sea, Lazio region. The *Neanderthals* likely would have performed ritualistic cannibalism from time to time. Some experts believe that the European species extinction was due to climate change.[37]

An Australian study of 2021 concluded that the 42-ka-old Earth's magnetic pole flip or 'Adams' event, had caused gradual extinction of the *Neanderthals* and megafauna starting around that time.[38a] Other studies reveal details of the eventual retreat and demise of the *Neanderthals*. Four *Neanderthal* caves in Gibraltar show that prolonged, unusual winter spells in northern Europe pushed them to warmer southern latitudes. They did not prosper there either, as their big game hunting skill was likely of no avail there. The *Homo sapiens* were much more flexible and resourceful in hunting smaller game. The lack of adequate food would have contributed to slowing fertility, dwindling population, and consequent loss of diversity. High inbreeding incidence would have also hastened extinction. The last *Neanderthals* in the Gibraltar caves lived ~33 ka ago. The walls of Gorham cave, one of the four caves in limestone hills right by the ocean, preserve valuable *Neanderthal* artwork.

Ample evidence exists of interbreeding among *Archaic H. sapiens*, i.e., *Neanderthals*, *Denisovans*, and *H. sapiens sapiens.* A computerized genome sequencing program revealed presence of genome and DNA of other species in *H. sapiens*, implying that *Homo sapiens*' interbreeding extended to a far wider group of extinct hominin species.[39] Experts opine that the inherited *Denisovan* gene equipped the Tibetans to withstand the harsh cold climate of Tibet. The mutual curiosity of *Neanderthals* and *H. sapiens* left an indelible mark on modern humans. *Neanderthal* genetic material is now common in non-African populations, and some modern human ailments are due to interaction with the *Neanderthals.*[38]

Homo sapiens sapiens, the 'Wise' People—Are We Sure?

"And God said, let us make man in our image, after our likeness: and let them have dominion over the fish of the sea, ... over all the earth, and over every creeping thing that creepeth upon the earth."— Genesis 1, Quote 26.[43] *Homo sapiens* arrived finally, after the exit of prehuman *Australopithecus*, and many of the other species of the *Homo genus*. Modern humans post-dated most ancestors but coexisted briefly with several: *H. erectus*, *H. heidelbergensis*, *H. florisiensis*, and *archaic H.*

sapiens, *i.e.*, the *Neanderthals* and *Denisovans*. Our direct ancestor *H. erectus*, held a record of longevity (about 1.8 Ma) but then succumbed.

Finally, the *H. sapiens* became the monarch of all they surveyed and surmounted all earthly barriers and traveled the globe from Pole to Pole. The species spread its wings and soared to heights beyond, delved into the bottomless oceans, becoming the unquestionable superpower. Humans put their unmistakable stamp everywhere on land, ocean, sky, even on the Moon, Mars, asteroids, and other heavenly bodies.

One may ask—what separates us from our forebears who trudged the same road for several million years with grit, ingenuity, and primitive technology amid many climatic and physical challenges, only to be overcome by natural causes. What stood us in good stead? The creativity and intelligence coming from a more varied omnivorous nutrient-rich diet fueling a larger brain, likely was the essential element in the equation. Several researchers concluded that without the abundant calories coming from meat-eating, the *Homo sapiens*' brain simply could not have developed so much.[19] Our ability to communicate through speech and scripts, our capacity to pass on our cumulative knowledge, ideas, and dreams from generation to generation by way of written words—all made us effective and productive as a group. But some of those who went before us, also tried to do all this in a more limited primal way. Today humans stand out because of our resourcefulness and adaptability, but also our yearnings for what is beyond our day-to-day physical existence.

Morning Shows the Day—Artistic Flair of the Early *Homo sapiens*

An eloquent example of abstract symbolic expression of our ancestors of the Middle Stone age, comes from the oceanside Blombos Cave, about 300 km from Cape Town, South Africa. It consists of a 73-ka-old ground stone flake with 6 subparallel red ochre lines, crossed by three painted oblique lines on the rock. The cross-hatched drawing, to date the oldest human creation, would have been meaningful to the cave dwellers. It highlights the symbolic expression of ideas of human ancestors that far back in time. Scientists also found nearby two red ochre painted abalone shells, a hearth, shells, bones, etc.[44]

In 2021, a limestone cave in the Sulawesi Island, Indonesia, yielded a startling secret of a figurative art painted by hunter-gatherers. The archeological report of a 45-ka-old red ochre-painted warty pig may be the world's oldest portrayal of an animal.[46] Cave dwellings aside, natural shelters with walls started appearing in the last 20 ka.

In Europe, the oldest figurative art is a 40-ka-old, 30-cm-high, ivory figurine 'lion-man' ('Loewen-mensch') from the Hohlenstein—Stadel Cave in Germany.[48] It depicts a human body and a lion head, delicately carved on a mammoth tusk. In 2013, experts painstakingly reconstructed it from 200 retrieved broken fragments. The artwork speaks volumes about an advanced culture, probably of a shamanic religion. The sculpture revealed creativity to conceive, visualize and make something imaginary. Mitochondrial DNA analysis of a femur sample in the cave revealed that it was earlier home to the *Neanderthals*, who interbred with *H. sapiens*.[47]

More sustained expression of artistry is later than 40 ka, distantly resembling modern art. Probably, human cognitive ability had a spurt around this time. A 32-ka-old French musical instrument is a testimony to early human creativity.

The Czech Republic holds some rare treasures of prehistoric human ingenuity. 'Dolni Vestonice' is the oldest Cro-Magnon human settlement ever found.[49] The site boasts of a 26-ka-old, 31-cm-tall, nude ivory figurine of Venus, with fine facial details—either a symbol of fertility, an idol or a goddess. Also, noteworthy is the relic of a high temperature (>400°C) kiln, amid scattered ceramic/fired-clay figurines.

Göbekli Tepe, Turkey—Hunter-Gatherers Built a Magnificent Temple

Foragers erected the monumental structure of the ~10,000-year-old (radiocarbon dated) edifice of pre-pottery Neolithic vintage, in southeast Turkey. It is about 10 km from the ancient city of Urfa, the 'city of prophets,' located near the northern edge of the Fertile Crescent; it probably predated human settlements anywhere (Fig. 12.8). Picto-

12-8 Aerial view of ancient, ~10-ka-old ruins of the Göbekli Tepe Complex, near Sanliurfa, southeastern Anatolia, Turkey.

gram, sculpture, and megalithic architecture embellished the world's earliest temple dedicated to an unknown deity. It created a buzz in the world's archeological scene when discovered in the 1990s. Some even likened it to the biblical Garden of Eden. It precedes many modern human inventions—pottery, metallurgy, written scripts, the wheel, and the dawn of agriculture.[50] Göbekli Tepe or the 'potbelly hill' is a 308-m-across, rock-cut mound perched atop a ridge—older than the Stonehenge (U.K.), Egyptian pyramids, etc. Geophysical probing suggests that only 5% of the structure is visible, the rest (~250 megaliths) is still buried. The T-shaped megalith pillars of massive limestone, up to 5- m-high, came from nearby limestone formations below. How did they transport them? The pillars, shaped by chisel-like flint tools, show refined artwork with carved animal forms that used to roam around locally. Prominent vulture images probably denote the ritual cleaning of corpses. The practice was common in communities across Iran and the Middle East. Even now, some Asian communities follow the Sky Burial custom.[51] It's astonishing that religious, nomadic hunter-gatherers had built the magnificent shrine at the dawn of civilization. The newly discovered Karahantepe settlement there is likely even older.

The Hands Cave, Patagonia, Argentina

The 13–9-ka-old, hunter-gatherer handprints in the "Cueva de las Manos" in Patagonia, Argentina, is a UNESCO World Heritage Site.

It depicts an assemblage of rock art of multicolored painted hands and hunting scenes on cave walls. The colorful, cleverly done, stenciled 'silhouettes of hands,' display something beyond the concern of daily subsistence. They used different mineral compounds—iron oxide, kaolin, manganese oxide, etc., to produce red, white, yellow, and black colors respectively.[52]

What Made Humans Special: The Role of an Enriched Diet

The *Australopithecus genus* possessed a bigger gut than the *Homo genus.*[19] Its dentition was also larger than *Homo*. Once human ancestors took to eating meat, they no longer needed large teeth. Particularly the molar size reduction was significant.[22] The use of tools and fire for cooking a variety of food also would have contributed to smaller tooth size. A comparative brain size study of the vegetarian *Australopithecus/Paranthropus boisei* (500 cc), the carcass scavenger *Homo habilis* (650 cc), and the hunter/meat-lover *H. erectus* (1000 cc) shows a direct relation between meat eating and brain size increase.[20] The *Homo* individuals outlasted *Australopithecus* probably because they ate cooked high protein diet.

The Use of Fire Was a Game Changer for Human Ancestors

Primatologist R. Wrangham argued that eating cooked nutrient-rich food is what made our brains so much larger than other non-human species. The 0.78-Ma-old Gesher-Benot Yaakov site in Israel contained scorched seeds, and remains of edible plants—wild grapes, barley, olive, etc. The locals used to eat a protein-rich diet of elephant, deer, hippo, among others.[21] Archeologists discovered a dozen hearths in the area, the first recorded habitual meeting place among human ancestors—a site for cooking, shelter, and perhaps social gatherings. The earliest evidence of fire use comes from the 2-Ma-old, ancient hearths of the Wonderwerk cave site, South Africa, and the million-year-old, Swartkran fossil site. For some, the fire would have come from natural sources like lightning.[25] In Asia, a circular layer of ashes along with animal bones confirms that Peking Man—a *H.*

erectus, had used fire in his Zhoukoudian cave in China. It is not clear whether Peking Man could control fire for regular use, or they used fire opportunistically, when available.[20]

Some experts now argue that the use of fire was not uncommon during the early Stone Age, 1.5 Ma ago or before. Domestic fire use would have come in handy for myriad tasks other than cooking—lighting cave dwellings, warding off predators at night, warmth on cold winter nights, etc. *H. erectus* had smaller mouth, teeth, and digestive system, probably due to eating cooked food by controlled fire.[25] The use of fire likely developed in stages over time. First was opportunistic use of natural sources like lightning strikes. Thereafter, the species learnt a limited conservation of natural fire by use of animal dung or charcoal. Finally came the mastery of controlled/kindled fire.[26]

We are the Most Hairless of All Species—Why So?

Another fact of far-reaching implication is that humans have been the most hairless creature of all land animals for a long time. Just compare us, or our direct forbears *H. erectus*, with our primate cousins. It did not happen overnight that we are practically hairless. It might have been a process that started ~2 Ma back when *H. habilis* transitioned into *H. erectus*. What triggered our hair loss? No one knows for sure, but things happen in nature because of adaptation. Earlier, Charles Darwin also contemplated this. When the Pleistocene global cooling took hold, some of the forest habitat gave way to expansive savannas. Right about then our arboreal ancestors were also becoming upright bipeds. They traversed large open territories for food and big game hunting. We recall bipedalism happened long before *H. erectus* came along.[20] The earliest *Homo* ancestors *S. tchadensis*, *O. tugenensis*, *A. ramidus*, were both arboreal and bipedal. A rudimentary tool making ability would have helped them to this end. Trudging, even running long hours in pursuit of game meant *H erectus* had to endure the heat of the glaring sun overhead. Their hunting technique likely would have included outrunning games they were hunting. Some African tribes still hunt by outrunning preys. Furry non-human primates are ill equipped to lose body heat fast enough to survive and function normally. The seminal idea of P. Wheeler and others was that the genus *Homo* had to lose its furry coat and develop sweating glands to combat heat and stay cool.[27]

Tools of the Homo erectus— The Signature Acheulean Hand Axes

Homo erectus opportunistically radiated toward areas of greater natural resources and easy access to food source—both plants and meat. They developed sea faring skills and populated countless islands in Java, Indonesia.[28] The Java Man (1.8-Ma-old) and other *H. erectus* fossils from Java (1.6–0.6-Ma-old) are older than those from China. Stone tools left behind are the only direct evidence of their presence, mostly near sites of bountiful food sources like river valleys and lakes.

The signature Acheulean hand axe always accompanied *H. erectus* wherever they lived in Africa, Asia, and Europe (Fig. 12.9). They also used elaborate sharp stone cleavers and handy scrapers for cutting bite-size meat for chewing. The stone tools were useful in butchery and perhaps big-game hunting. The open-air Spanish sites, Ambrona and Torralba, about 2 km apart by the Ambrona River, were originally swampy terrains. Bones and skulls of elephant, horse, deer, and cattle abound in the area, amid some hand axes and other stone tools.[29]

Stone tools aside, they also used wooden artifacts both in Asia and Europe. The best example is the Schoningen lignite mine, Germany, where researchers discovered eight Paleolithic 'throwing spears' in excellent shape. Thermoluminescence dating gave an age of 0.337–0.300 Ma for the 2-m-long spruce spears. A cache of about 10,000 bones of horse, deer, and bison, etc., lay scattered along with stone tools.

The Evolution of Language: Was *Homo erectus* the First to Talk?

Earlier, the idea prevailed that language is a recent development perhaps not going far beyond the last 500,000 years. But experts argued that *H. erectus*—our direct ancestor, was probably the first to use language in social groups 1.5 Ma back.[12] A larger brain capacity and com-

12-9 A representative signature bifacial Acheulian hand axe of *Homo erectus*, displayed at the Saint-Remi Reims Museum, France.

munication skill may explain *H. erectus*' singular success in settling in varied geographic areas and environment. The bulge of Broca's area ('frontal lobe area of dominant brain hemisphere with speech function') in a *H. habilis* brain cast made experts think that the species could speak a rudimentary language.[30] The precursors of speech and gestures may be older (~7 Ma), when the *Homo genus* split from the *genus Pan*.[31] Laughter among great apes may date back farther.

Hot Debate on Human Origin—'Out-of-Africa' or 'Polygenetic Candelabra'?

Earlier, we considered the sole possibility of *Homo genus* originating in Africa; all other species lasted much longer than *H. sapiens* yet did not make the cut. Where did the final change occur? What made us who we are, though looking somewhat distinct in different regions of the globe? Two schools of thought have emerged on the topic (Fig. 12.10).[30]

The unitary Noah's Ark (Out-of-Africa) model suggests modern humans originated in Africa, and spread across Asia and Europe, and replaced the pre-humans. A 2021 study argues that 'Modern human lineages' developed from a small African group of 10,000 to 50,000

HUMAN EVOLUTION

Unitary Noah's Ark (Out-of-Africa) model
or
Complete Replacement Model

Humans originated in Africa and later radiated across the globe and replaced local pre-humans

Polygenetic Candelabra model
or
Multiregional Model

Humans developed independently, interbred with archaic groups everywhere, creating distinct geographic populations in all old-world continents

12-10 The debate on the origin of *Homo sapiens*—Africa or Multiregional?

individuals ~200 ka ago.[35] The Eurasian lineage supposedly emerged from this group less than 100 ka ago. The fossil of *Cro-Magnon1* man, an early modern human, came from a cave shelter of the same name in France. Its similarity with specimens from Africa may suggest a relatively recent migration of *H. sapiens* from Africa to Europe.[33] This model argues that differences among modern humans across the globe are superficial. The expert majority probably tilt towards the Out-of-Africa/Complete Replacement Model.

The Polygenetic Candelabra model, in contrast, holds that modern humans developed independently through interbreeding with different archaic groups.[40] They interbred in widely separated areas of the world at the same time, giving rise to today's varied geographic, physically distinct populations. The DNA evidence of several other hominin species in humans tends to support the polygenetic model.[39]

Sang-Hee Lee, a paleoanthropologist, investigated human origin as a species. She highlighted the dissonant voice of the second school, supporting a multiregional origin of our species. She argues that *H. erectus* would have originated in Asia and migrated to Africa—giving rise to later *H. sapiens* there.[20, 41] For her, Asia was equally a cradle of humankind—the ~1.75-Ma-old *dmanisi* hominin (*H. erectus*) of Georgia is as old as the oldest *Homo* fossil in Africa. Her '*Neanderthals* of Asia,' i.e., the *Denisovans*, from the Denisova cave, Siberia, overlapped with *Neanderthals* and *H. sapiens*.

Wu Xinzhi, another proponent of Asian provenance, made a clear distinction between the Recent Multiregional Evolution concept that he defended, and the Out of Africa Model. He argues that *H. erectus* evolved into *H. sapiens* in Asia, and that Asia should have an equal billing with Africa as the birthplace of *Homo sapiens*. The morphological similarity between *H. erectus* and *H. sapiens* makes him believe in the continuity of evolution in China.[42] The arguments of both Sang-Hee Lee and Wu Xinzhi merit farther discussion.

The Quaternary Fauna and Flora: Many Perished, Some Thrived

The Quaternary was a time of numerous glacial events alternating with interglacial warmer interludes. Atmospheric cooling created wide savannas replacing forests during the past 2 Ma. The world's soil cover is largely of Quaternary origin. Rivers transported glacial rock flour and loess deposits to deltas and floodplains where trees and grasslands thrived on fertile soil.

Presently, Earth has been basking in the 11.7-ka-old, interglacial warm period. The prediction is that the planet may enter a future deep freeze again, perhaps in several thousand years or later. Snow blanketed nearly one-third of land surface during the Last Glacial Maximum (LGM), in tandem with a marked sea level drop. Conversely, warmer interglacial spells earlier had caused ice to melt, sea level to rise and submerge land areas. All this instability would have seriously impacted the animal kingdom. Also, the relative shortness (2.58 Ma) of the Quaternary Period made the force of disturbance more palpable. Glacial cycles and sea level variation aside, vast arid plains during LGM would have caused some scarcity of plants, and the consequent herbivore food chain collapse. Other natural events like super-volcano eruptions, asteroid impacts, and diseases like pandemics, etc., also could have triggered an ecological disturbance.

Many animals that roamed around Africa a million years back are extinct mainly due to climate change. Species like zebra, buffalo, lion, hyena are later arrivals from Europe and Asia.[2] About 90% of more recent Quaternary animals are like the modern ones. The vast, wide steppe grasslands were ideal grazing ground for horse, antelope, bison, and their predators. The Quaternary was also the era of gigantism. Many land mammals attained enormous size during the time.[53] Some argue that larger body size is an adaptation to colder climates. While that may be true of Eurasia and North America, it fails to explain gigantism in Africa and Australia.

Many megafaunas would have perished due to Earth's short-lived (800-year-long) magnetic pole flip 42,000 years back.[38a] The straight tusked elephant, bison, rhino, mammoth, mastodon, musk oxen, camels, species of horse, etc., used to roam around in North America. The large predators were sabre-tooth cat, cave lion, dire wolf, cheetah, etc. It's surprising that native horses and camels once galloped across North America. Luckily, some of these animals escaped via Bering Land Bridge before their extinction in North America.

Cave Paintings and Extinct Animals

The contemporary cave paintings in several continents depict some of the extinct animals. About 600 wall paintings depict ~6,000 figures

12-11 The amazing, up to 20-ka-old rock paintings on the vertical rock shelter walls of the Paleozoic table mountains in Colombia. The Chiribiquete National Park, a UNESCO World Heritage Centre, showcases scenes of dancing, hunting, battles, ceremonies, and flora and fauna. Faint images of jaguars are present next to the handprints near the top.

in the 17-ka-old Lascaux cave's wall and ceilings, showing once native animals including rhino, and a 5.2-m-tall bull.[55]

The amazing Chiribiquete rock paintings shine among ancient paintings, now duly recognized by UNESCO (Fig. 12.11). These embellish the miles-long, imposing vertical walls of 60 rock shelters in the Paleozoic sandstone table mountains right in the heart of the dense Colombian rainforests. The Chiribiquete National Park boasts of a staggering 75,000 rock paintings up to 20-ka-old, and many rock shelters.[56] The motifs include routine social themes—animals, humans, some extinct plants, and flowers. The images of jaguar feature prominently there. The artists belonged to the Carizona tribe, extinct since the early 20th century. However, some reclusive tribes still live in typical native 'malokas' (house) in the park area.[57]

Similar artwork in red ochre, probably of a later age (~12.5-ka-old), also beautifies the walls of the Paleozoic Serranía La Lindosa Table Mountain in the Amazon rainforest, Colombia.[56] Some paintings depicting megafauna on the Cerro Azul peak, highlight hunter-gatherers killing giant mastodons. The images of mastodon and horse, etc., suggest that the Amazon rainforest climate was probably unlike today. Was a rainforest environment a congenial habitat for mastodon—an ice age megafauna, or for horse that is basically a grass land creature?

The Flora Adapted Well to Climate Change

It is perhaps safe to say that most plants we see today survived the Quaternary Period (2.58-Ma-old), marked by the oscillating glacial cold and interglacial warm spells. Ground-hugging plants like moss, sedge, lichen, shrub, and low grass, covered more land surface during the cold period, while forests and woodlands thrived during the interglacial. The conifer, cypress, oak, flowering plants, and prairie grass—all survive to this day.

The global carbon and hydrological cycle depended on the Pleistocene Amazon ecosystem and forest cover, just as today. The rainforests stood their ground even during the Last Glacial Maximum (LGM), though the climate got dry. The average air temperature was ~6°C during LGM (now 14–15°C). The most pronounced vegetation shift occurred about ~16 ka ago when significantly open, savanna vegetation encroached even into the Amazon territory. The rainforests proved resilient and weathered well the bitterly cold ice-house spells of the Quaternary.[65] However, human activity is a serious challenge for the Amazon ecosystem today.

The Tragedy Down-Under—The Same Story, Many Animals Perished

Things were no different in Australia with all the giant marsupial varieties.[60] The supersized herbivore wombat *Diprotodon* was the largest ever, weighing ~2 tons, and displaying teeth like knife blades. The predator animals were also predictably with a marsupial appendage. The meter-long *Thylacoleo* is called a marsupial lion, a misnomer. The wolf-like Tasmanian tiger fell prey to the newly arrived biped islanders.[61] The 5.5-m-long lizard *Megalania* with curved lethal teeth, a cousin of the northern Komodo Dragon, also became extinct. Australian birds attained super-size like the 3-m-long *Dromornis;* it became extinct naturally. The 2.5-m-long, and 2-m-tall bird *Genyornis* had the dubious distinction of coexisting with the first islanders, implicated in its demise. Incidentally, the European settlers' arrival in Australia coincided with the disappearance of 27 mammals, and 52 floral species.

Asphalt Seep Graveyards of the Quaternary—Buried the Predator and Prey Alike

Many Quaternary giant animals and birds became extinct—the stork-like, great Teratorn birds with a wingspan of 5 m that used to terrorize their prey. The Teratorn fossils remain preserved in the La Brea Tar Pits, California. The Rancho La Brea property of Los Angeles represents the world's largest repository of fossils in asphalt. The 40–12-ka-old tar pits preserve intact specimens of mammoth, dire wolf, ground sloth, bison, horse, turtle, fish, gopher, American lion, and sabre-tooth cat (*Smilodon*). Dead birds, insects, snails, plants, and pollen grains are also present. The entrapped birds were mostly carnivorous including some condor, luckily the species is still present in the Americas.[58]

In South America, the 25–30-ka-old, 0.5 km^2 Inciarte 'Mene' (seep), Zulia State, Venezuela, is ten times larger than the La Brea. It contains many fossils of mammal, bird, frog, and lizard. The other Venezuelan Orocual tar seep (0.5–1-Ma-old), Monagas State, shows rare fossil snakes (Squamata—Serpentes). The site is significant—the entombed animals confirm the beginning of the 'Great American Biotic Interchange' during the 2.5–1-Ma interval, after the Isthmus of Panama formed. The snakes appeared in North America in the Eocene, migrated to Panama in Miocene, and finally crossed over to South America in the Pleistocene.[59]

So Many Quaternary Animals and Birds Gone—Why?

A brewing debate is about the real cause(s) of the end of many Pleistocene animals. Climate change, among several probable causes, would have been crucial in some cases. It is possible that loss of flora because of adverse climate led to food chain collapse for the herbivores. Disease and predators also could have been probable causes for some. One even wonders if any cosmic invader could have caused the demise of these animals.

A serious reason might be the "Prehistoric overkill hypothesis" that implicates omnipotent humans. A strong correlation exists between human arrival (40 ka or later), and disappearance of several

12-12 Size of some extinct and modern giant birds and dinosaurs. Elephant bird (*Aepyornis maximus*) marked 3, ostrich bird (*Struthio camelus*) marked 4, the rest are non-avian theropod dinosaurs. Grid spacing: 1m.

local faunas.[63] In Madagascar, extinction of giant lemurs, 3-m-tall elephant birds and others, coincided with the arrival of humans just 1,500 years ago [64] (Fig. 12.12).

The Toll of Exotic Species on Native Fauna—A Misfit, Aquatic Leviathan in Colombia

A recent example of introduction of exotic species, and the toll it takes on the local ecosystem, comes from Colombia, South America.[68] In the 1980s, a local drug lord introduced hippos, 1 male, 3 females, in his ranch near the Magdalena River valley. The giant quadrupeds were super comfortable in their new menagerie. However, when the estate went unattended after the drug lord's death, the hippos found a cozier habitat in the nearby Magdalena River. By early 2023, the 4 giants had multiplied into a large herd of 130 animals in 40 years. They are destroying crops, obstructing navigation, scaring away native animals like monkeys, etc. Hippo waste causes bacteria to thrive in water, making it oxygen-poor, especially affecting the fish population. They pose a menace to villagers, including some who were seriously hurt. If unchecked, Colombia may have over 400 hippos in the next 8 years.

A similar new millennium sad tale comes from the Florida Everglades where thousands of invasive Burmese pythons (pets—released or escaped) have decimated the population of wild bobcat, racoon, opossum, and rabbit.

Mass Extinction—Nature's Renewal or Survival of the Fittest?

Five major mass extinctions across land and sea happened during the 539-Ma-old Phanerozoic Earth history. Table 12.2 shows details of each extinction event—age, biota loss and ranking as per its severity. It is curious that all mass extinctions occurred at or near geological period boundaries. Earlier, we discussed the possible causes, in the absence of a conclusive one. Of the five, the K/Pg extinction is rather contentious. We recall from Chapter 8, the father-son Alvarez and Alvarez team, along with other scientists, favored a cosmic body im-

Table 12-2: Mass Extinction Through Geologic Time

Age of Extinction	Biota Loss %	Fossil Variety Extinction/Loss	Probable Causes/ Severity Rank
End-Cretaceous ~66 Ma	76%–80%	Dinosaur, ammonite, many plankton, and orbitoid foraminifer. Bivalves, including rudist, gryphaea, exogyra, and inoceramid.	• Asteroid impact • Deccan Trap eruption and release of toxic gases • Natural climate change • Sulfate aerosols from impacted rocks • Collapse of photosynthesis and the food chain • Impact-related energy release and earthquakes • *5th ranked extinction*
End-Triassic ~210–201Ma	80%	Brachiopod, conodont, many gastropod and bivalve—all extinct. Marine reptiles—all extinct except ichthyosaur and plesiosaur.	• Central Atlantic Magmatic Province volcanism • CO_2-related global warming • Methane release from permafrost melting • Possible Manicouagan asteroid impact • *4th ranked extinction*
2. End-Permian 1. Capitanian (Upper Permian) ~266–252 Ma for both phases	70% Land life 95% Marine life	Trilobite, some echinoderm, two-thirds of the amphibian population, and large herbivores.	• Siberian magmatism causing global warming • Major sea-level fall • No seafloor spreading • Oxygen depletion • Bedout meteorite impact • Temperature rise due to high CO_2 levels and locked methane release • Known as the "Great Dying" • *Worst extinction event in Earth's history*
End-Devonian ~360–375 Ma	75% General 97% Vertebrate	Some rugose and tabulate coral, many algae, acritarch, stromatoporoid, conodont, jawed vertebrate, and placoderm fish.	• Vascular plants and CO_2-related cooling • Siljan and Woodleigh meteorites • Siberian Viluy trap eruption • *3rd ranked extinction*
End-Ordovician ~441–444 Ma	86%	Many graptolite, trilobite, bivalve, brachiopod, bryozoan, coral, and other species extinct or affected.	• Glaciation • 50–100 m sea level drop • Lower ocean temperature • Volcanic activity • Asteroid impact • Meteor showers • *2nd ranked extinction*

Rank of extinction highlights degree of biota loss and severity. [69, 70]

pact model.[66] However, some experts, including paleontologists, argued that eruptive processes like the Deccan Trap had a hand in the K/Pg, and other mass extinctions.[67]

The Sixth mass extinction may be currently in progress as many mammals, birds and flora have become extinct or are facing existential threat. The Smithsonian National Museum, Washington D.C., displays interesting statistics linking the extinction percentage of animal species (>50 kg weight) in different continents during the last thousands of years: South America 82% (in 4 ka), North America 74% (14 ka), Australasia 97% (50–40 ka), Eurasia 18% (45–4 ka), and Africa 17% (50–10 ka). Everywhere, humans and climatic downturn contributed to the unfolding disaster. The U.S. holds the record of the highest number (266) of extinct animal and plant species,[62] Humans are impacting the environment in myriad ways—industrial wastes, high level of atmospheric pollution, large scale encroachment on wildlife habitat, spread of cultivation, largescale logging, deforestation—the list continues.

Evolution: A Requiem for Life from Eternity to *Homo sapiens*

Chapter 3 (*Table 3.2*) gave us a sense of how and where humble life would have had started in the hazy Archean Eon, including five notable mass extinctions later. But that was just the beginning. Life evolved continuously in the long Proterozoic Eon, and then through the Paleozoic, Mesozoic, and Cenozoic eras of the Phanerozoic Eon. Table 12.3 summarizes the important biotic milestones on Earth. The extinction events, triggered by natural causes like climate change, volcanic eruption, asteroid impact, among others, took a toll on the existing biota. Luckily, they proved to be temporary (geologically speaking) dampers. The remarkable parade of diverse life marched on till the recent debut of *Homo sapiens* on the unfolding scene.

It is amazing that life evolved into the astonishingly complex reptiles and mammals of later times from the lowly archaea and cyanobacteria of the Archean Eon (Figure 12.13). It shows the breathtaking evolution of life, though punctuated by five mass extinction events. The change from the Proterozoic to the Cambrian may seem

Table 12-3: Evolution of Life and Earthly Markers

Era	Time Period	Key Features
Quaternary	0.0-2.58 Ma	Most living species appear. *Homo genus* emerges, leading to the crown jewel: *Homo sapiens*.
Neogene	2.58-23 Ma	**Terrestrial life:** Panthera family (big cats), woolly mammoth, grazing animals. **Marine life:** Dugongs, megalodon sharks, kelp forests. **Flora:** Deciduous trees, grasslands. **Primates & Hominins:** Great apes (orangutans, gorillas, *Pan genus*), transitional hominins, first hominin (*Australopithecus*).
Paleogene	23-66 Ma	**Mammals:** Lemurs, tarsiers, whales (evolved from land animals), horses, elephants (Paleomastodon), rhinos, Old World monkeys. **Flora:** Angiosperms such as cashew, lychee, roses, and legumes.
Cretaceous	66-145 Ma	**Faunal peak and mass extinction:** Dinosaurs, pterosaurs, plesiosaurs, ichthyosaurs, ammonites, rudist clam. **Marine life:** Giant Archelon turtles, ammonites. **Others:** Unhatched dinosaur eggs, diatoms, coccoliths, foraminifera. **Flora:** Angiosperms, grasses, shrubs, flowers evolving color, nectar, and scent.
Jurassic	145-201 Ma	**Dinosaurs dominate**: Sauropods (e.g., Titanosaur), theropods, early birds (*Archaeopteryx*). **Marine life:** Crocodiles, mosasaurs, amphibians (frogs, toads). **Flora:** Gymnosperms.
Triassic	201-252 Ma	**Rise of archosaurs:** Ancestors of birds, dinosaurs, and pterosaurs. **First dinosaurs; early mammals**: small, rodent-like. **Marine reptiles:** Ichthyosaurs, plesiosaurs.
Permian	252-299 Ma	Marine diversity and tetrapod expansion. Amphibians thrive. **Flora:** Gymnosperms emerge (cycads, ginkgo, yews, conifers). Glossopteris plants dominate.
Carboniferous	299-359 Ma	Early reptiles and land vertebrates develop amniotic eggs. **Flora:** Giant ferns, scale trees, lycopods.
Devonian	359-419 Ma	**Marine life:** Stromatoporoid reefs, corals, bony fish, sharks, and early jawed fish (Acanthodians). Tetrapods begin moving onto land. **Flora:** Plants.
Silurian	419-444 Ma	**Jawless fish** (Agnatha, lamprey-like). **Large marine arthropods** ('sea scorpions'). **First land arthropods** (insects, arachnids, centipedes, millipedes).
Ordovician	444-485 Ma	**Marine life:** Bryozoans, early sharks. **Land evolution:** First plant microfossils, spores, cuticles. Possible first land arthropods.
Cambrian	485-539 Ma	**Rapid diversification of life.** **First hard-shelled animals:** Arthropods, brachiopods, pelecypods, gastropods, trilobites, corals, sponges, graptolites, and more.
Proterozoic	539 Ma-2.5 Ga	**Soft-bodied animals:** (630–539 Ma). **Flora & Microbial Life:** Algae, fungi (1200 Ma); Acritarchs (1800 Ma); cyanobacteria (2500–539 Ma).
Archean	2.5-4.0 Ga	3.8 Ga: oldest fossil of Cyanobacteria.

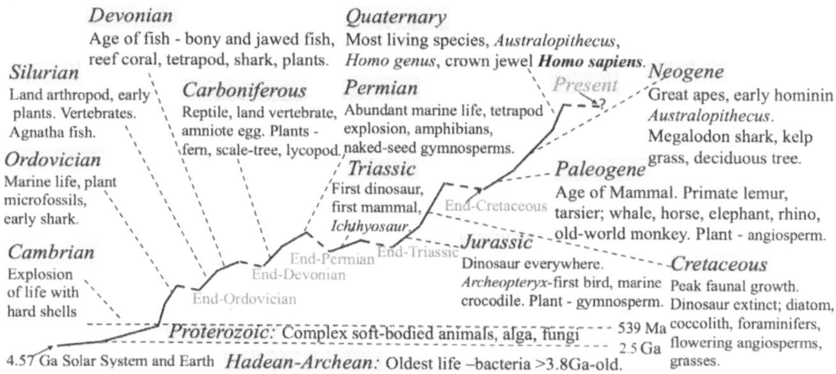

12-13 A schematic representation of evolution of life through multiple periods of mass extinction of species during the Phanerozoic Eon. The upward trend of the solid curve marks evolution of new species, while the downward trend (short, dashed lines) represents setbacks through mass mortality. The modern loss of wildlife and flora is worrisome. No scale is implied.

very abrupt, but in reality, the last phase of the Proterozoic already hinted at the trend toward more complex hard-shelled fauna of the Cambrian Period. The Precambrian Ediacaran fauna (630–539 Ma) in different continents yielded fossils of complex soft-bodied animals. The Cambrian Sea had more available calcium and oxygen for the organisms to form mineralized calcite skeleton.

* * *

Changing the Subject—Life Beyond Earth?

No discussion on mundane life can end without a brief reference to our quest for the possibility of alien life. Many centuries ago, Greek philosophers wondered about the possibility of life elsewhere in the Universe. Democritus in the fifth century BCE, the proponent of 'atomism,' first mooted the point that atoms of only a few elements make up life; the same atoms equally pervade the universe. These magical atoms may give rise to life elsewhere. Later, Epicurus (371–270 BCE) concurred that life might exist in many other worlds.[71a] Fast forward to the last century when astronomer Frank Drake notably led the search for radio signals from alien life.

Scientists have been actively probing the clue to alien life, aided by the gift of advanced telescopes and instruments during the many preceding decades. As of April 2021, the United States authorities released videos by pilots of "unidentified aerial phenomena" (UAP), perhaps tacitly hinting at the possibility of the existence of aliens. Simply, UAP is what we knew as UFO or 'unidentified flying object' for many dozens of years. Earlier, the U.S. govt. set up a taskforce to study the issue in Aug. 2020.[71] In July 2023, three former U.S. military officials testified before the U.S. Congress about the reported proof of crashed alien spacecrafts and remains of 'non-human biologics' of extraterrestrial pilots (NPR, July 27, 2023).

Planet Mars—We May Find Out Soon if Life Ever Existed There

Earth may be the only solar planet with conducive physical milieu apt for sustaining life, especially liquid water. Mars resembles our blue planet in many ways, and the presence of Martian ice and methane may be suggestive of once favorable conditions there. Scientists have always considered the possibility of primitive life on Mars since the 19th century. Some NASA scientists interpreted wiggly nano-scale tubular images in Martian meteorite samples from Antarctica, as suggestive of primitive life on Mars. However, doubts persist if these are true microbes. The 2018 NASA Curiosity rover's discovery of complex organic molecules in Martian samples heightens the possibility of life on Mars. Hopefully, NASA's Perseverance rover exploring the Jezero crater since 2021, as well as the Chinese Zhurong, and rovers from several other countries, would finally provide a definitive proof of biosignatures in Martian soils/rocks, if any.

The geological setting on Mars may be ideal for preserving clues to ancient life. Photos by spacecrafts and rovers point to mounded small structures that resemble earthly stromatolite, a product of cyanobacteria. Earth's Precambrian limestones are often replete with stromatolites. Rarely, modern seas like Shark Bay, Australia, and the Caribbean Bahamas, also host stromatolites.

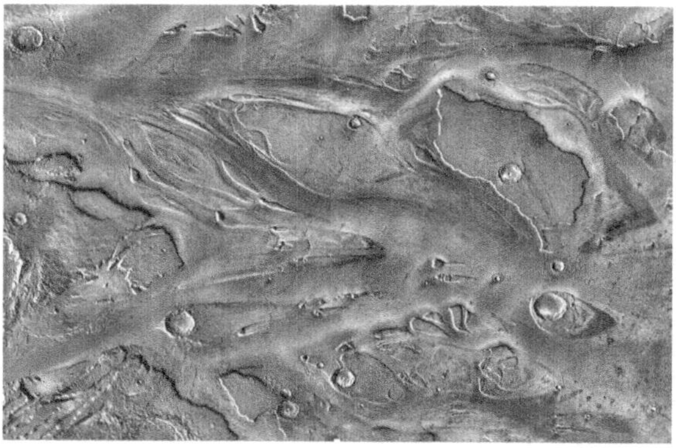

12-14 Clear evidence of water flow in the Kasei Valles on Mars. It shows how a crater (extreme left center, barely visible) divides the flow of a catastrophic flood into two streams entering at upper and lower left. Circular areas—Martian impact craters.

Live primitive bacteria in Martian soil today are unlikely because the atmosphere is too harsh for organic presence now. However, the earlier Martian magnetic field would have likely maintained an earth-like atmosphere with rivers and oceans. Many sedimentary structures in Martian sandstones imply the earlier presence of flowing water on Mars (Fig. 12.14). The Martian magnetic field stopped working later because its liquid outer core turned cold and solid. The absence of a magnetic field allowed solar wind and ultraviolet radiation to strip away most of the atmosphere and water resources, according to NASA's MAVEN probe results.

Life Elsewhere in the Solar System?

In Chapter 3, we mentioned the prevalent idea that life had an abiotic origin in deep oceans near hydrothermal vents. We know that proton gradients exist between deep-ocean alkaline vents and ocean water. The unique happenstance could have triggered the burst of microbial life in the nascent Earth around 3.5 Ga ago.[54] Similar deep liquid oceans and hydrothermal alkaline vents may be common in other planets, and especially, in some of their moons. The U.S. space missions reported that icy moons of distant gas-rich solar planets may hold such vents in the oceans. Especially, two moons—Jupiter's *Europa* and Saturn's *Enceladus* are promising; they contain vast oceans of liquid water underneath an icy surface cover.

Simulation exercises imply presence of liquid water on Venus about 3 Ga ago, at a temperature of 50–20°C. Later, a massive 'outgassing' of volcanic CO_2 would have created a runaway hothouse effect, as the present Venetian unearthly 462°C temperature confirms.[81] Will there be traces of 3-Ga-old primitive life on Venus?

Any Sign of Life in Exoplanets? Let's Keep Our Fingers Crossed

Does life exist anywhere else beyond our solar system? Are we alone in the universe, studded with billions of stars and their planets? The answer perhaps would be—it is up in the air. However, research shows probably that may not be the case. Some form of life, unlike what we know on Earth, may exist elsewhere. NASA's Telescopes (Hubble, and James Web Space Telescope) detected atmospheric water vapor, an essential ingredient of life, in a few exoplanets, raising the prospect of life elsewhere.

The University of Chicago researchers used a NASA software ROCKE-3D to simulate oceanic circulation in different types of exo-planets. Oceanic life depends largely on upwelling of nutrients from deeper ocean to shallow waters where organisms thrive. It revealed a favorable milieu for high upwelling rates, a necessary condition for life elsewhere.[73]

In 1995, two Swiss astronomers discovered the first exo-planet—'51 Pegasi.'[72] Since then, NASA has compiled a roster of more than 5,600 documented exo-planets in the Universe. Belgian astronomers made an important discovery of a red dwarf star in the Aquarius constellation—'Trappist-1.' The powerful Spitzer Space Telescope (NASA) and the Very Large Telescope at Paranal (ESO) in the Atacama Desert[75] identified seven orbiting temperate terrestrial exoplanets of the ultracool star 'Trappist-1'.[74] The earth-size exo-planets are orbiting around the star about 40-light-years away. Trappist-1 is very small and has only 9% the mass of our Sun. Six of its planets fall within its habitable zone—a zone defined by the distance of a planet from its star, and the prevalent temperature. It is a goldilocks-like zone where temperature is mild enough to hold water on a temperate surface.[82] Three of the exoplanets (*e*,*f*, *g*) likely contain water, and another three (*c*, *d*, *h*) may be possibly water-bearing. The planets look promising as vigorous research is in progress. However, exoplanet life may be of a totally different kind than what we know on Earth.

Recently NASA scientists discovered a new exoplanet—TOI-1231b, about 90-light-years away from Earth, that could contain water clouds.[76] The planet shows an ambient temperature of 60°C. Soon NASA's James Webb Space Telescope will be able to observe this planet closely.

DEBATE ABOUT THE FINAL EPOCH: 11,700 ka TO PRESENT

Anthropocene or Holocene: Which Better Reflects the Reality?

How should we call this razor-thin time slice? The International Commission on Stratigraphy (ICS) and International Union of Geological Societies (IUGS) have traditionally defined the post-glacial—11.7-ka interval of Earth as the Holocene Epoch. However, many geoscientists are urging the two entities to formalize the name 'Anthropocene' for the last bit of the Holocene to reflect the grave human impact on the slim time interval. Both societies reconsidered the issue and set up a special committee for a final recommendation.

The impact of our footprint is undeniable as simple data confirms.[77] About 10 ka ago, humans made up ~1% of living vertebrate fauna, and 99% were wild animals. Now fast forward to 2020: Humans 32%, wild animals 1%, livestock 67%. Human impact on the climate and environment is disproportionate, as our numbers are somewhat out of balance. Since the 1750s, the volume of atmospheric CO_2 level, eco-system modification, freshwater use, marine fish capture, animal husbandry, fertilizer consumption, etc., have increased perilously. At the same time ~20% of Amazon tropical rainforest cover has disappeared.[78]

An earlier proposed date for the Anthropocene was the beginning of the Neolithic civilization and dawn of agriculture about 12,000 years ago.

Fast forward to 2000, Nobel laureate Paul Crutzen, and chemist E. Stoermer assessed the severity of human impact on Earth's environment and proposed the term 'Anthropocene' for the period starting from 1784 CE, the year when James Watt invented the steam engine.[83] Their activism proved to be a catalyst, it galvanized the geoscience and environmental communities. The term caught on, appeared in >450 publications by 2016,[79] and a new journal "Anthropocene" debuted in 2013.

The flood of publications highlighted the impact of human activities, including energy consumption on the biosphere and atmosphere, especially since ~1945.[80] This is the unique moment when humans made unprecedented technological progress, yet altered the environment, leaving behind unsightly, perhaps irreparable scars.

After all the commotion, the ICS-IUGS body formalized the special committee recommendation of a 3-fold subdivision of the 'Holocene' in 2018 (Table 12.4), bypassing the Anthropocene issue.[45] To be fair, at least 50 glacial-interglacial cycles marked the past 2.58 Ma. The traditionalists argue that why then should we single out part of the current interglacial as 'Anthropocene.' While it's true that ~200 years (1784 onward) is a mere blink of an eye in the 4.57-Ga-old Earth

Table 12-4: Holocene 3-Fold Subdivisions

Stage name	Age (ka)	Holocene Division	Remarks
Meghalayan*	0–4,250	Upper	Global drought condition (4.2 ka BP)[16]
Northgrippian	4,250–8,236	Middle	Sudden cold spell in northern hemisphere (Greenland ice core data)
Greenlandian	8,236–11,700	Lower	Warmer climate (Greenland ice core data)

**Results of samples from Mawmluh limestone cave deposits, near Cherapunji, Meghalaya, India*

history, who can ignore the disproportionate degradation of the geo-environment during the short period.

The very latest 2021 proposal by the Anthropocene Working Group (AWG), was to name the period from 1950 onward (a Post-World War II period coinciding with the Atomic Age) as the Anthropocene, based on evidence of sediments from the meromictic (the top and bottom layers don't mix) Crawford Lake, Milton, Ontario (Figure 12.15).[85]

In 2024, the Quaternary Stratigraphy Subcommittee (QSS) of the ICS studied the proposal and voted against the adoption of the Anthropocene nomenclature. Soon, the IAS-IUGS formally rejected the AWG proposal. However, the entities fully recognized the value of the continued use of 'Anthropocene' for the last bit of Holocene as a testimony to the negative human impact on the Earth eco-system.[84]

12-15 The Holocene sediments of Crawford Lake, Milton, ON, Canada, have been proposed as evidence of human impact during the suggested Anthropocene time period.

CHAPTER HIGHLIGHTS

- Review of the Quaternary flora and fauna, especially the great ape *Homo sapiens* of the *Homo genus.*
- The hominin *Australopithecus* evolved into the *Homo genus* ~2.5 Ma back.
- Simplified *Homo genus* evolution: *H. habilis—H. erectus—H. neanderthalensis* (*'Archaic Homo sapiens'*), and finally *H. sapiens sapiens* (barely 0.2–0.3 ka old).
- Discussion on human ancestor's diet, use of fire, lack of body hair, tools and weapons.
- Evolution of language, and examples of early human artistry.
- Debate on the place of origin of *H. sapiens*—Africa or Multiregional.
- Deciduous forests in a cooler climate are becoming common.
- Analysis of asphalt seeps, significance of naturally trapped animals and birds in them.
- Examples of extinction of many Quaternary fauna and flora. Reviewing the impact of humans on the possibility of a sixth mass extinction.
- Possibility of life beyond Earth—Mars, exoplanets, and moons of other solar planets.
- Debate about the name of part of the last 11.7-ka-old Interglacial period—*Holocene* or *Anthropocene Epoch.*

WHAT'S NEXT?

Finally, the travelogue of Earth's long journey comes to a wayside stop. It unfolded the complex, yet steady evolution of an initial glowing fire ball into the mature habitable 'Blue Marble'– our home. It laid out countless earthly treasures—mineral, rock, plant, and animal, crowning with the debut of *Homo sapiens.*

The final chapter will be somewhat speculative, dwelling on Earth's nebulous future path. Will the past be the key to the future?

Star coral polyps (Montastraea) show central openings, and a delicate radial fabric. Playa Cocos, Cuba. Specimen width – 10 cm.

Chinese Money Tree made of semiprecious Carnelian (SiO_2) symbolizes wealth and prosperity, often used in New Year celebrations. Its distinctive flesh-like color, derived from iron traces, reflects its name's origin, Carneus (Latin for flesh-colored). The tradition dates back 2,000 years. India celebrates a similar wish-fulfilling tree festival, Kalpataru, on New Year's Day.

"There is grandeur in this view of life [...] having been originally breathed into a few forms or into one; and [...] from so simple a beginning, endless forms most beautiful and most wonderful have been, and are being,

evolved."

Charles Darwin
On the Origin of Species (1859)

PART THREE

THE FUTURE *of* EARTH

13

The Past: Illuminating the Mysteries of the Future

We Are Finally Here

As one arrives at the last visible milepost on Earth's epic journey route, one looks back on the many veritable ups and downs, fireworks, celestial bombardments, and extensive dying of many species. The ever-changing climatic swings have been the hallmark—from infernal heat to long, repeated bone-chilling cold spells. Through it all, life marched on and vanquished all odds. Some genera or species fell by the wayside. Natural selection dictated the fate of those that survived in the dynamic geologic milieu. Especially, the arrival of *Homo sapiens* has been a game changer.

Planet Earth Has Seen It All—A Key to the Future?

At a bird's eye level, we ask ourselves—which way will Earth be heading? A geologist examines the mysteries of the past and tries to make sense of it with specific tools in his/her toolkit. Can they predict the future—a million years or farther down the road? One may think there probably needs to be a role reversal—from Sherlock Holmes

13-1 Depiction, in watercolor, of the 2-headed Roman God Janus, who had the ability of dual visions of the past with one head, and the future with the other.

to an astrologer. Or one may beseech the two-headed Roman God Janus. The mythical Janus used to see the past with one head, while he could foresee the future with the other, while facing the opposite side (Fig. 13.1). However, it is not so offbeat to speculate where the planet may be heading. Our old planet has seen it all—major earthly milestones from the nascent stage up until today. Can the past be the crystal ball to foretell the future? To that end, let us very briefly recapitulate the past.

Earth Story in a Nutshell

The narrative started with the hellish heat of the Hadean-Archean, lasting 2 Ga during Earth's nascent stage. The ~3.8-Ga-old primitive bacteria—the first ever sign of life on Earth, became palpable even in that inferno. The reprieve came with the Snowball Earth (717–635 Ma) when a thick icy blanket extended from the Poles to the Equator. Several landmarks in the ensuing short-lived Phanerozoic Eon (539 Ma-Present) stand out—the Cambrian eruption of marine life with novel hard shells, and the highest ever sea level in the Ordovician (>220 m above the modern level). Also, the Ordovician (first) mass extinction, the Carboniferous-Permian icehouse-greenhouse cycles, and later lush coal forests from pole to pole, were noteworthy. The supercontinent Pangea emerged soon. The most severe mass extinction ever of the End-Permian preceded the Triassic lowest sea level and aridity when the dinosaur made its debut. The story continues unfolding through the benign Jurassic, and especially the Cretaceous with its warm subtropical balmy climate, and second highest sea level ever. The Pangea supercontinent went asunder in the Jurassic-Cretaceous. Dinosaurs and many marine animals flourished, yet the dinosaurs and most other animals met their abrupt end because of cosmic intervention—the Chixulub meteorite impact. The Deccan lava eruption on the Indian Plate was also an accomplice. Next, the still warm Paleogene ushered in the 'Age of Mammals.' Antarctica was the harbinger of the future—it donned a snow cap in the Oligocene. Soon the Paleogene gradually transitioned into the cooler Neogene that climaxed into the Quaternary icehouse. The Neogene global event of the Panama Isthmus connected North and South America, while permanently separating the Atlantic and Pacific oceans. Fast forward to the 20-ka-old, Last Glacial Maximum, when sea level dropped a whopping 125 m below the present level. However, it was not all chill and gloom. Numerous geologically short-lived, congenial warmer interludes—the 'interglacial,' punctuated the inconvenience and boredom of prolonged Pleistocene icehouse climate. The *Homo genus*, and importantly, our species, was an eyewitness to many of the latest climatic cycles. We are now enjoying one such fleeting warm interglacial for the last ~11,000 ka.

The debut of *Homo Sapiens* was undoubtedly the most transcendental Quaternary event. Before us, biped hominins—the *Australopithecus* and pre-human species of the *Homo genus* toiled and slogged, and for a much longer period. In course of time, our distant ancestors came down from their arboreal habitat as the colder climate created savannas replacing forests. They started living on level ground and in caves. Our direct ancestor *Homo erectus*, was the longest-lived species (1.9 Ma–0.1 Ma). One of our close kins, the *H. neanderthalensis*, became extinct rather recently, <40 ka ago, while *H. florisiensis* may not be extinct, as an expert opines.[1]

Table 13.1, "Earthly Milestones," highlights all the major events of the marvelous time-ladder of Earth. This Period-wise geologic summary will give readers an integrated, bird's-eye view of our ever-evolving 4.57-Ga-old planet.

Table 13-1: Earthly Milestones

Geological Division	Geological Time Span	Major Events and Characteristics
Holocene Epoch	Quarternary Period *2.58 Ma-Present*	Interglacial reprieve. *H. sapiens* rule. Human numbers explode & wildlife dwindles. CO_2 level rising. Scientists prefer to call the period from 1950-present as Anthropocene due to significant human impact. Sixth Mass Extinction?
Pleistocene Epoch		Active hotspots, mid-ocean volcanoes, Ring of Fire, San Andreas Fault. Over 1,200 volcanoes. Icehouse climate with short warm periods. Noted reef development. Last glacial event (125–11.7 ka), sea fell 125m. Great Lakes formed. Land bridges appeared. Faunal gigantism. Many extinctions. *Homo genus, H. sapiens* emerged.
Pliocene Epoch	Neogene Period *23-2.58 Ma*	Himalayas, Rockies, Andes, Tibetan Plateau set. Global cooling (~15°C). Arctic snow cap ~3 Ma. Panama Isthmus connects Americas (2.8 Ma). Pacific Ring of Fire more active. Neogene markers—Antarctica isolated, Hawaiian & Caribbean islands, African rift lakes. Amazon, Missouri, Indus-Ganges basins. Bengal Fan forms. Mediterranean dries up. Columbia River Basalt erupts. Mammals thrive—whales, dugongs, Megalodons, hippos. Kelp forests spread. Great apes emerge, leading to Homo ancestors.
Miocene Epoch		
Oligocene Epoch	Paleogene Period *66-23 Ma*	Antarctic snow cap forms, sea falls 50m. Andes rise, Alpine orogeny. Southern Sea fully open. Lake Baikal forms. Paleomastodon, old world monkeys emerge. Angiosperm flowering plants. Major European extinction.
Eocene Epoch		Thermal max ~30°C. Pacific Ring of Fire active. Birth of the Rockies. Drake Passage opens. India collides with Eurasia. Eocene brings horses, camels, rhinos. K/Pg survivors—snakes, crocs, turtles—thrive. Birds rule the sky. Coal forests widespread.
Paleocene Epoch		Dinosaurs gone. Mammals and plants rebound in 0.7 Ma. Hotter climate than today. Continents fully separate. Rockies rise. Western Interior Seaway vanishes. India nears Asia. Coal forests flourish. Birds soar. Early primates emerge.
Cretaceous Period	Mesozoic Era *252-66 Ma*	Pangea breaks up. Warm Earth (~25°C), sea 220m higher. Atlantic forms, continents drift. Deccan eruptions, Rockies rising. Dover chalk, oil-rich shale. Yucatán asteroid impact. Dinosaurs, ammonites peak, then **Fifth Mass Extinction.**
Jurassic Period		Pangea breaking up, salt in rifted Gulf of Mexico. Dinosaurs thrived. *Archeopteryx*, a reptile-bird transition, in Solenhofen limestone. Marine crocodile in Jurassic Coast, UK.
Triassic Period		Pangea Supercontinent. Lowest sea level ever. Hot, av. 25°C. Red beds in major rift-basins. Zechstein salt movements. Corals in Italian dolomite. Meteorite impact. Archosaur, dinosaur & marine Icthyosaur. **Fourth Mass Extinction.**

Geological Division	Geological Time Span	Major Events and Characteristics
Permian Period	Paleozoic Era *539-252 Ma*	Pangea buildup. Icehouse - Greenhouse cycles. End-Permian very dry, drastic sea level fall. Zechstein salt strata formed. Siberian lava eruption, largest ever. Gymnosperm, conifer dominated. Amniotes thrived. Mammal-like large reptile Dimetrodon reigned. **Third & Worst Mass Extinction**.
Carboniferous Period		Gondwana & Laurasia draw closer. Hercynian orogeny hints at Pangea. Coal forests spread, even in Antarctica. Greenhouse world dominates. Protective shelled eggs evolve. Ferns, dragonflies thrive. Amphibians, reptiles roam.
Devonian Period		Acadian Orogeny. Age of the fish - Acanthodian with jaw & spine. Sharks evolved. Land 'tetrapod', insect, plants, arachnid appeared. Prolific coral reefs. First coal deposits. **Second Mass Extinction**.
Silurian Period		Orogeny in full swing. Goldilocks climate. Prolific reef builder - special sea sponges, corals common. Jawless earliest fish 'Agnatha'. Plants, centipedes appeared.
Ordovician Period		First Paleozoic glaciation. Highest sea level ever. Fauna - starfish, graptolite, crinoid. **First Mass Extinction**.
Cambrian Period		Explosion of life with hard skeletons. Fauna - brachiopod, trilobite, arthropod, graptolite, 2m 'abnormal shrimp'. More ambient O_2. 'Great American Carbonate Bank' flourished.
Neoproterozoic Era	Precambrian *4.57-0.539 Ga*	Rodinia break up. Snowball Earth event. Icehouse-Greenhouse. Abundant O_2 dependant soft-bodied Ediacaran fauna. Ozone (O_3) appears.
Mesoproterozoic Era		Birth of supercontintent Rodinia. Abundant algal stromalotites. Low O_2
Paleoproterozoic Era		Glacial deep freeze. 'Great Oxygenation Event' ~2.4Ga. Low O_2. Cyanobacterial stromalotite.
Archean Eon		Supercontinent cycle begins. Cratons form. Very low O_2. Early sulfate-reducing bacteria. Archaea. Microbial banded iron formations.
Hadean Eon		Hot Earth, meteorite impacts, oldest rock (4.375 Ga), water from space/volcanoes, earliest microbes(?).
Big Bang, Solar System	Big Bang: 13.77Ga Solar System: ~4.6 Ga	Big Bang: Zero hour of Universe. Galaxies, stars, and planets follow soon after. The Solar System appears about 9.1 Ga post-Big Bang.

The Future—"Tomorrow and Distant Tomorrow"

What can one say of the future, equipped with the knowledge of the past? Today, it depends on two mighty powers—the colossal forces of nature as ever, and the immense techno-scientific might of humankind. More than 99.9% of species that ever lived became extinct through natural causes. It will happen again, likely faster, aided by our indiscretion. Posterity may not totally absolve our species of the 'sixth mass extinction' now in progress. In addition, the world's nuclear stockpile and biological weapons are enough to destroy the living world many times over if human imprudence prevails.

Here we will ponder future scenarios of our planet at different details: very short term, and short term of plus/minus million to several million years respectively. We will later reflect on the long-term geologic future spanning many dozen million years or more, that involves inevitable changes beyond humans. Finally, we will speculate upon the even more hazy distant future—billion years and beyond.

Human Impact on the Future Earth—A Short-Term Perspective

No other species has wielded such power to compromise the planet's future, its wildlife and habitat. What will happen if we consume resources indiscriminately, oblivious of the ecosystem upheaval we are causing? In the very short term, the impact of human excesses on Earth is ominous. In contrast, wildlife and plants basically take from nature what they need for their daily need, living in perfect harmony in a sustainable way.

The 2023 world population of >8 billion may become 11 billion by the year 2100. The number of farmed animals and poultry will increase manifold to keep up with this scenario. More poultry and livestock means more greenhouse gases, and potential contamination through animal wastes. The Covid-19 is a reminder that such future hazards may occur more often through zoonotic transfer and also chance lab leaks.

The Geologic Future—Inevitable Changes, Some Beyond Us

No one has a crystal ball to know how it will pan out. We can only speculate, using our geological hindsight. Short-term natural changes of tens of thousands to several million years, will include recurrence of sea level oscillation, icehouse-greenhouse cycle, supervolcano eruption, Earth's magnetic pole reversal, asteroid impact, etc. Plate tectonic acrobatics, massive hurricane and flood, extreme heat and drought, solar variation, wildfires, etc., will also continue.

Sea-level Rise Impacted the Holocene Climate

The main cause of Pleistocene-Holocene Sea level rise was the melting of glaciers and ice sheets at the polar high latitudes. A change of Earth's orbit and rotation would have caused increased solar incident light, melting of ice and sea level rise. An excellent example of recent sea level rise comes from the Holocene of North America. About 8.4 ka ago after the glacial thaw, the sea-like proglacial Agassiz, and Ojibway lakes emptied 105 km^3 of freshwater into the ocean, raising global sea level by 1–3 m. The melt water reaching the North Atlantic Ocean triggered the abrupt 8.2-ka-old Holocene cooling by 3°C. "NOAA" estimates any future sea level rise of 2m will affect 13 million people only on the US east coast, and millions more worldwide.

Global Warming—A Looming Risk

The other cause of sea level rise is global warming. Greenhouse gases (GHG)—CO_2, CH_4, and N_2O, are contributing significantly to the present warming and expansion of seawater volume, inundating many coastal areas worldwide. NH_3 gas is also toxic for the ecosystem, especially for the ozone layer. The culprit gases may come from fossil fuel use, deforestation, and industrial and domestic effluents, etc. Frequent flooding of coastal areas causes loss of swamps and marshlands, mudslides, loss of farmlands, properties, etc. Especially, flooding, salination, and soil deterioration may uproot the livelihood of millions of people. The tropical, topographically low islands and countries—Maldives, Bangladesh, and many others, may face existential problems, triggering massive human migration. The highest elevation in Maldives is only two meters. Many climate-refugees are fleeing from developing nations even today.

The Arctic Sea and Greenland will be even more affected by global-warming gases. The decadal loss of sea ice has been staggering. The melting of Antarctic Thwaites or 'doomsday glacier' may lead to substantial sea level rise and a catastrophe of global proportion. The unusual warming may release methane hydrates from high latitude permafrost and deep seas, worsening the situation. Environmentally, methane is 30 times more damaging than carbon dioxide.

Sea Level Fall, Nothing New—Testimony: A Perched, 150–200-m-high Paleo-Coastline

At the peak of the Last Glacial Maximum, glaciers covered >25% of Earth's surface (only 10% today), as the sea level fell drastically. The sea level rose after glaciers melted and the 13–9-ka-old Champlain and Tyrell seas of North America inundated elevated inland areas up to 300 km from the modern Atlantic coast. Raised parallel beaches by the ancient Tyrell Sea more than 100 kilometers from the present shorelines is evidence of the receding sea, due to isostatic rebound of land. Likewise, the ancient shoreline of the Champlain Sea, 150–200 m above the present sea level, was preserved as the sea level fell. Many marine fossils including Beluga and other whales remain buried in the mud of the ancient Champlain Sea in eastern Ontario and Quebec, Canada, and Vermont, New York, U.S.A.

The Mississippi River advanced nearly 200 km, deep into the Gulf of Mexico forming a new shoreline and a shelf-edge delta, when sea level fell 125 m due to the 20-ka-old Last Glacial Maximum. The next icehouse following the present balmy interglacial period may pose similar challenges for humankind in the future.

If Sea Level Drops in Future, What Will Happen?

Sea level fall may cause some other drastic changes in the future. A visible effect will be the emergence of a large swath of marine shelf as dry land in the process isolating some bay, lagoon or estuary. Once the restricted sea becomes cut off from the main ocean during sea level drop, it may become a lake; it may eventually dry up, and precipitate thick salt deposits. The Late Miocene Mediterranean Sea went dry and precipitated a 3-km-thick salt column. Similarly, the stark Danakil depression, Ethiopia, is now a hot, dry, salty expanse, though lying 150 m below modern sea level. Earlier, the Red Sea rift basin between Africa and the Arabian Peninsula, became a dry salt bed upon isolation from the Gulf of Aden.[2] The modern Lake Nicaragua with frolicking sharks, once an arm of the Pacific Ocean, morphed into a vast lake because of tectonic uplift, and lava flows.

Post-Glacial, On-going Isostatic Rebound—May Risk the existence of the Niagara Falls

The weight of large ice sheets depresses land surface that slowly bounces back later with evanescing glaciers. Scientists fear that some famous natural wonders like the Niagara Falls, may or may not exist for too long. The ongoing isostatic rebound of land may erase the American Niagara Falls in the next 2 ka or so. Another scenario is that the Canadian Horseshoe Falls may erode back and morph into a river with rapids in about 15 ka.[3]

What Happens After the Current Interglacial? Deep Freeze or a Greenhouse?

We are now enjoying a congenial interglacial spell within the icehouse. It will be transitory, until the onset of the next icy spell. A Quaternary glacial stage may last about 80,000 years in a 100,000-year glacial-interglacial cycle. The past average interglacial stage lasted about 10–20 ka. So, the present 11.7-ka-old interglacial may continue for several thousand years more. However, short chills like the historic 'Little Ice Age' (~1300–1850 CE) of Europe and North America may visit us again.

Some authors paint a grim apocalyptic scenario, in case of a future ice age.[4] The boreal North America, Europe and Asia will be buried under ice, leaving minimal agricultural land for subsistence. An ice age dawns when summer temperature in the northern hemisphere stays below freezing point for many years. However, a dubious silver lining is that anthropogenic global warming may slightly prolong the current balmy spell. This may open a Pandora's box, full of intractable issues.

The Quaternary icehouse of rhythmic glacial-interglacial cycles may persist indefinitely as in the past ~2.58 Ma. In Earth's 4.57-Ga history, icehouse climate had occurred much less (~25%) than the greenhouse (~75%). The icy chill will depend on the intensity of the Sun's incident rays on the planet. Earth's axial tilt angle, and variable orbital path, are crucial and may conspire to initiate a new icehouse cycle.

However, an icehouse condition does not last forever either. Changing external and earthly conditions may eventually reinvite the next interglacial, or better still a warmer, long-term, greenhouse stage like before. Earth's previous greenhouse stage lasted from the Triassic to End-Eocene—226 Ma (260–34 Ma), when global warming gases—CO_2, CH_4, N_2O ('laughing gas') were abundant. The gases enveloped the globe keeping it warm and cozy. The long warm climate triggered an explosion of life when the poles were free of snow. Other causes of greenhouse effect are volcanic activity, solar activity, Earth's shorter orbit path and higher axial rotation angle, etc. The past greenhouses coincided with abnormal CO_2 values—up to 20 times higher than today's 420 ppm. The preindustrial age CO_2 value was 300 ppm.[5]

In summary, the current icehouse-interglacial cycle may persist 100,000 years or so, though no one knows for sure. Alternatively, Earth may bask in a prolonged greenhouse climate after the present interglacial.

SHORT TERM NATURAL CALAMITIES—THE LIKELIHOODS

A Supervolcano Eruption—An Existential Threat or Hype?

A supervolcano registers the maximum volcanic explosivity index.

NASA reports about 50 supervolcanoes on Earth, any of which can seriously impact the living world by belching out >1000 km^3 of lava, gas, and pyroclastic debris. To date, the worst case is the Toba crater lake volcano in Sumatra that erupted 2,800 km^3 of lava 74,000 years ago, causing the greatest mass extinction in recent history. It is a more likely threat than any wayward cosmic visitor.

Indeed, a supervolcano eruption is a more real threat than an asteroid impact for some. The exhaled gas from the catastrophic eruption will induce a volcanic winter and cause collapse of agriculture, leading to famine and countless deaths. One of the last supervolcano eruptions happened in New Zealand 26,500 years ago. NASA estimated the supervolcano disaster frequency to be 1.4 eruption in a million years.[6] However, it is simply a statistical probability. It may happen sooner than the New Zealand event, or not in a million years to the relief of the living world. No doubt, it will happen in the geological very short-term as we recount below.

Mount Tambora's Fury, Indonesia, 1815—The Worst Eruption in Recorded History

Mount Tambora on the Pacific Ring of Fire, created a global catastrophe, though it was not ranked as a supervolcano. The 4,000-m-high stratovolcano on Sumbawa Island (between Bali and Flores) was grumbling from 1812 onward, till it unraveled on April 10, 1815 (Fig. 13.2). To compare historically, the Tambora event was 10 times more powerful than the well-known 1883 Krakatoa, the recent 1991 Pinatu-

13-2 Aerial view of Mt. Tambora caldera (6-km diameter) on the Sumbawa Island, Indonesia, 2011.

bo eruption, or the Vesuvius disaster of Pompeii fame, Italy (79 CE). The apocalyptic event expelled an estimated 150km^3 of tephra (ash, lapilli, bombs) and many megatons of sulfur aerosols into the stratosphere, the ash reaching 1,300 km away.[18]

Torrents of lava cascading down from the summit buried foothills villages and killed nearly 10,000 people on the island. Massive tsunamis magnified the damage. The aerosols traveled far and wide and created a volcanic chill, weather extremes, harvest loss, and a spike in crime for several years in the area, and in distant North America, Europe and elsewhere. The year 1816 was appropriately dubbed as the 'year without a summer' in the West. The post-eruption crop loss, famine, and epidemic diseases killed many tens of thousands of people in the affected area. Imagine the effect of such an eruption today. The one billion world population of 1800 CE is now eight-fold in 2024.

Lastly, of note the Tambora eruption also helped to create a literary masterpiece that regaled readers ever since. The 19-year-old Mary Shelley (wife of poet P. B. Shelley) wrote her memorable novel *Frankenstein* during the dismal, gloomy, rainy, and frosty Post-Tambora days in Geneva.

Geomagnetic Polarity Shifts Are Frequent—Should We Worry?

The magnetic poles flip every 200–300 ka, so pole reversal is a given for the future Earth. The *Homo* genus witnessed the late Stone age geomagnetic reversal event 780 ka ago when the process continued for many thousand years. Humans will witness the North Pole literally becoming the South Pole in a few hundred thousand years or sooner. Field geologists, if still in business that far into the future, will have to get used to the new reality—the compass needle pointing south. During reversal, geomagnetic fields weaken, allowing harmful cosmic rays to bombard our planet. The spectacular *Aurora borealis* display is a colorful evidence of Earth's magnetic field. The kaleidoscopic northern light proves Earth is valiantly combating the injurious charged particles of solar winds and cosmic rays in the high magnetosphere. Cosmic rays are currently affecting the Antarctic ozone layer and have a harmful effect on the whole biosphere.

Need any proof of past magnetic reversal? High ^{14}C isotope content in a fossilized 1,600-year-old Kauri tree (diameter >2.5m), New Zealand, confirmed that the 42,000-yr-old pole reversal would have impacted *Neanderthal* and megafauna extinction.[7] The Atlantic seafloor basalts also bear testimony to numerous magnetic reversal stripes (Fig. 4.2). The weakening of Earth's magnetic strength for the last 200 years is likely foreshadowing the onset of a future reversal event. Magnetic instability compromises satellite communication and electric grids. It disorients animal, bird, and aquatic life that depend on the geomagnetic field for navigation. The harmful radiation may damage cells and cause cancer.

Space Intruders in the Atmosphere—How Often Do They Visit?

Any out-of-the-blue, unwelcome celestial visitors striking Earth have a chance of occurring, though unlikely soon. In 1908, the Tunguska meteor impact in an uninhabited region of Russia, caused 2,000 km^2 of forest land to burn down. The Feb 15, 2013, Chelyabinsk meteor or super-bolide explosion in the Siberian sky is another instance, suggesting that a cosmic stranger may not be just a science-fiction theme line. A total of 1,491 non-fatal injuries and more than 7,200 damaged buildings was the toll of that meteor that didn't even hit

13-3 The sky above Siberia lit up in the glowing trail of the Chelyabinsk meteor that exploded in air, causing over a thousand non-fatal injuries, and damaging 7,200 buildings.

the ground, as it exploded in the sky (Fig. 13.3). NASA has flagged[8] a near-Earth asteroid *Apophis* that has a remote chance of hitting Earth many years from now, though probably it will pass by harmlessly, far above Earth in 2029.

Asteroid fragments enter Earth's orbit regularly as meteor, varying in size from a basketball to much larger. On December 2, 2020, a fireball (0.8–0.9-m-across) trailed through the midday sky, about 35 km directly above Syracuse and Rochester, upstate New York. Hurtling down at an incredible speed, it broke into pieces in the air creating a blinding flash over Lake Ontario. The sonic boom was audible over New York State, and adjacent Ontario, Canada. Cameras atop Toronto's 560-m-high CN Tower captured the flash of the meteor burst.

NASA gives some interesting statistics about space bodies showing up in Earth's atmosphere or even hitting it.[9] Our planet receives over 100 tons of sand-sized particles every day. The meteorites, usually ~25-m in size, burn down upon entering Earth's atmosphere. A football-field-size asteroid may inflict serious damage every couple of thousand years. Rarely, a large enough asteroid may sneak in once every few million years and cause an existential threat on Earth.[9] Well, the 1.4-Ma-old, Pleistocene Pingualuit meteorite was large enough to punch a circular crater hole with the force of 8,500 nuclear bombs.[10] The crater is now a lake, 3.5-km-across and 400-m-deep in northern Quebec, Canada. Our ancestor *Homo erectus* would have felt the reverberations everywhere. The pink Lonar crater-lake in western India (1.8-km-across, 290-m-deep) is the product of a more recent ~50-ka-old, meteorite impact. The 1.2-km-across Berringer Crater, Arizona, USA, is of a similar vintage.

The possibility of a cosmic intruder remains, even though slim. The chance of a 5-km-across asteroid impact is 6 Ma, while that of a km-size asteroid is 500,000 years.[11]

Earth's Incessant Move—
Plate Tectonic Acrobatics Will Continue

The long-term changes of tens to hundreds of millions of years are closely related to plate tectonic movements, causing reconfiguration of Earth's continents and oceans. Earth is like a living entity, always mobile, though imperceptible visually. Our planet (circumference ~40,000 km) rotates on its axis every 24 hours, at a speed of 460 m/sec. Do we perceive mobility while merrily riding on it? The crustal plates are likewise always on the go because of the inner workings within the deep mantle. The data from some crustal plates give us a sense of their respective mobility rate per year—Pacific subduction–8 cm, Mid-Atlantic spreading–2 cm, India's push into Asia's belly–3.7 cm, and Australia's headlong rush toward Asia–5.6 cm. The rates may not seem much but are geologically significant.

Indeed, let us review how some on-going large-scale geological activities will evolve in the future. The Pacific Plate and the ocean above are slowly subducting, i.e., disappearing, beneath the Americas and Asia. The plate may lose 800 km of crust in 10 million years, as it dives beneath the continents. Eventually, the Americas and Asia will become one landmass, as the Pacific plate disappears in the mantle.

In another scenario, the Atlantic Sea floor, spreading laterally since the Jurassic, has been moving apart the Americas and Europe/Africa. It will likely push aside the continents an additional ~250 km in the next 10 Ma (2.5 cm/year). Even tiny Iceland is literally splitting apart because the Atlantic mid-oceanic rift passes right through it. However, this is only part of the story. As it happens with plates, one or both of the drifting landmasses (Americas or Africa/Europe) will change course, and eventually morph into colliding plate(s). Both plates will cease to be idle drifters, they will approach, collide, and the Atlantic Ocean will slowly disappear, as had happened earlier. The Paleozoic Iapetus Ocean, in the same location as the Atlantic, had literally disappeared due to colliding plates. The Jurassic opening of the Atlantic Ocean pushed aside the deformed rocks of the vanished Iapetus Ocean. The outcropping Paleozoic rocks in eastern Canada and Spain/Portugal, are testaments of the vanished Iapetus Ocean.

A New Supercontinent in the Works?

India is an integral part of Eurasia for the last 40 Ma, after the fateful collision that obliterated the intervening Tethys Ocean. Australia is sailing northward and currently approaching the east Asian plate margin. In about 50–60 Ma, Australia may smash against the Asian Plate. The encounter will create an Everest-size summit in eastern Asia, and Australia becomes part of the America-Eurasia continent, i.e., 'Amasia.' In the process, the Indian Ocean will disappear, like the Tethys earlier. A new Pangea-like successor supercontinent may debut someday, fulfilling the tenets of the famous 'Wilson Cycle.' It is unclear how the Antarctic plate may fit into this scenario.

A new study by several experts in 2023 provided the clue as to how a new supercontinent may come into being, Antarctica included, in about 250 Ma from now.[19] It will be a uniform supercontinental landmass completely surrounding a land-locked small ocean. By that time or long before, run-away heat will cause total mammal extinction on Earth.

The End of *Homo Sapiens*—When It Will Happen?

Scientists accept the Copernican principle that we are neither unique nor special. We are like all other species, extinct and extant. Darwin thought that extinction is a natural process. We will also become extinct someday in the future, as new species will emerge. However, *H. sapiens* is the only species that can tweak its own destiny to a degree to prolong its stay on Earth. Unlike other animals, humans will never passively submit to overwhelming natural calamities. They will utilize ingenuity and resources to their advantage and prolong the journey in the face of challenging situations. The opposite is also true. Humans may hasten their own demise if they adopt a 'golden goose' mentality and fail to take care of the unique earthly home and ecosystem.

Many experts have debated about the longevity of humankind, some even thinking the end will be soon in a few centuries to many thousand years. The potential natural agents of human extinction are—supervolcano eruption, climate change, pandemics, asteroid/comet, solar flare, stellar explosion (gamma-ray bursts, supernovae), etc. Prof. J. Richard Gott calculated *Home sapiens* may live 8 million years at the 95% confidence level.[12] A 2019 study estimated an extinction probability of <1 in 870,000.[11] However, an inherent problem

13-4 A sprightly sunflower, Riverside Park, Mississauga, Canada.

besets all predictions. Experts normally extrapolate the likelihood of a future event based on past examples. But human extinction had never happened before. The fossil record shows that an average mammalian species goes extinct in about 1–2 million years. Our longest-lived ancestor *Homo erectus* valiantly braved physical challenges for 1.8 Ma, yet ultimately succumbed. The *H. sapiens neanderthalensis*, our fellow traveler for most of our journey, trudged about 400 ka, before giving in. During the last 42-ka-old polarity reversal, the process lasted only 800 years, an unusually short time for polarity reversal, yet it proved lethal for Neanderthals and megafaunas. Earth's geomagnetic field weakens to about one-tenth of its full-strength during polarity transition and fails to combat the harmful solar cosmic rays from entering Earth.[13] Humans may face the risk of extinction sooner because the average polarity transition time is only ~7000 years.

In earlier chapters, we reviewed that several mundane or cosmic factors, especially climate change, might have played a role in species collapse. Climate change was a crucial factor in four mass extinctions in geologic history—Ordovician, Permian, Triassic, and Cretaceous. It didn't matter if the main cause be cataclysmic volcanic eruption, an asteroid impact, or of any other origin, all compromised the reigning climate equilibrium.

The Holocene climate change caused widespread demise of pre-historic or later cultures.[14] The Sumer, and Hittite civilizations, and probably one Egyptian kingdom perished due to climatic down-

13-5 Photosynthesis—Plants and other organisms synthesize food from CO_2 and H_2O, aided by light, i.e., photo from the sun. The byproduct of pure oxygen is a gift for all life on Earth. The huge leaf plant (>1.5-m-across) grows in the cool tropical rainforest of Costa Rica by the Poás volcano.

turn. The abnormal monsoon and droughts in Angkor, Cambodia, led to its tragic end. The recurring droughts caused decline of the Indus Valley culture around 2000 BCE, Asia, and the Mesoamerican Mayan society in the 8th–9th century. The Pacific Easter Island (Rapa Nui) culture withered because of prolonged drought, population pressure, and total forest destruction during the 'Little Ice Age.'

THE END OF EARTH? MAYBE IN A FEW BILLION YEARS

What if Earth's Core Geodynamo Stops Working?

Earth's liquid outer core is vital, now protecting its unique atmosphere from the powerful solar winds and shower of charged particles. If ever the liquid core becomes solid through heat loss, Earth will face the same fate as Mars did 3.5 Ga ago, i.e., it will lose its precious magnetic field and the celestial colorful *Aurora borealis*, and *A. australis*. Indeed, the 2022 research data reveals that Earth's core may be cooling down slowly.[15] However, the doomsday scenario is unlikely to happen for about a billion years or so.

What If the Mundane Photosynthesis Factory Turns Off?

Indeed, what does the crystal ball predict about our Sun in the blurry billion-year territory? Above all else, the Sun holds the key to Earth's very existence, its well-being, and even in keeping the entire solar system ticking in the long-term. The vital heat and light energy for the vibrant biosphere emanates from the Sun. The source of energy are simply two elements—hydrogen 70% and helium 28%. Four hydrogen atoms undergo nuclear fusion in the Sun's core producing a helium atom and releasing energy. Our star has been burning its hydrogen fuel for the past 5 Ga to light up the solar system.

The entire earthly food chain enterprise runs on photosynthesis, where atmospheric carbon dioxide is converted into sugar and oxygen, mediated by the Sun's precious light, and water from the soil (Fig. 13.5).

What Else May Spell Doom for Earth?

The first remote scenario involves a distant galactic collision. NASA's powerful JWST has already captured similar galactic encounters. Our Milky Way Galaxy and closest neighbor Andromeda will be colliding someday—the D-Day being ~4 Ga from now. Sadly, no earthlings will be invited to enjoy the spectacular show (Fig. 13.6). The Sun's natural increase in heat and luminosity in about a billion years from now, may slowly vaporize all our oceans, and all life, if it still exists, will end.[16]

What happens if the collision scenario proves to be a false alarm and Gaia survives? By that time, the ~10-Ga-old Sol may run out of its precious hydrogen fuel. It will implode because hydrogen maintains its internal thermal pressure to combat the immense pull of gravity. It will likely collapse and contract inward, unleashing helium fusion at the core, with massive expansion of gases in the outer layer. The stony planets—Mercury, Venus, Earth, and Mars will become absorbed in the new avatar of the Sun—a red giant, and finally a white dwarf.[17] It will just vaporize, or be blown off into smithereens, spilling all primordial elements in the vast interstellar space. Enough raw ingredients for a new cosmic entity—a future sun, and a set of obedient offspring planets and satellites. It is back to square one, the cycle begins anew.

13-6 Illustration of the galactic proximity of the Milky Way and Andromeda (left). The latter's tidal pull affects the Milky Way image.

"Who divines the last word, who dreams of the end ...?
From the olden heart, the new sprouts again.
Life of a bud, one day soon be over.
A fruit will show up, behind the wilted flower." *

a translated song of **Rabindranath Tagore**
India's first Nobel Laureate

*Rabindranath Tagore (1861-1941), Song "Shesh nahi je....": Gita Bitan (song anthology vol.), Song 606 (Gitalekha), p.238, Visva-Bharati University Publisher, 1995. Public Domain. Translated by Santosh Ghosh.

CHAPTER HIGHLIGHTS

- Analysis of Earth's future under varied scenarios and time frame.
- A brief recount of Earth story, highlighting geologic events through time, some of them likely recurring in Earth's future.
- Earth's fate and future depending largely on our species, and on natural events—super-volcano eruption, geomagnetic polarity changes, and unexpected cosmic visitor, etc.
- Earth's ultimate climate driver is Nature.
- Global warming and sea level rise, creating existential problems for humanity.
- A future icehouse stage is likely to severely impact all life.
- Future active plate tectonic movements likely causing—disappearance of oceans, creation of a supercontinent or rise of new mountains.
- Expert opinions on the longevity of humankind on Earth.
- Finally, future scenarios of the eventual demise of Planet Earth.

The "Catedral del Sal" in the ~200-m-deep, Zipaquirá salt mine, is a famous tourist site in Colombia, in the Sábana de Bogota. The 70-Ma-old halite salt is the dried-up residue of an inland sea. The Andean orogeny uplifted the 2650-m-high Zipaquirá region from the Cretaceous Sea. **1** The Virgin Guasa with laminated salt backdrop. 'Guasa,' is the miner's angel, the name comes from agua (water) + sal (salt) **2** The 'Cardinal Cross' sculpted on bedded salt **3** The deep, underground Salt Cathedral and pews for Catholic church services.

The 2.5-meter-high Olmec colossal head (Monument 1) from 700–850 BCE, located in the La Venta Museum Park, Villahermosa, Tabasco, Mexico. It is astonishing to consider how the basalt sculpture, weighing several tons, was transported over 80 kilometers from the nearest quarry at Cerro Cintepec in the Tuxtla Mountains.

"Science is advancing at a galloping pace today. We cannot even imagine the many new discoveries to happen in the coming decades about the cosmos, the solar system and many aspects of Earth. Stay tuned."

Santosh Ghosh

EPILOGUE I

Symphony of Earth

We have finished the multi-tome travelogue of Earth's ongoing, exciting journey through several billion years, full of interesting tidbits of earthly genre. Geology brings together all the loose physical strings to blend them into the unique choral symphony of the natural world. An old philosophical concept, '*Música universalis*' imagines celestial music created by movements of stars, planets, and satellites. Johannes Kepler also spoke of the 'music of the spheres'—rhythmic orchestra of predictable movements of orbiting celestial bodies.

Human music has three main pillars—melody, rhythm, harmony. We imagine *Melody* in nature to be what we perceive as the visage of Earth—the lofty snow-clad mountains under a blue sky, meandering rivers and undulating verdant plains, vast blue oceans lapping the coast where a pale beach beckons us. We enjoy worldly melody with all our senses. We picture Earth's external beauty as melody, the focal point of earthly music. Melody is the face of music, just as a band singer carries the melody. In human music, melody consists of high and low pitches. Similarly, Earth's melody consists of contrasting vistas and sounds. The low pitches of tweeting birds, soft rolling waves, and soothing, serene, placid landscape alternate with high pitches of catastrophe, thunderous disaster, darkness, and desolation.

Rhythm, the second element of music, is easier to grasp in Earth's case. The musical rhythm consists of repetition of notes separated by silence. The seasonal rhythmicity is baked into Earth's natural laws operating at multiple scales. We regularly see small scale rhythms of waves, tides, and seasons. The spring flowers make a meadow smile again, after the winter snow buried the vibrant landscape. The lush green summer foliage wanes as the fall finally sends a colorful farewell, before the winter chill again descends. In the living world, the monarch butterflies make the 4,800-km-long, transoceanic journey from summer breeding grounds in North America to the hibernating destination in central Mexico every year. Take another example—the emergence of a cicada species every 17 years as it coincided with the Spring of 2021. They emerge only when the ground temperature is 17.8°C. Their noisy mating orchestra lasts only for a few weeks before the females deposit their eggs, and then all return to dust again. A new brood of millions of cicadas will emerge exactly 17 years later. The rhythmicity is even more striking when we think of our planet's celestial journey. It revolves around its axis every 24 hours bringing days and nights unfailingly. It travels around the Sun, traversing one complete orbital journey in 365 days, and bringing the four seasons alternately in the two hemispheres. Rhythm is pervasive and omnipresent in nature as in music.

Harmony is the third pillar of music. It reveals when two or more instruments play simultaneously, it complements and builds upon the existing melody. Simply put, harmony is composite in nature. The harmony on Earth has many invisible moving parts which are working in unison to heighten the effect of the melody we observe in nature. Earth's core, mantle and crust, restless tectonic plates, ocean currents, to name a few, work ceaselessly to make it a benign goldilocks place of harmony where life can blossom and prosper.

Earth's all-important stratospheric ozone layer protects all living creatures from the harmful ultraviolet radiation of the Sun. The geomagnetic field, generated in Earth's deep, liquid outer core, neutralizes the dangerous charged particles of solar winds from the Sun's corona. The harmony of different levels of the atmosphere works in concert with Earth's geomagnetic field to play the unearthly visual melody of the *Aurora borealis/australis*. Geology is the master conductor of the unique symphony orchestra of Planet Earth.

EPILOGUE II

Facing the Music

The Symphony of Earth would have been an ideal ending of the book; but we need to mention a few inconvenient facts before we sign off—the effect of our collective actions or inactions vis-à-vis the blue planet. Some big problems are directly human-generated, others involve human impact on the geo-environment.

The direct human-centered problems are many—destruction of wildlife habitat and alarming biodiversity loss, population explosion, displacement of migrants from developing nations, the ever-increasing number of farmed animals to feed growing population, etc. Our propensity to tinker with nature, and consume exotic animals, may unleash a future pandemic worse than the one of 2020–2022. Our numbers ballooned, while wildlife and forest cover dwindled alarmingly.

Also, the latent danger of a possible nuclear confrontation is real. The humongous number of nuclear warheads (12,500 as of early 2023, data—*Federation of American Scientists*) already stockpiled by various countries has the capacity to destroy the world many times over. It may create a nuclear winter, widespread fire, total crop loss and famine, and a human toll in billions. In the case of nuclear war in this century, the survivors will think the dead fared better.[6]

Another looming problem is *H. sapiens*' brainchild 'AI or artificial intelligence,' based on computer algorithms and a large amount of data. AI applications may be both salutary and risky. A computer machine learns to do the same tasks that we do habitually, if empowered with artificial intelligence. The potential benefit of AI is limitless, while its dark side is alarming. A real danger is that AI may prove to be like the fictional Victor Frankenstein's creation—a nameless monster[1] who played god. Many experts are warning us that AI is no substitute for truth, it needs strict regulations by appropriate agencies. Many AI company executives and researchers (350 in total) warned heads of states that AI poses an extinction risk same as a global pandemic or nuclear war. The AI generated language models can spread biased half-truths or worse, creating a narrative of false news and views for political or other purposes. Similarly, vision models can generate lifelike images or videos of imaginary events. Some of the interpersonal or international conflicts are now moving from the traditional arenas to the dark corners of the internet where its tentacles are everywhere. Cyber-attacks involving ransomware can interrupt our power grid, water supply, educational system, travel industry, medical facilities, etc.

'Deepfake,' a troublesome recent phenomenon, uses artificial intelligence tools to alter visual or audio content of a video, where unsuspecting viewers may not perceive it as fake. One normally does this by digitally altering the face or body of someone to make him/her appear like someone else, any famous person, for example. In the movie industry, stuntpersons routinely impersonate main actors in violent movie scenes. However, deepfake is more serious than that. It can generate fake news, pernicious videos, hoaxes, and different types of frauds.[2]

As far back as July 2017, the University of Washington researchers showcased the potential deceptive power of the technology; they produced a convincing fake portrayal of President Obama giving a speech, applying artificial intelligence. They took one of his years-old audio speeches and successfully lip-synced it to his more recent video giving a speech in the White House.[3] The deepfake technology and refined deep learning can create serious headaches at all levels. As of March 2024, a disturbing deepfake video of the Italian President was circulating on the internet. The deepfake cases are more commonplace in 2023/2024 than before.

Changing the topic, another menace may be biological/chemical weapons that some countries unleash to punish enemies. We know of chemical weapon attacks in the past used by some nations against adversaries. Many state-run biological research labs exist to fight epidemics and pandemics. What if one of these labs were to leak deadly microbes or toxins accidentally or intentionally unleashing horror and havoc in future?

Two other human inventions with lots of promise but also serious potential downside are nanotechnology and solar geoengineering to combat droughts. The former has many beneficial applications like energy generation and diagnostic care and treatment; yet nano particles, being nano (1 nanometer = one-billionth/meter) may enter the blood stream causing vital organ problems. In solar geoengineering, injected aerosols in the stratosphere may reflect some of the sunlight and heat back into space and thereby making Earth cooler. But it may open a Pandora's box of destabilized global climate, ecosystem, etc. (information from Global Challenges Foundation).

Now let us focus attention on how human activities are affecting Earth's geo-environment, i.e., air, water, land, and the vibrant ecosystem. We are depleting nature fast, destabilizing the natural equilibrium that has always reigned in Earth's integrated complex environment. Is the vast Amazon rainforest, holding the precious oxygen reservoir on Earth, gradually losing the battle?

The world population will be about 11 billion by the year 2100, thus needing more cattle, poultry, fertilizer for agriculture, etc., to meet the needs of this increased population. These may generate more greenhouse gases and potential contamination through manure, animal wastes and gases. Lately, some emerging economies are aggressively burning fossil fuels to run their economic engines.

The present global warming trend from high atmospheric CO_2 and CH_4 is having many unusual repercussions—frequent flooding, sea level rise in coastal areas, abnormal heat and drought, wildfires, mudslides, hurricanes, and untimely tornadoes, among other natural events. N_2O and NH_3 are also potential GHG gases. Global warming mostly happens in the troposphere[4] (Figure below). As greenhouse gases go up unchecked and raise ambient temperature to a critical level, both sea level rise and farmland deterioration will uproot the livelihood of millions of people.

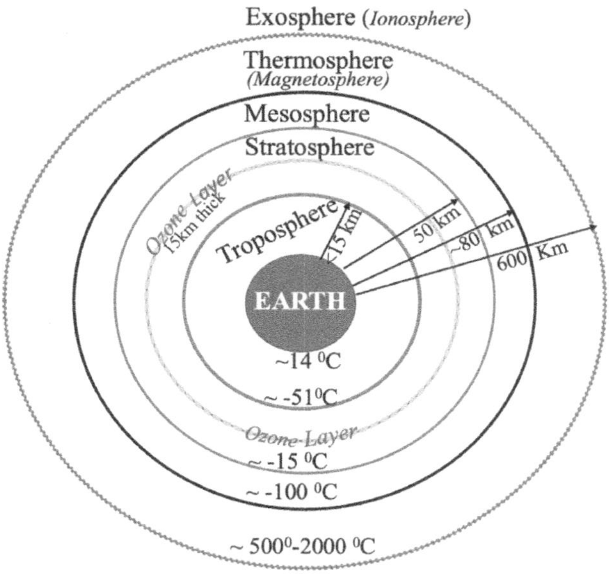

Earth's layered atmosphere (Not to scale). Temperature or distance range approximate. Upper stratosphere hotter because of ozone formation. The Aurora happens in the Magnetosphere which is part of the red-hot Thermosphere. The Exosphere includes the Ionosphere and extends up to 10,000 km in space.

The staggering, steady loss of sea ice is significantly affecting global wind patterns. Global warming is particularly taking an unprecedented toll on the Arctic Sea ice and Greenland. At the same time, many tropical island nations, and topographically low regions may face existential problems. The unusual warming will also release copious toxic methane, locked up in high latitude permafrost and in deep seas as hydrates, further worsening the situation.

Carbon dioxide and methane emissions aside, other urgent environmental problems are also staring us in the face. Clean water supply, waste management, natural hazard prediction and prevention, an eco-friendly way of mineral resource exploitation, etc., are some immediate issues. The continent-size plastic garbage piles like the GPGP or 'Great Pacific Garbage Patch' in distant oceans, the countless overly polluted rivers, and the over-use of fossil fuels may irreversibly alter our ecosystem. In particular, the disastrous level of plastic pollution requires concerted remedial efforts by public and private sectors alike.[5] Hopefully, our collective sanity will prevail, and we will be prudent enough to protect our only 'home.' Planet Earth is resilient and has the capacity to heal itself if we allow it to have a fighting chance. Let us strive so that the Earth-human relationship does not turn into a zero-sum game.

[1]Mary Shelley, 1818—"Frankenstein; or the Modern Prometheus", https://en.wikipedia.org/wiki/Frankenstein

[2]Wikipedia, Deepfake, https://en.wikipedia.org/wiki/Deepfake

[3]J Langston, July 11, 2017, Lip-syncing Obama: New tools turn audio clips into realistic video: UW News, https://www.washington.edu/news/2017/07/11/lip-syncing-obama-new-tools-turn-audio-clips-into-realistic-video/

[4]NASA, Jan 21, 2013, Earth's atmospheric layers, https://www.nasa.gov/image-article/earths-atmospheric-layers-3/

[5]AGI, Sept 22, 2020, AGI releases critical needs document for 2020: https://www.american-geo-sciences.org/news/agi-releases-critical-needs-document-2020

[6]Sean Illing, June 23, 2024, 'What nuclear annihilation could look like': podcast (interview of Annie Jacobsen, the writer of 'Nuclear War A Scenario' Mar 26, 2024, 361p, Penguin Random House LLC).

Afterword

It has been my privilege to review the book *Deep Time: Earth's Geologic Journey from Eternity to Homo Sapiens* by Dr. Santosh Ghosh. I truly enjoyed his engaging narrative that combines precise technical descriptions with depictions and anecdotes that make the topic under discussion more enticing to the readers, irrespective of their degree of command of the Earth Sciences. His chronicles intertwine scientific facts with storytelling, providing a sense of awe at the complexity and beauty of Earth's geologic history.

Deep Time takes us on a journey that starts at the birth of the universe 13.8 billion years ago and parades us through the formation of the Solar System and the emergence of Earth as a unique physical entity; then it leads us through the main geological events, i.e., the inception of plate tectonics and the evolution of the continents. It reveals the secrets of the origin and changes of life until the ascent of humankind at the top, and how they interrelate. All in all, it gives the reader a sense of the immensity of geologic time and how difficult it is for us to truly understand its limits and its transformative impact on the planet. Dr. Ghosh also speculates what the future holds for Earth as a geological being.

In *Deep Time*, Dr. Ghosh not only explores geological events, but also examines the reason why they happened and by doing so, he offers explanations on how the changing landscapes of Earth, climate and life forms are intertwined and influence one another. He describes life evolution and mass extinctions in light of plate tectonics and climate evolution. He traces the development of early life forms, the emergence of complex organisms later, and the eventual rise of *Homo sapiens*. This general approach provides readers with a deeper understanding of how interconnected Earth's history is with the story of life itself.

Dr. Ghosh also addresses the impact of human activity on the planet making special emphasis on the importance of understanding our geological past to better set a course for our future. His insights into the proposed Anthropocene, the current geological epoch characterized by significant human influence, are particularly thought-provoking and relevant.

As a geologist, I was completely drawn by *Deep Time* but you don't have to be an expert to access the information it offers and enjoy the narrative of Dr. Ghosh. Both casual non-specialist readers, as well as more serious earth science researchers will find valuable insights into the geohistory of Earth, life evolution and how human beings fit into the larger historical context of our planet, and ultimately, of the universe.

Bernabe Aguado
Former Senior Research Geoscientist,
Saudi Aramco, Petronas Carigali & Petróleos de Venezuela
M.Sc., *University of Wyoming*

Acknowledgments

I thank Dr. Robert Hickman, Andrew Cross, Sarmistha De, Dr. Maria A. Lorente, and Dr. Carlos Molinares, as they read the manuscript, partly or otherwise, and made useful suggestions. I wish to single out my former colleague Bernabe Aguado, for his careful technical editing of the entire book and numerous constructive suggestions. Dr. Suhas Talukdar provided useful comments on Chapter 1. Dr. Abhijit Basu – Herman B. Wells Professor of Sedimentary Geology, Indiana University, Bloomington, USA, read the manuscript and offered valuable inputs and editorial suggestions.

My son Robin helped me throughout the process, including while writing the Epilogue—I. Robin did the excellent cover design, and final editing and formatting of the book for the publication platform. My wife Nina kept me grounded, and patiently let me pursue my seemingly endless, more than 5-year-long effort.

I thank Upendra Pai, Ritwik Ray, Arnav Adhikari, Saurav Adhikari, Marianto Castro, Drs. Omar Colmenares, Ani Adhikari, for helpful suggestions.

Finally, I thank all my friends, ex-colleagues, and students in different continents for their support and camaraderie throughout my working years.

Chapter 1

1-2: https://commons.wikimedia.org/wiki/File:WMAP_2012.png. *Attribution:* NASA/WMAP Science Team, Public domain, via Wikimedia Commons.

1-3 left: https://commons.wikimedia.org/wiki/File:NASA%E2%80%99s_Webb_Reveals_Cosmic_Cliffs,_Glittering_Landscape_of_Star_Birth.jpg *Attribution:* Image: National Aeronautics and Space Administration (a U.S. federal government agency; https:www.nasa.gov), European Space Agency (https://www.esa.int/), Canadian Space Agency (https://www.asc-csa.gc.ca), Space Telescope Science Institute (https://www.stsci.edu/, science operations center for the Hubble Space Telescope), Public domain, via Wikimedia Commons.

1-3 right: https://commons.wikimedia.org/wiki/File:Crab_Nebula_imaged_using_James_Webb_Space_Telescope.png *Attribution:* NASA, ESA, CSA, STScl, T. Temim (Princeton University), CC BY 2.0, via Wikimedia Commons.

1-4: https://commons.wikimedia.org/wiki/File:Solar_system.jpg *Attribution:* NASA/JPL, public domain, via Wikimedia Commons.

1-5: https://commons.wikimedia.org/wiki/File: Periodic table.jpg *Attribution:* LeVanHan, CC BY-SA 3.0, via Wikimedia Commons.

1-6: https://commons.wikimedia.org/wiki/File:Corundum-27472.jpg *Attribution:* Rob Lavinsky, iRocks.com—CC-BY-SA 3.0, CC BY-SA 3.0, via Wikimedia Commons.

Chapter 2

2-2: https://commons.wikimedia.org/wiki/File:Moon_rock_2, JSC.jpg *Attribution:* Shannon Moore, CC BY 2.0, via Wikimedia Commons.

Chapter 3

3-8 img 5: Zipaquirá salt mine view.

3-12: https://commons.wikimedia.org/wiki/File:Hierve_el_Agua.jpg *Attribution:* Adrián Cerón, CC BY-SA 3.0, via Wikimedia Commons.

3-24: https://commons.wikimedia.org/wiki/File:Mars_(51175398512).png *Attribution:* Raziel Abulafia, CC0, via Wikimedia Commons.

Chapter 4

4-1: https://commons.wikimedia.org/wiki/File:Plates_tect2_en.svg *Attribution:* Map: USGS Description: Scott Nash, Public domain, via Wikimedia Commons.

4-3: https://commons.wikimedia.org/wiki/File:Plate-tectonics.png. *Attribution:* Pavel Bokr, CC BY-SA 4.0, via Wikimedia Commons.

4-4: Styled after https://commons.wikimedia.org/wiki/File:Plate_Tectonics_ru.png *Attribution:* Jose F. Vigil, USGS, Public domain, via Wikimedia Commons.

4-6 img 2: https://commons.wikimedia.org/wiki/File:Santorini_Caldera_ASTER.jpg *Attribution:* Santorini_Landsat.jpg: NASA derivative work: Giorgostr, Public domain, via Wikimedia Commons.

Chapter 5

5-3: https://commons.wikimedia.org/wiki/File:Angel_Falls_from_our_camp_-_Flickr_-_davidkjelkerud.jpg *Attribution:* David Kjelkerud from Stockholm, Sweden, CC BY-2.0, via Wikimedia Commons.

5-4: https://commons.wikimedia.org/wiki/File:Rodinia_reconstruction.ar.png *Attribution:* Original: John Goodge Derivative, Public domain, via Wikimedia Commons.

5-8: https://commons.wikimedia.org/wiki/File:Inzeria_tjomusi_(stromatolite)_(Neoproterozoic,_850_Ma_to_1_Ga;_Bashkiria,_Russia)_(41559659220).jpg *Attribution:* James St. John, CC BY 2.0, via Wikimedia Commons.

5.9: https://commons.wikimedia.org/wiki/File:Ediacaran_fossils_Mistaken_Point_Newfoundland.jpg *Attribution:* Alicejmichel, CC BY-SA 4.0, via Wikimedia Commons.

Chapter 6

6-3: https://commons.wikimedia.org/wiki/File:Pangaea_continents.png *Attribution:* en: User: Kieff, CC BY-SA 3.0, via Wikimedia Commons.

6.9: https://commons.wikimedia.org/wiki/File:Sydney_Mines_Point_Aconi_Seam_038.JPG *Attribution:* Rygel, M. C., CC BY- SA 3.0, via Wikimedia Commons.

Chapter 7

7-1: Simplified from https://commons.wikimedia.org/wiki/File:Outcrops_of_Newark_Supergroup_Lutrell_(1989)_USGS_Bulletin_1572.png *Attribution:* Gwendolyn Lewise Werth Lutrell, Public domain, via Wikimedia Commons.

7-7: https://commons.wikimedia.org/wiki/File:STS009_Manicouagan.jpg *Attribution:* National Aeronautics and Space Administration, Public domain, via Wikimedia Commons.

7-9: Skull of ichthyosaur fossil in Museo Minas Nemocon, Colombia.

Chapter 8

8-1: https://commons.wikimedia.org/wiki/File:Pangaea_to_present.gif *Attribution:* Kious, Jacquelyne, Tilling, Robert I., Kiger, Martha, Russel, Jane, Public domain, via Wikimedia Commons.

8-2: https://commons.wikimedia.org/wiki/File:Chicxulub-Anomaly-Grav-3.jpg *Attribution:* See page for author (unknown), Public domain, via Wikimedia Commons.

8-3: https://commons.wikimedia.org/wiki/File:Arnager-kalk-bornholm_11_hg.jpg *Attribution:* Hannes Grobe/AWI, CC BY 3.0, via Wikimedia Commons.

8-4: Global geologic temperature, modified from "Climate History"—ICEHOUSE or HOTHOUSE? Copyright © Christopher Scotese, 2020. https://www.scotese.com/climate.htm

8-6: https://commons.wikimedia.org/wiki/File:Bikini_2244493428_13643c6505_o.jpg *Attribution:* Peter Minton (EVS Island), CC BY 2.0, via Wikimedia Commons.

8-13 left: https://commons.wikimedia.org/wiki/File:Indroda_eggs.JPG *Attribution:* Sballal, CC BY- SA 3.0, via Wikimedia Commons.

8-13 right: https://commons.wikimedia.org/wiki/File:11-Figura4_Bernardo.jpg *Attribution:* Titan2155, CC BY-SA 4.0, via Wikimedia Commons.

8-14 img 1: https://commons.wikimedia.org/wiki/File:Sanajeh_fossil.png *Attribution:* Jeffrey A. Wilson, Dhananjay M. Mohabey, Shanan E. Peters, Jason J. Head, CC BY 2.5, via Wikimedia Commons.

8-14 img 2: https://commons.wikimedia.org/wiki/File:20170806_Bolivia_1268_Sucre_sRGB_(24128480908).jpg *Attribution:* Dan Lundberg, CC BY-SA 2.0, via Wikimedia Commons.

8-15: Marine Pliosaur reptile—110-Ma-old Kronosaurus, 10m long, with a missing right flipper and tail. Museo Villa de Leyva, near Bogota, Colombia.

8-16: Cretaceous ammonites on the entrance wall, Villa de Leyva Museum, Colombia.

Chapter 9

9-3: Simplified sketch from Fig. 1 of P. H. Mattson, 1984, Caribbean structural breaks and plate movements, p131-152. Geol. Soc. America Memoir 162, Copyright © GSA.

9-5: https://commons.wikimedia.org/wiki/File:Kanchenjunga_from_Sandakphu.jpg *Attribution:* Anirban Biswas, CC BY-SA 2.0, via Wikimedia Commons.

9-12 bottom: https://commons.wikimedia.org/wiki/File:Nummulites.jpg *Attribution:* Wilhelms, CC BY-SA 3.0, via Wikimedia Commons.

9-14: https://commons.wikimedia.org/wiki/File:Great_Artesian_Basin.png *Attribution:* Tentotwo, CC BY-SA 3.0, via Wikimedia Commons.

9-15: https://commons.wikimedia.org/wiki/File:Pacific_Ring_of_Fire.png *Attribution:* USGS, Public domain, via Wikimedia Commons.

9-16: https://commons.wikimedia.org/wiki/File:Wadi_al_hitan.jpg *Attribution:* Mohammed ali Moussa, CC BY-SA 3.0, via Wikimedia Commons.

Chapter 10

10-1: https://commons.wikimedia.org/wiki/File:Ocean_circulation_conveyor_belt.jpg *Attribution:* Thomas Splettstoesser, Public domain, via Wikimedia Commons.

10-2: https://commons.wikimedia.org/wiki/File:Panama_Topography.png *Attribution:* Sadalmelik, CC BY-SA 3.0, via Wikimedia Commons.

10-8: Simplified from USGS, https://commons.wikimedia.org/wiki/File:Tectonical_map_of_East_Africa.png *Attribution:* USGS, Public domain, via Wikimedia Commons.

10-12: https://commons.wikimedia.org/wiki/File:Qapaq_Nan_at_Mosollaqta_lake_DSC_2822.jpg *Attribution:* Aga Khan (IT), CC BY-SA 4.0, via Wikimedia Commons.

10-13 img 1: https://commons.wikimedia.org/wiki/File:Trapiche_emerald.jpg *Attribution:* Luciana Barbosa. CC BY-SA 3.0, via Wikimedia Commons.

10-15: https://commons.wikimedia.org/wiki/File:Amazonriverbasin_basemap.png *Attribution:* Kmusser, CC BY-SA 3.0, via Wikimedia Commons.

10-19: https://commons.wikimedia.org/wiki/File:ZagrosMtns_SaltDome_ISS012-E-18774.jpg *Attribution:* NASA, Public domain, via Wikimedia Commons.

10-21: Gomez-Cabrera and Jackson, 2009, AAPG Mem. 90. AAPG Copyright © 2009. Reprinted by permission of the AAPG whose permission is required for further use.

10-22: Chavez et al., 2009, AAPG Memoir 90. AAPG Copyright © 2009. Reprinted by permission of the AAPG whose permission is required for further use.

10-23: Left: https://commons.wikimedia.org/wiki/File:Topographic_map_of_Central_America.jpg *Attribution:* NASA/JPL/NIMA, Public domain, via Wikimedia Commons.

10-24: https://commons.wikimedia.org/wiki/File:Megalodon.jpg *Attribution:* Lonfat, Public domain, via Wikimedia Commons.

10-27: https://commons.wikimedia.org/wiki/Commons:Valued_image_candidates/Sahelanthropus_tchadensis_-_TM_266-01-060-1_Global_fond.jpg
Attribution: Didier Descouens, CC BY-SA 4.0, via Wikimedia Commons.

Chapter 11

11-1: https://commons.wikimedia.org/wiki/File:Sea_level_temp_140ky.gif *Attribution:* NOAA U.S., Public domain, via Wikimedia Commons.

11-2: https://commons.wikimedia.org/wiki/File:Glacial_Lake_Agassiz_and_Glacial_Lake_Ojibway,_7,900_BPE.png *Attribution:* See page for author (U.S. NOAA), Public domain, via Wikimedia Commons.

11-4: https://commons.wikimedia.org/wiki/File:Rama%27s_bridge.jpg *Attribution:* THK, CC BY-SA 3.0, via Wikimedia Commons.

11-5: https://commons.wikimedia.org/wiki/File:Danakil.jpg *Attribution:* sharanbhurke, CC BY-SA 4.0, via Wikimedia Commons.

11-7: https://commons.wikimedia.org/wiki/File:Georges_island.jpg *Attribution:* Thparkth, CC BY-SA 3.0, via Wikimedia Commons.

11-8: https://commons.wikimedia.org/wiki/File: New_York%27s_Finger_Lakes.jpg *Attribution:* NASA, Public Domain, via Wikimedia Commons.

11-9: https://commons.wikimedia.org/wiki/File:Oak_Ridges_Moraine_map.png *Attribution:* Norman Einstein, Public Domain, via Wikimedia Commons.

11-10: https://commons.wikimedia.org/wiki/File:Bigrock.jpg Attribution—Coaxial at English Wikipedia, CC BY 3.0, via Wikimedia Commons.

11-11: https://commons.wikimedia.org/wiki/File:ISS-30_Perito_Moreno_Glacier_near_Lake_Argentino,_Argentina.jpg *Attribution:* NASA, Public domain, via Wikimedia Commons.

11-13: Backward erosion data from Parks Canada falls area display.

11-16: https://commons.wikimedia.org/wiki/File:Lava_fountain_dome.jpg *Attribution:* J. B. Judd—USGS, Public domain, via Wikimedia Commons.

11-18: https://commons.wikimedia.org/wiki/File: Eje_Neovolc%C3%A1nico_Mexico.jpg *Attribution:* Soleincitta, CC BY-SA 3.0, via Wikimedia Commons.

11-20: https://commons.wikimedia.org/wiki/File:STAPP_106_La_Brea_Pitch_Lake.jpg *Attribution:* Grueslayer @Wikipedia, CC BY- SA 4.0, via Wikimedia Commons.

11-22: https://commons.wikimedia.org/wiki/File:Cosecha_de_sal_en_el_Salar_de_Uyuni.JPG *Attribution:* Alfredobi, CC BY-SA 3.0, via Wikimedia Commons.

11-24: https://commons.wikimedia.org/wiki/File:Miss_lobe.jpg *Attribution:* Patcharaporn Ch., CC BY 3.0, via Wikimedia Commons.

11-32: https://commons.wikimedia.org/wiki/File:GreatBarrierReef-EO.JPG *Attribution:* NASA, by MISR, Public domain, via Wikimedia Commons. Modified: partial view of original.

Chapter 12

12-2: https://commons.wikimedia.org/wiki/File:Australopithecus_sediba_and_Lucy.jpg *Attribution:* Profberger, CC SA-BY 3.0, via Wikimedia Commons.

12-3: https://commons.wikimedia.org/wiki/File:NHM_-_Laetoli_Fu%C3%9Fspuren.jpg *Attribution:* Wolfgang Sauber, CC BY-SA 4.0, via Wikimedia Commons.

12-4: https://commons.wikimedia.org/wiki/File:Homo_habilis_-_forensic_facial_reconstruction.png *Attribution:* Cicero Moraes, CC BY-SA 4.0, via Wikimedia Commons.

12-5: https://commons.wikimedia.org/wiki/File:Turkana_Boy.jpg *Attribution:* Claire Houck from New York City, USA, CC BY-SA 2.0, via Wikimedia Commons.

12-7: https://commons.wikimedia.org/wiki/File:Sapiens_neanderthal_comparison_en_blackbackground.png *Attribution:* hairymuseummatt (original photo), DrMikebaxter (derivative work), CC BY-SA 2.0, via Wikimedia Commons.

12-8: https://commons.wikimedia.org/wiki/File:The_archaeological_site_of_G%C3%B6bekli_Tepe_-_main_excavation_area_(rotated_270_deg).png *Attribution:* German Archaeological Institute, photo E. Kücük, CC BY 2.5, via Wikimedia Commons.

12-9: https://commons.wikimedia.org/wiki/File:Biface_Acheul%C3%A9en_Mus%C3%A9e_Saint-Remi_Reims_21092016_03.jpg *Attribution:* Vassil, CC0, via Wikimedia Commons.

12-11: https://commons.wikimedia.org/wiki/File:Chiribiquete_AJ11calabazos.JPG *Attribution:* Carlos Castaño Uribe, CC BY-SA 3.0, via Wikimedia Commons.

12-12: https://commons.wikimedia.org/wiki/File:Giantbirds.png *Attribution:* Matt Martyniuk, CC BY-SA 4.0, via Wikimedia Commons (modified - birds are numbered and legend modified to reflect numbering).

12-14: https://commons.wikimedia.org/wiki/File:Kasei_Valles_THEMIS_flow_features.jpg *Attribution:* NASA/JPL- Caltech/ Arizona State University, Public domain, via Wikimedia Commons.

Chapter 13

13-1: https://commons.wikimedia.org/wiki/File:Janus_.jpg *Attribution:* Tony Grist, CC0, via Wikimedia Commons.

13-2: https://commons.wikimedia.org/wiki/File:Caldera_Mt_Tambora_Sumbawa_Indonesia.jpg *Attribution:* Jialiang Gao (peace-on-earth.org), CC BY-SA 3.0, via Wikimedia Commons.

13-3: https://commons.wikimedia.org/wiki/File:2013_Chelyabinsk_meteor_trace.jpg *Attribution:* Alex Alishevskikh, CC BY-SA 2.0, via Wikimedia Commons.

13-6: https://commons.wikimedia.org/wiki/File:Andromeda_Collides_Milky_Way.jpg *Attribution:* NASA; ESA; Z. Levay and R. van der Marel, STScI; T. Hallas; and A. Mellinger, Public domain, via Wikimedia Commons.

Chapter-end image 1: Zipaquirá salt mine, Colombia.

Chapter-end image 2: Olmec head: La Venta Museum Park, Villahermosa, Tabasco, Mexico.

Front Cover

Fossil Image: Modified from: Animals in Prehistoric Exhibit – Photo by Ajale, via Pixabay. *Attribution:* https://pixabay.com/photos/animals-prehistory-museum-history-3530629/

REFERENCES

Chapter 1

[1] **C. Sagan,** 1996, Cosmos: 365p., Random House.

[1a] **C. Sagan,** 1977, The Dragons of Eden: 263p. Random House.

[2] **J. Mark,** Feb 16, 2022, Aristarchus of Samos: World History Encyclopedia. www.worldhistory.org/Aristarchus_of_Samos

[3] **V. J. Stenger,** Sept 9, 2014, God and the Multiverse–Humanities expanding views of the Cosmos: 447p. Prometheus Books, NY.

[4] **Creation Myths,** The Creation in Rig Veda 10.129: creationmyths.org. https://www.creationmyths.org/rigveda-10-129-indian-creation/

[5] **Wikipedia,** Religious interpretations of the Big Bang theory: https://en.wikipedia.org>wiki>Religious_interpretation....

[6] **Swami Mukundananda,** Bhagavad Gita—Commentary: How old is the universe: BG 8.17. Chapter 8, Verse 17. Holy-bhagavad-gita.org.

[7] **P. J. Steinhardt and N. Turok,** , June 2002, A Cyclic Model of the Universe: Science, v296(5572): 1436-1439. DOI:10.1126/science.1070462.

[8] **S. Hossenfelder,** Sept 11, 2022, The Trouble with "The Big Bang": Nautilus Magazine.

[9] **Wikipedia,** Observable universe: https://en.wikipedia.org/wiki/Observable_universe

[10] **NASA-WMAP,** Jan 24, 2014, Shape of the Universe: https://wmap.gsfc.nasa.gov/universe/uni_shape.html

[11] **A. Robinson,** April 30, 2018, Did Einstein really say that? Nature, v557(7703):30-30. doi:10.1038/d41586-018-05004-4.

[12] **Wikipedia,** George Le Maître: https://en.wikipedia.org/wiki/Georges_Lemaître

[13] **P. de Felipe,** Aug 23, 2017, "Georges Lemaitre, the Scientist and Priest Who "Could Conceive the Beginning of the Universe": https://biologos.org/articles/georges-lemaitre-the-scientist-and-priest-who-could-conceive-the-beginning-of-the-universe

[14] **J. Horgan,** April 7, 2020, Remembering Big Bang basher Fred Hoyle: https://blogs.scientificamerican.com>cross-check>remembering-big-bang-basher-fred-hoyle/

[15] **Hubble Space Telescope,** Edwin Powell Hubble— The man who discovered the cosmos: https://www.space-telescope.org/about/history/the_man_behind_the_name

[16] **J. Bennett & S. Shostak,** 2007, Life in the Universe: 485p. Pearson Addison Wesley Publisher.

[17] **American Museum of Nat. History,** Georges Lemaitre, Father of the Big Bang: https://www.amnh.org/learn-teach/curriculum-collections/cosmic-horizons-book/georges-lemaitre-big-bang

[18] **Pope Pius XII,** Nov 22, 1951, The Proofs for the Existence of God in the light of the modern Science: Papal Encyclicals https://www.papalencyclicals.net/pius12/p12exist.htm

[19] **S. Hawking,** 1998, A Brief History of Time: 240p. Bantam Publisher.

[20] **Wikipedia,** Chronology of the universe: https://en.wikipedia.org/wiki/Chronology_of_the_universe

[21] **Univ. of Oregon,** May 23, 2017, The Early Universe toward the beginning of Time: https://pages.uoregon.edu/jimbrau/astr123/Notes/Chapter27.html

[22] **Wikipedia,** Timeline of the universe: https//:en.wikipedia.org/wiki/Timeline_of_the_early_universe

[23] **NASA- WMAP,** May 09, 2016, Tests of Big Bang: The CMB: https://wmap.gsfc.nasa.gov/universe/bb_tests_ cmb

[24] **K. Tate,** April 03, 2013, Cosmic Microwave Background: Big Bang relic explained (Infographic): Space.com. https://www.space.com>20330-cosmic-microwave-bac...

[25] **N. deGrasse Tyson,** 2017, Astrophysics for people in a hurry: 222p. W. Norton & Co, Inc.

[26] **F. Hoyle and W. A. Fowler,** Nov 1960, Nucleosynthesis in supernovae: Astrophysical Jour., v132:565-590. Abstract. https://ui.adsabs.harvard.edu/abs/1960ApJ...132..565H/abstract

[27] **European Space Agency,** July 22, 2019, Hubble spots a stunning spiral: https://science.nasa.gov/missions/hubble/hubble-spots-a-stunning-spiral/

[28] **A. Strickland,** July 24, 2019, Stunning spiral galaxy spotted by Hubble more than 70 million light years away from our solar system: CNN news. https://www.cnn.com/2019/07/24/world/spiral-galaxy-hubble-scn-trnd/index.html

[29] **J. Choi, C. Conroy, J. Moustakas, G. Graves, B. P. Holden, M. Brodwin, M. J. I. Brown, P. G. van Dokkum,** Aug 20, 2014, The assembly histories of quiescent galaxies since z= from absorption line spectroscopy: The Astrophysical Jour., v792(2): https://arxiv.org/abs/1403.4932

[30] **M. Carlos, P. E. Nissen, J. Melendez,** Mar 2016, Correlation between lithium abundances and ages of solar twin stars: Astronomy Astrophysics, v587, A100. Abstract.

[31] **Sci. Ref. Section, Library of Congress,** Nov 19, 2019, What does it mean when they say the universe is expanding: https://www.loc.gov/everyday-mysteries/astronomy/item/what-does-it-mean-when-they-say-the-universe-is-expanding/

[32] **ESO Osterreich,** Aug 17, 2004, How old is the Milky Way? https://www.eso.org/public/news/eso0425/

[33] **R. Roy Britt,** Aug 17, 2004, Milky Way's age narrowed down: SPACE. www.space.com/263-milky-age-narrowed

[34] **A. Harvey and E. Howell,** Feb 01, 2022, How many galaxies are there? SPACE.com. https://www.space.com/25303-how-many-galaxies-are-in-the-universe.html

[35] **B. Jackson,** Sept 2021, Astro for kids: How many stars are there in space? https://www.astronomy.com/science/astro-for-kids-how-many-stars-are-there-in-space/

[36] **J. Brau,** Nov 03, 2016, Star formation—a traumatic birth: Univ. of Oregon. https://pages.uoregon.edu/jimbrau/astr122/Notes/Chapter19.html

[37] **Wikipedia,** Star Formation: https://en.wikipedia.org/wiki/Star_formation

[38] **H. E. Bond, E. P. Nelan, D. A. VandenBerg, G. H. Schaefer, D. Harmer,** 2013, HD 140283: A star in the solar neighborhood that formed shortly after the Big Bang: The Astrophysical Jour., v765(1): L12. DOI:10.1088/2041-8205/765/1/L12.

[39] **STScI,** March 07, 2013, Hubble finds 'birth certificate of oldest known star': Science Daily. https://www.sciencedaily.com/releases/2013/03/130307145103.htm

[40] **M. Williams,** June 01, 2016, How was the solar system formed? The Nebular hypothesis: Universe Today. https://www. universetoday.com /how-was-the-solar-system-formed

[41] **K. Schrock,** July 24, 2006, Planet-forming disks slow the spinning stars: Scientific American. https://www.scientificamerican.com/article/planet-forming-disks-slow/

[42] **NASA,** How does our sun compare with other stars? https://spaceplace.nasa.gov/sun-compare/en/

[43] **Cool Cosmos,** What is the Sun made of? https://coolcosmos.ipac.caltech.edu/ask/4-what-is-the-sun-made-of-

[44] **B. Dorminey,** April 22, 2018, Why Mars turned into a planetary runt: Forbes. https://www.forbes.com/sites/brucedorminey/2018/04/22/why-mars-turned-into-a-planetary-runt/

[45] **F. Cain,** April 15, 2013, How long does it take sunlight to reach the earth? Phys.org. https://phys.org/news/2013-04-sunlight-earth.html

[46] **P. Strathern,** 2000, Mendeleyev's Dream: 308p. Berkley Books, NY.

[47] **D. Brown and E. Berger,** Oct 24, 2017, Cosmic alchemy: Colliding neutron stars show how the universe creates gold: The Conversation. https://theconversation.com/cosmic-alchemy-colliding-neutron-stars-show-us-how-the-universe-creates-gold-86104

[48] **LeVanHan,** 2008, File: Periodic table.jpg: https://commons.wikimedia.org/wiki/File:Periodic-table.jpg. CC BY-SA 3.0 license.

[49] **E. Lagatta,** July 14, 2023, How old is our universe? Big Bang may be 26.7billion years old: USA Today. https://www.usatoday.com/story/news/nation/2023/07/14/universe-may-older-than-thought-study-shows/70411343007/

[50] **N. T. Tillman,** Jan 21, 2022, Red Giant Stars: facts, definition and the future of the Sun: SPACE. https://www.space.com/22471-red-giant-stars.html

[51] **Science Daily** (Univ. of Bristol source), Sept 9, 2011, Where does all the gold comes from? Precious metals the result of meteorite bombardment, rock analysis finds: Science Daily. https://www.sciencedaily.com/releases/2011/09/110907132044.htm

[52] **M. Willbold, T. Elliott, S. Moorbath,** Sep 7, 2011, The tungsten isotopic composition of the Earth's mantle before the terminal bombardment: Nature 2011, v477(7363): p195-198. https://pubmed.ncbi.nlm.nih.gov/21901010/

[53] **NASA,** July 12, 2022, First images from the James Webb Space Telescope | NASA: https://science.nasa.gov/mission/webb/multimedia/images/#First-Images

[54] **S. Gontcho A. Gontcho, J-B Kikwaya Eluo, P. Gabor**, Jan 17, 2023, Resurfaced 1964 VRT video interview of Georges La Maitre: Jour. For the History of Astronomy, XX(X):p1-7 (English transcript of French interview).

[55] **G. W. Bowersox, E. E. Foord, B. M. Laurs, J. E. Shigley, C. P. Smith**, 2000, Ruby and sapphire from Jegdalek, Afghanistan: Gems and Gemology, Summer 2000, p110-126. https://www.gia.eduWWW/doc/Ruby-and-Sapphire-from-Jegdalek-Afghanistan.pdf

Chapter 2

[1] **B. Hodge,** May 30, 2007, How Old is the Earth: https://answersingenesis.org

[2] **Bible-Science Guy,** 26. Age of the Earth—Kepler and Newton: https://biblescienceguy.wordpress.com/2008/09/11/26-age-of-the-earth-kepler-newton/

[3] **American Institute of Physics,** How old is the universe? https://history.aip.org./history/exhibits/curie/age-of-earth.htm

[4] **Wikipedia,** Moon rock: https://en.wikipedia.org/wiki/Moon_Rock

[5] **M. Norman,** April 21, 2004, The oldest Moon rocks: Planetary Sci. Research. http://www.psrd.hawaii.edu/April04/lunarAnorthosites.html

[6] **J. Hrala,** Jan 12, 2017, Moon rocks reveal that we were very wrong about the age of the Moon: Science Alert. https://www.sciencealert.com/we-finally-know-the-precise-age-of-the-moon-thanks-to-apollo-14-moon-rocks

[7] **A. Olvera,** Oct 10, 2016, The legend of Coatlicue and Coyolxauhqui—of how the moon and stars were formed: Inside Mexico. https://www.inside-mexico.com/the-legend-of-coatlicue-coyolxauhqui/

[8] **NASA Science,** Overview | Inside & Out—Moon: https://science.nasa.gov/moon/composition/

[9] **Phys.org.,** Sept 17, 2009, New NASA temperature maps provide 'whole new way of seeing the moon': https://phys.org/news/2009-09-nasa-temperature-moon.html

[10] **J. Green,** June 13, 2019, Gravity Assist Podcast: mapping the Moon, with Noah Petro: https://www.nasa.gov/podcasts/gravity-assist/gravity-assist-podcast-mapping-the-moon-with-noah-petro/

[11] **N. DeGrasse Tyson,** 2017, Astrophysics for people in a hurry: 222p. W. Norton & Co, Inc.

[12] **M. Weissberger,** Aug 6, 2019, Thousands of Tardigrades stranded on the Moon after Lunar Lander crash: LiveScience. https://www.livescience.com/66109-tardigrades-moon-israeli-lander.html

[13] **N. Taylor Redd,** Jan 06, 2011, Details of the Moon's core revealed by 30-year-old data: SPACE. https://www.space.com/9710-details-moon-core-revealed-30-year-data.html

[14] **Wikipedia,** Geology of the Moon: https://en.wikipedia.org>wiki >Geology_of_the-Moon

[15] **NASA,** Antarctic Meteorites: https://curator.jsc.nasa.gov/antmet/

[16] **Wikipedia,** Clavius (crater): https://en.wikipedia.org/wiki/Clavius_(crater)

[17] **C. Li, D. Liu, B. Liu, X. Ren, J. Liu, Z. He, W. Zuo, X. Zeng, R. Xu, Xu Tan, X. Zhang, W. Chen, R. Shu, W. Wen, Y. Su, H. Zhang, Z. Ouyang,** May 15, 2019, Chang'e 'E-4 initial spectroscopic identification of lunar far-side mantle derived materials: Nature, v.569:p378-382.

[18] **Purdue University,** May 30, 2013, Team solves the origin of the Moon's 'mascons' mystery: Phys.org. https://phys.org/news/2013-05-team-moon-mascons-mystery.html

[19] **NASA,** Exploring the Moon pdf: https://ares.jsc.nasa.gov/interaction/lmdp/documents/58199main_exploring_the_moon.pdf

[20] **E. C. Robertson,** 1966, The Interior of the Earth: USGS. https://pubs.usgs.gov/gip/interior/

[21] **NASA,** Earth's Moon – NASA Science: https://science.nasa.gov/moon/

[22] **T. Machemer,** July 14, 2021, Moon's Wobbly Orbit and Rising Sea Level Will Cause Record Flooding in the 2030s: Smithsonian Magazine. https://www.smithsonianmag.com/smart-news/moons-orbit-and-rising-sea-levels-will-cause-record-flooding-2030s-180978180/

[23] **ABC Science,** Nov 28, 2012, How many hours were in a dinosaur's day? (Based on interview of Prof. R. Mardling). https://www.abc.net.au/science/articles/2012/11/28/3642932.htm

[24a] **M. Zimecki,** 2006, The lunar cycle: effects on human and animal behavior and physiology: National Institute Health. https://pubmed.ncbi.nlm.nih.gov/16407788/

[24] **L. Poppick,** July 2013, 6 Wild ways the moon affects animals:https://www.livescience.com/37928-ways-the-moon-affects-animals.html

[25] **P. Kaneiwska, S. Alon, S. Karako-Lampert, O. Hoegh-Guldberg, O. Levy,** Dec 15, 2015, Signaling cascades and the importance of moon light in coral broadcast mass spawning: eLife, 2015, v4. https://elifesciences.org/articles/09991

[26] **G. Budde, C. Burkhardt, T. Kleine,** May 20, 2019, Molybdenum isotopic evidence for the late accretion of outer solar system material to Earth: Nature Astronomy, v3:p736-741. Abstract. https://ui.adsabs.harvard.edu/abs/2019NatAs...3..736B/abstract

[27] **The Economic Times,** May 22, 2019, Formation of the Moon brought water to Earth: Study: https://economictimes.indiatimes.com/news/science/formation-of-moon-brought-water-to-earth-study/articleshow/69440284.cms?from=mdr

[28] **S. Wolpert,** Jan 28, 2016, Moon was produced by a head-on collision between Earth and a forming planet: UCLA Newsroom: https://newsroom.ucla.edu/releases/moon-was-produced-by-a-head-on-collision-between-earth-and-a-forming-planet

[29] **R. L. Korotev,** Feb 2000, The great lunar hotspot and the composition and origin of "LKFM" impact-melt breccias: Jour. Geophysical Res. Atmospheres. v105(E2):p4317-4346. Abstract.

[30] **SpaceRef,** Oct 30, 2012, Lunar scientists develop new theory on Earth and Moon formation: https://spacenews.com/nasa-lunar-scientists-develop-new-theory-on-earth-and-moon-formation/

[31] **National Space Society,** Aug 15, 2006, Book review: Return to the Moon (by Harrison Schmitt): https://nss.org/book-review-return-to-the-moon-by-harrison-schmitt/

[32] **W. A. Ambrose, J. F. Reilly II, D. C. Peters,** 2013, Earth Resources for Human Sett ement in the Solar System and Earth's future in space: AAPG Mem. v101:p1-7.

[33] **H. S. Schmitt,** 2013, Lunar Helium-3 energy resources: AAPG Mem. v101:p33-52.

[34] **W. A. Ambrose,** 2013, The significance of lunar water and other mineral resources for rocket propellants and human settlement of the moon: AAPG Mem. v101:p7-31.

[35] **A. Voiland,** Jan 29, 2021, Violent formation of Mistastin Lake: https://earthobservatory.nasa.gov/images/148491/violent-formation-for-mistastin-lake

Chapter 3

[1] **J. Bennett and S. Shostak,** 2007, Life in the Universe: 485p. Pearson Addison Wesley Publisher.

[2] **The Geol. Society,** Plate Tectonics, Continental/Continental: The Himalayas: https://www.geolsoc.org.uk/Plate-Tectonics/Chap3-Plate-Margins/Convergent/Continental-Collision

[3] **USGS,** May 5, 1999, Inside the Earth: https://pubs.usgs.gov/gip/dynamic/inside.html

[3a] **National Geographic Society,** Sep 27, 2022, Mantle: https://education.nationalgeographic.org/resource/core/

[4] **P. C. Rickwood,** 1981, The largest crystals: American Mineralogist, v66:p885-908. http://www.minsocam.org/ammin/am66/am66_885.pdf

[5] **A. P. Nuttman, V. C. Bennett, C. R. L. Friend, M. T. Rosing,** Sept 25, 1997, ca.3710 and >= 3790 Ma volcanic sequence in the Isua (Greenland) supracrustal belt, structure, and Nd isotope implications: Chemical Geology, v4(3-4):p271-287. https://doi.org/10.1016/S0009-2541(97)00084-3

[6] **NOAA,** Sept 4, 2020, How does sand form: NOAA National Ocean Service. https://oceanservice.noaa.gov/facts/sand.html

[7] **National Park Foundation,** Sea of Sand—White Sands National Park: https://www.nationalparks.org/explore/parks/white-sands-national-park

[8] **E. Wurgaft, Z. Steiner, B. Luz, B. Lazar,** Sept 2016, Evidence for inorganic precipitation of $CaCO_3$ on suspended solids in the open waters of the Red Sea: Marine Chemistry, 2016, v186: p145-155.

[9] **Atlas Obscura,** Fly Geyser, Gerlach, Nevada (Added By Tre): https://www.atlasobscura.com/places/fly-ranch-geyser

[10] **Wikipedia,** Hierve el Agua: https://en.wikipedia.org/wiki/Hierve_el_Agua

[11] **Wikipedia,** Laterite: https://en.wikipedia.org/wiki/Laterite

[12] **P. L. Binda & J. G. Moltzer,** Dec 1979, Origin of oolitic iron formation: a discussion: Jour. Sed. Petrology, v49: p1351-1362.

[13] **D. P. Bhattacharyya and P. K. Kakimoto,** 1982, Origin of ferriferous ooids; an SEM study of ironstone ooids and bauxite pisoids: Jour. Sed. Research, v52(3): p849-857. Abstract.

[14] **J. C. Moore,** 2001, Accretionary Prisms: Encyclopedia of Ocean Sciences. Abstract. www.sciencedirect.com/topics/earth-and-planetary-sciences/accretionary-prism

[15] **J. Morrison,** Aug 29, 2016, The blasphemous geologist who rocked our understanding of Earth's age: Smithsonian Magazine. https://www.smithsonianmag.com/history/father-modern-geology-youve-never-heard-180960203/

[16] **Encyclopedia.com,** May 23, 2018, 'James Hutton': https://www.encyclopedia.com/people/science-and-technology/geology-and-oceanography-biographies/james-hutton

[17] **Wikipedia,** 'William Smith' (geologist): https://en.wikipedia.org/wiki/William_Smith_(geologist)

[18] **USGS,** June 13, 2001, Radiometric Time Scale: https://pubs.usgs.gov/gip/geotime/radiometric.html

[19] **F. Dorey,** Nov 19, 2018, Homo habilis: Australian Museum. https://australian.museum/learn/science/human-evolution/homo-habilis/

[20] **C. Paterson,** 1956, Age of the Meteorites and the Earth PDF: Geochimica et Cosmochimica Acta. v10(4): p230-237.

[21] **K. M. Cohen, S. C. Finney, P. L. Gibbard, J.-X. Fan,** 2013, The ICS Chronostratigraphic Chart: Episodes, v36:p199-204.

[22] **G. M. Cooper,** 2000, The Cell: A Molecular Approach. The origin and evolution of cells. 2nd edition. Sunderland (MA): (Sinauer Associates, 2000). https://www.ncbi.nlm.nih.gov/books/NBK9841/

[23] **W. F. Martin, S. Garg, V. Zimorski,** Sept 26, 2015, Endosymbiotic theories of eukaryote origin: Phil.Trans. Royal Soc. B. Biol Sci, v370(1678). https://pmc.ncbi.nlm.nih.gov/articles/PMC4571569/

[24] **N. Lane,** 2010, Why are cells powered by proton gradients? Nature education, v3(9):18. https://www.nature.com/scitable/topicpage/why-are-cells-powered-by-proton-gradients-14373960/

[25] **U. College of London,** Nov 4, 2019, Deep Sea vents had ideal conditions for origin of life: https://www.ucl.ac.uk/news/2019/nov/deep-sea-vents-had-ideal-conditions-origin-life

[26] **D. S. Kelley, J. Karson, G. Fruh-Green et al.**, 2005, Life is full of surprises: NOAA Ocean Explorer. https://oceanexplorer.noaa.gov/explorations/05lostcity/background/overview/overview.html

[27] **NASA's Mars Exploration program,** Mars Facts | All about Mars: https://mars.nasa.gov/all-about-mars/facts

[28] **M. Williams,** Dec 7, 2015, Mars compared to Earth, Universe Today: Phys.org. https://phys.org/news/2015-12-mars-earth.html

[29] **D. Cook,** Jan 2021, The Search for Life on Mars: AAPG Explorer, p6-8.

[30] **K. M. Stack,** Feb 2021, Destination Jezero Crater: Landing site of the Mars 2020 Perseverance Rover: AAPG Explorer. https://explorer.aapg.org/story/articleid/58913/destination-jezero-crater-landing-site-of-the-mars-2020-perseverance-rover

[31] **D. Cook,** Feb 2021, Mars 2020, Perseverance Rover, and International Mars Fleet Arrival: AAPG Explorer, p18-19.

[32] **D. Cook,** Oct 2020, Mars resources for human settlement: AAPG Explorer, p6-8.

[33] **National Geographic Society,** Sep 27, 2022, Core: https://education.nationalgeographic.org/resource/core/

[34] **D. Strain,** April 27, 2020, Researchers dig into case of geologic amnesia: Phys.org. https://phys.org/news/2020-04-case-geologic-amnesia.html

Chapter 4

[1] **R. W. Wilson, G. A. Houseman, K. J. W. McCaffrey, A G. Doré, S. J. H. Buiter,** July 25, 2019, Fifty years of the Wilson Cycle concept in plate tectonics: An overview, Lyell Collection, Geol. Soc. London, Sp. Pubs, v470: p1-17. https://sp.lyell-collection.org/content/470/1/1#ref-134.

[2] **C. L. Johnson, C. G. Constable, L. Tauxe, R. Barendregt, L. L Brown, R. S. Coe, P. Layer, V. Mejia, N. D. Opdyke, B. S. Singer, H. Staudigel, D. B. Stone,** April 23, 2008, Recent investigations of the 0-5 Ma geomagnetic field recorded by lava flows: Geochemistry, Geophysics, Geosystems, v9(4). https://agupubs.onlinelibrary.wiley.com/doi/full/10.1029/2007GC001696

[3] **D. Forsyth & S. Uyeda,** Oct 01, 1975, On the relative importance of driving forces of plate motion: Geophysical Jour. International, v43(1): p163-200. https://www.researchgate.net/publication/229699636_On_the_Relative_Importance_of_the_Driving_Forces_of_Plate_Motion

[4] **S. Marshak,** 2009, Essentials of Geology: 518p, W.W. Norton Pub.

[5] **P. G. DeCelles, D. M. Robinson, G. Zandt,** Dec 2002, Implications of shortening in the Himalayan fold-thrust belt for uplift of Tibetan Plateau: Tectonics, v21(6): p12-25.

[6] **R. E. Reilinger, S. C. McClusky, M. B. Oral, R. W. King, M. N. Toksoz, A. A. Barka, I. Kinik, O. Lenk, I. Sanli,** May 1997, Global positioning system measurements of present-day crust movements in the Arabia-Africa-Eurasia plate collision zone: Jour. Geophysical Res, v102(B5): p9983-9999. Abstract. https://ui.adsabs.harvard.edu/abs/1997JGR...102.9983R/abstract

[7] **N. Eyles and A. Miall,** 2007, Canada Rocks—The Geologic Journey: 512p. Fitzhenry & Whiteside, Canada.

[8] **Wikipedia,** Mariana Trench: en.wikipedia.org/wiki/Mariana_Trench

[9] **NOAA,** Ocean Exploration and Research, How deep is the ocean: https://oceanexplorer.noaa.gov/facts/ocean-depth.html

[10] **C. Conrad and C. Lithgow-Bertelloni,** 2004, The temporal evolution of plate driving forces: Importance of slab suction versus "slab pull" during the Cenozoic: Jour. Geophysical Res.—Solid Earth, v109(B10).

[11] **J. F. Casey, J. A. Karson, D. Elthon, E. Rosencrantz, M. Titus,** Dec 1983, Reconstruction of the geometry of accretion during formation of the Bay of Islands ophiolite complex: Tectonics, v2(6): p509-528. https://www.researchgate.net/publication/248815872

[12] **M. P. Searle, R. I. Corfield, B. J. Stephenson, J. McCarron ,** May 01,1997, Structure of the North Indian continental margin in the Ladakh-Zanskar Himalayas: implications for the timing of obduction of Spontang ophiolite, India-Asia collision and deformation events in the Himalaya: Geol. Magazine, v134(3): p297-316. https://www.researchgate.net/publication/231889494

[13] **P. E. Bartok, O. Renz, G. E. G. Westermann,** Aug 01,1985, The Siquisique Ophiolites, Northern Lara State, Venezuela: A discussion of Middle Jurassic ammonite, and tectonic implications: Bull. Geol. Soc. Am., v96(8): p1050-1055.

[14] **Wikipedia,** Ordovician: en.wikipedia.org/wiki/Ordovician

[15] **V. Gornitz,** Jan 2007, Sea level rise, after the ice melted and today: https://www.giss.nasa.gov/research/briefs/archive/2007_gornitz_09/

[16] **T. E. Tornqvist, S. R. Wortman, Z. R. P. Mateo, G. A. Milne, J. B. Swenson,** Oct 14, 2006, Did the last sea level low stand always lead to cross-shelf valley formation and source-to-sink sediment flux? https://www.researchgate.net/publication/242232405

[17] **M. F. Coffin and O. Eldholm,** 2005, Large Igneous Provinces: Encyclopedia of Geology. https://www.sciencedirect.com/topics/earth-and-planetary-sciences/large-igneous-province

[18] **B. Taylor,** Jan 31, 2006, The single largest oceanic plateau: Ontong Java-Manihiki-Hikurangi: Earth and Planetary Sci. Letters, v241(3-40): p372-380.

[19] **Wikipedia,** Ontong Java Plateau: https://en.wikipedia.org>wiki>Ontong_Java_Plateau

[20] **Woods Hole Oceanographic Institute,** Seamounts. https://www.whoi.edu/know-your-ocean/ocean-topics/how-the-ocean-works/seafloor-below/seamounts/

[21] **M. Coffin,** 3.3.3 Large Igneous Provinces: https://www.ldeo.columbia.edu/~polsen/nbcp/lipmc.html

[22] **L. L. Sloss,** 1963, Sequences in the cratonic interior of North America: Bull. Geol. Soc. America, v74: p93-113.

[23] **G. P. Eaton,** 2008, Epeirogeny in the Southern Rocky Mountain region: Evidence and origin: Geosphere, v4(5) p764-784. https://pubs.geoscienceworld.org/gsa/geosphere/article/4/5/764/31206/

[24] **P. Perreault,** Dec 29, 1997, Are the Earth's magnetic poles moving? How do navigators adjust to the change? https://www.scientificamerican.com/article/are-the-earths-magnetic-p/

[25] **J. Boyd,** Nov 19, 2018, 'True polar wander' may have caused ice age: Science News. https://news2.rice.edu/2018/11/18/true-polar-wander-may-have-caused-ice-age/

[26] **D. Woodworth and R. G. Gordon,** Oct 2018, Paleolatitudes of the Hawaiian hot spot since 48Ma: Evidence for a Mid-Cenozoic True polar stillstand followed by Late Cenozoic True Polar Wonder coincident with northern hemisphere glaciation.: Geophysical Res. Letters, v45. 10.1029/2018GL080787.

[27] **NASA Earth Observatory (caption—M.J. Wilkinson),** Sep 3, 2016, Santorini, Greece: https://earthobservatory.nasa.gov/images/88926/santorini-greece

[28] **T. Kokkonidis,** Jan 8, 2024, Growing magma under Santorini volcano a 'Real Threat': Greek Reporter.com, https://greekreporter.com/2024/01/08/growing-magma-under-santorini-volcano-real-threat/

[29] **IUGS | International Commission on Geoheritage,** The Quaternary Santorini Ca dera: https://iugs-geoheritage.org/geoheritage_sites/quaternary-santorini-caldera/

Chapter 5

[1a] **Harvard University,** Origin of Water on Earth: https://courses.seas.harvard.edu/climate/eli/Courses/EPS281r/Sources/Origin-of-oceans/1-Wikipedia-Origin-of-water-on-Earth.pdf

[1] **S. Nadeau, W. Chen, J. Reece, D. Lachhman,** 2013, In Guyana: The lost Hadean crust of South America: Brazilian Jour. of Geology, v43(4): p601-606.

[2] **J. O'Neil, R. W. Carlson, D. Francis, R. K. Stevenson,** Sept 26, 2008, Neodymium-142 Evidence for Hadean Mafic Crust: Science v321(5897): p1828-1831. DOI:10.1126/science.1161925.

[3] **CBC News,** Sept 25, 2008, Canada home to the world's oldest rocks, researchers say: https://www.cbc.ca/news/science/canada-home-to-world-s-oldest-rocks-researchers-say-1.759567

[4] **National Science Foundation,** Sept 25, 2008, Oldest known rock on Earth has been discovered. Eureka Alert: https://www.eurekalert.org/news-releases/604451

[5a] **N. Sleep,** 2010, Hadean-Archean environment: National Institute of Health (.gov). https://pmc.ncbi.nlm.nih.gov/articles/PMC2869525/

[5] **B. Oskin,** Feb 23, 2014, Confirmed: Oldest fragment of Early Earth is 4.4 billion years old: https://www.livescience.com/43584-earth-oldest-rock-jack-hills-zircon.html

[6] **K. Spuller,** Feb 27, 2018, Geol. Soc. Am. Guest Blog, Rethinking the Hadean: Evidence of Earth's earliest life: https://speakingofgeoscience.org/2018/02/27/rethinking-the-hadean-evidence-of-earths-earliest-life/

[7] **B. Rizk,** Aug 08, 2024, The search for the earliest traces of life on Earth: Univ. of Ottawa, https://www.uottawa.ca/about-us/news-all/search-earliest-traces-life-earth

[8] **J. Bennett and S. Shostak,** 2007, Life in the Universe: 485p. Pearson Addison Wesley Publisher.

[9] **W. Bleeker and B. W. Davis,** 2004, What is a craton? How many are there? How do they relate? And how did they form? Abstract. https://ui.adsabs.harvard.edu/abs/2004AGUSM.T41C..01B/abstract

[10a] **J. Leanuik,** April 22, 2016, Earth's Supercontinents: The Dialogue. www.the-dialogue.com/en/en33-earthhttps://www.the-dialogue.com/en/en33-earths-supercontinents/

[10b] **Wikipedia,** Supercontinent: https://en.wikipedia.org/wiki/Supercontinent

[10] **N. Eyles and A. Miall,** 2007, Canada Rocks—The Geologic Journey: 512p. Fitzhenry & Whiteside, Canada.

[11] **D. Catling and K. J. Zahnle,** Feb 26, 2020, The Archean Atmosphere: Science Advances, 2020, v6(9) eaax1420. Abstract.

[12] **S. M. Awramik,**1992, The oldest records of photosynthesis, 1992: Photosynthesis Res., v33: p75-89.

[13] **J. Dostal,** Mar 01, 2008, Igneous rock associations 10. Komatite: Geoscience Canada, v35(1). Abstract. https://journals.lib.unb.ca/index.php/GC/article/view/11074

[14] **AMNH,** NY, Banded Iron Formation: https://www.amnh.org/exhibitions/permanent/planet-earth/how-has-the-earth-evolved/banded-iron-formation

[15] **E. Chi Fru, M. Ivarsson, S. P. Kilias, A. Bengtson, V. Belivanova, F. Marone, D. Fortin, C. Broman, M. Stampanoni**, June 2013, Fossilized iron bacteria reveal a pathway to the biological origin of Banded Iron Formation: Nature Communications 4, Article No.2050 (2013). https://www.nature.com/articles/ncomms3050

[16] **S. Marshak,** 2009, Essentials of Geology: 518p, W.W. Norton Pub.

[17] **M. Woo,** July 30, 2018, Oldest evidence for life on land unearthed in South Africa: LiveScience. https://www.livescience.com/63199-oldest-life-on-land.html

[18] **Y. Shen, R. Buick, D.E. Canfield,** Mar 1, 2001, Isotopic evidence for microbial sulphate reduction in the early Archean era: Nature, v410(6824): p77-81. https://pubmed.ncbi.nlm.nih.gov/11242044/

[19] **N. Noffke, D. Christian, D. Wacey, R. M. Hazen,** 2013, Microbially induced sedimentary structures recording an ancient ecosystem in the ca.3.48billion-year-old Dresser Formation, Pilbara, Western Australia: Astrobiology, 2013, v13(12): p1103-1124. https://pmc.ncbi.nlm.nih.gov/articles/PMC3870916/

[20] **J. W. Schopf,** June 29, 2006, Fossil evidence of Archean life: Phil.Trans. Royal Soc. London B. Biol. Sci., v361(1470): p869-885. https://pubmed.ncbi.nlm.nih.gov/16754604/

[21] **Palaeos,** The Paleoproterozoic Era: http://palaeos.com/proterozoic/paleoproterozoic/paleoproterozoic.html

[22] **C. Brewer-Carias,** 1978, La Vegetacion del Mundo Perdido, 223p, Fundación E. Mendoza, Venezuela,

[23a] **NSF News Release** 08-119, July 17, 2008, A single boulder may prove that Antarctica and North America were once connected: Science Daily: https://www.sciencedaily.com/releases/2008/07/080717134558.htm

[23] **D. Praetorius,** Oct 9, 2011, Rodinia evidence suggests Supercontinent linked North America and Antarctica: Huffington Post, https://www.huffpost.com/entry/ancient-supercontinent-evidence-pangea_n_922366

[24] **Snowball Earth,** When did the snowball earths occur? https://www.snowballearth.org/when.html

[25] **L. Poppick,** April 5, 2019, Snowball Earth: The times our planet was covered in ice: https://www.astronomy.com/science/snowball-earth-the-times-our-planet-was-covered-in-ice/ (Poppick's interview of Prof. P. F. Hoffman).

[26] **P. F. Hoffman,** May 2019, Big Time: Annual Review of the Earth and Planetary Sci., v47:p1-17. Abstract. https://www.annualreviews.org/content/journals/10.1146/annurev-earth-053018-060145

[27] **J. P. Pu, S. A. Bowring, J. Ramezani, P. M. Myrow,** Nov 1, 2016, Dodging snowballs: Geochronology of the Gaskiers glaciation and the first appearance of Ediacaran biota: Geology, v44(11): p955-958. Abstract.

[28] **D. Brassen,** Jan 27, 2022, New study suggests 'Missing billion years in rock record carved away by 'Snowball Earth' glaciers: Forbes Magazine. https://www.forbes.com/sites/davidbressan/2022/01/27/new-study-suggests-missing-billion-years-in-rock-record-carved-away-by-snowball-earth-glaciers/

[29] **J. E. Kamis,** Aug 4, 2015, Ancient ice ages related to volcanic activity—A major confirmation of plate climatology theory: Plate Climatology. https://www.plateclimatology.com/ancient-ice-ages-related-to-volcanic-activity-a-major-confirmation-of-plate-climatology-theory

[30] **N. J. Planavsky,** Oct 31, 2014, Low Mid-Proterozoic atmospheric oxygen levels and the delayed rise of animals: Science, 2014, v346(6209): p635-638.

[31] **Understanding Evolution,** May 2017, The bacteria that changed the world: https://evolution.berkeley.edu/evo-news/the-bacteria-that-changed-the-world/

[32] **Univ. of Albany,** Ozone: https://www.albany.edu/faculty/rgk/atm101/ozone.htm

[33] **A. Buis,** Feb 27, 2020, Milankovitch (Orbital) cycles and their role in Earth's climate: NASA Global Climate Change. https://science.nasa.gov/science-research/earth-science/milankovitch-orbital-cycles-and-their-role-in-earths-climate/

[34] **Wikipedia,** Pole Star: https://en.wikipedia.org/wiki/Pole_Star

[34a] **S. R. Meyers,** June 4, 2018, Proterozoic Milankovitch cycles and the history of the Solar System, PNAS, v115 (25). https://www.pnas.org/doi/10.1073/pnas.1717689115

[35] **Énergie et Ressources Naturelles,** Aug 14, 2018, Grenville Province: Lexique Stratigraphique .gq.mines. https://gq.mines.gouv.qc.ca/lexique-stratigraphique/province-de-grenville_en/

[35a] **R. P. Tollo,** 2005, Grenvilllian Orogeny, Encyclopedia of Geology: https://www.sciencedirect.com/topics/earth-and-planetary-sciences/grenvillian-orogeny

[36] **A. Indares,** Feb 2025, The Grenville Province: revisiting the orogenic framework and integrating recent findings: Can. Jour. Earth Sciences, v62(2):99-118. Canadian Science Publishing. http://dx.doi.org/10.1139/cjes-2023-0143

[37] **Wikipedia,** Banded Iron Formation: en.wikipedia.org/wiki/Banded_Iron_Formation

[38] **A. H. Knoll, E. J. Javaux, D. Hewitt, P. Cohen,** June 29, Eukaryotic organisms in Proterozoic oceans: Phil.Trans. Royal Soc. London B Biol Sci., v361(1470): p1023-1038.

[39] **Wikipedia,** Ediacaran biota: en.wikipedia.org/wiki/Ediacaran_biota

[40] **Newfoundland & Labrador Tourism,** Canada, Oct 11, 2017, Mistaken Point Ecological Reserve—A story 565 million years in the making: https://www.newfoundlandlabrador.com/trip-ideas/travel-stories/mistaken-point-ecological-reserve-a-story-565-million-years-in-the-making

[41] **M. M. Anderson and S. B. Misra,** Nov 1968, Fossils found in the Pre-Cambrian Conception Group of South-eastern Newfoundland: Nature, v 220(5168): p680-681. Abstract. https://www.nature.com/articles/220680a0

Chapter 6

[1] **Geological Society,** Siccar Point: The Scottish Geology.com. https://www.geolsoc.org.uk/GeositesSiccarPoint

[2] **In-C. Ryu, C. W. Oh, S. W. Kim,** Oct 2005, A Middle Ordovician Drowning Unconformity on the northeastern flank of the Okcheon (Ogcheon) Belt, South Korea: Gondwana Research, v8(4): p511-528.

[3] **D. R. Kolata, W. Huff, S. M. Bergström et al.,** 2001, The Ordovician Sebree Trough: An oceanic passage to the Mid-Continent United States: Bull. GSA 2001, v113(8): p1067-1078.

[4] **J Derby, R. Fritz, S. Longacre, W. Morgan, C. Sternbach (Editors),** 2012, The Great American Carbonate Bank—The Geology and Economic Resources of the Cambrian-Ordovician Sauk Megasequence of Laurentia: AAPG Mem. 98, USA:504p.

[4a] **AAPG Wiki,** Great American Carbonate Bank. https://wiki.aapg.org/Great_American_carbonate_bank

[5a] **Wikipedia,** Late Ordovician Glaciation: https://en.wikipedia.org/wiki/Hirnantian_glaciation

[5] **W. B. N. Berry, M. S. Quinby-Hunt, P. Wilde,** 1995, 2. Impact of Late Ordovician glaciation-deglaciation on marine life: In 'Effects of Past Global Change on Life', National Research Council (US) Panel on Effects of Past Global Change on Life, Washington (DC): National Academic Press (US), 250p. https://www.ncbi.nlm.nih.gov/books/NBK231940/

[6] **J. Paterson,** June 15, 2018, Earth's first giant predators produced killer babies: University of New England (UNE). https://www.une.edu.au/connect/news/2018/06/earths-first-giant-predators-produced-killer-babies

[7] **M. Barash,** Nov 2014, Mass Extinction of marine biota at the Ordovician-Silurian transition due to environmental changes: Oceanology, v54(6): p780-787. https://www.researchgate.net/publication/269398569_Mass_Extinction_of_the_Marine_Biota_at_the_Ordovician-Silurian_Transition_Due_to_Environmental_Changes

[8] **C. P. T. O'Dale,** Glasford Impact Structure: Crater Explorer. https://craterexplorer.ca/glasford-impact-structure/

[9a] **J. Ormö, E. Sturkell, C. Alwmark, J. Melosh,** 2014, First known terrestrial impact of a binary asteroid from a main belt breakup event: Sci Rep 4 (6724). https://www.nature.com/articles/srep06724

[9] **N. Eyles,** Mar 15, 2013, Road Rocks of Ontario: 570p. Fitzhenry & Whiteside Pub.

[10] **M. Bagley,** Feb 20, 2014, Silurian Period Facts: Climate, Animals & Plants Facts: Live Science. https://www.livescience.com/43514-silurian-period.html

[11] **S. C. Finney,** Stratigraphy, and structure of the Roberts Mountains Allochthon: Cal. State Univ, Long Beach, California. https://www.csulb.edu/geological-sciences/stratigraphy-and-structure-of-the-roberts-mountains-allochthon

[12] **J. R. Beerbower,**1968, Search for the Past: 512p, Prentice-Hall.

[13] **M. E. Zorn, M. W. Caldwell, M.V.H. Wilson,** Feb 2005, Lithological analysis of the Lower Devonian vertebrate-bearing beds at the MOTH locality, N.W.T., Canada: Insights to taphonomy and depositional setting: Can Jour. Earth Sciences, v42(2):1-13. Abstract. https://cdnsciencepub.com/doi/abs/10.1139/e05-015

[14] **P. E. Playford,** 1980, Devonian 'Great Barrier Reef' of Canning Basin, W. Australia: Bull. AAPG, v64(6): p814-840.

[15] **M. H. Monroe,** Sept 30, 2011, Gogo Reef Formation: https://austhrutime.com/gogo_formation.htm

[16a] **Oil, Gas and Salt Resources Library,** Petroleum Industry—Ontario: https://www.ogsrlibrary.com/industry_statistics_ontario_petroleum

[16] **Science-News Staff,** Mar 16, 2017, New study confirms sharks arose from Acanthodian fishes: Sci-News. https://www.sci.news/paleontology/sharks-acanthodians-04706.html

[17] **T. J. Algeo, R. A. Berner, J. B. Maynard, S. Scheckler,** Jan 1995, Late Devonian oceanic anoxic events, and biotic crises: "rooted" in the evolution of vascular plants: GSA Today, v5(3):45:64-66. https://www.researchgate.net/publication/230891610

[18] **N. Eyles & C. H. Eyles,** 1992, 'Glacial Models', p73-100. In GeoText1:Facies Models. Response to Sea Level Change. 454p. (Eds. R.G. Walker and N.P. James), Publisher Geol. Assoc. Canada.

[19] **N. Eyles, A. B. Franca, G. Gonzalez Bonorino, C. H. Eyles, O. López Paulsen,** Jan 1, 1995, Hydrocarbon-Bearing Late Paleozoic Glaciated Basins of Southern and Central South America: AAPG Mem. v62: p165-182. In Petroleum basins of South America: 800p. (Eds T. Tankard, R. Suárez Soruco, H. J. Welsink).

[20] **P. E. Potter, A. B. Franca, C.W. Spencer, M. V. Caputo,** 1995, Petroleum in glacially—related sandstones of Gondwana: A Review: Jour. Pet. Geol., v18(4): p397-420. Abstract. https://www.researchgate.net/publication/227894082

[21] **UCMP Berkeley,** The Carboniferous Period: https://ucmp.berkeley.edu/carboniferous/carboniferous.php

[22] **The Carboniferous Period**: Plants covered the Earth: Fossils, Facts and Finds. https://www.fossils-facts-and-finds.com/carboniferous_period.html

[23] **Bio-Libre Texts,** Mar 05, 2021, 26.1B: Evolution of Gymnosperms: https://bio.libretexts.org/Bookshelves/Introductory_and_General_Biology/General_Biology_(Boundless)/26%3A_Seed_Plants/26.01%3A_Evolution_of_Seed_Plants/26.1B%3A_Evolution_of_Gymnosperms

[24] **E. Glerlowski-Kordesch and C Cassie,** Sept 2015, The 'Spirorbis' problem revisited: Sedimentology and Biology of Microconchids in marine-nonmarine transition: Earth-Science Reviews v148, p209-227.

[25] **K. S. Brink, H. C. Maddin, D. C. Evans, R. R. Reisz.,** Nov 23, 2015, Reevaluation of the historic Canadian fossil Bathygnathus borealis from the Early Permian of Prince Edward Island: Can. Jour. Earth Sci., v52(12):1-12.

[26] **H. Martin Gottingen,** 1981, The late Paleozoic Gondwana glaciation: Geologische Rundschau, v70(2): p480-496.

[27] **G. Feulner,** Oct 2017, Formation of most of our coal brought Earth close to global glaciation: Abstract. PNAS, USA, v114(43): p11333-11337.

[28] **PBS.org.,** Evolution change Deep Time—Paleozoic Era—Permian Period: https://www.pbs.org/wgbh/evolution/change/deeptime/permian.html

[29] **H. Hickey,** Dec. 6, 2018, What caused Earth's biggest mass extinction? https://sustainability.stanford.edu/news/what-caused-earths-biggest-mass-extinction

[30] **W. Parry,** Mar 31, 2011, Ancient Ocean 'Dead Zone' delayed life after mass extinction: Live Science, https://www.livescience.com/13504-permian-mass-extinction-anoxic-volcanoes-carbon-dioxide.html

[31] **L. Becker, R. J. Poreda, A. R. Basu, K. O. Pope, T. M. Harrison, C. Nicholson, R. Lasky,** June 4, 2004, Bedout: A possible end-Permian impact crater offshore of Northwestern Australia: Science, 2004, v304(5676): p1469-1476. Abstract.

[32] **C. Q. Choi,** Oct 18, 2012, Extreme global warming may have caused largest extinction ever: Live Science. https://www.livescience.com/24091-extreme-global-warming-mass-extinction.html

[33] **J. L. Penn, C. Deutsch, J. L. Payne, E. A. Sperling,** Dec 7, 2018, Temperature-dependent hypoxia explains biogeography and severity of end-Permian marine mass extinction: Science, v362(6419) .eaat1327. https://www.science.org/doi/10.1126/science.aat1327

[34] **E. Blanco,** Aug 13, 2013, First appearance datum at Green Point. https://agilescientific.com/blog/2013/8/13/first-appearance-datum-at-green-point.html

[35] **Bruce Trail Conservancy,** The Niagara Escarpment Conservation: https://brucetrail.org/what-we-do/

Chapter 7

[1] **C. R. Scotese,** PaleoMap Project: http://www.scotese.com/

[2] **A. Embry & B. Beauchamp,** 2008, Chapter 13 Sverdrup Basin, p451-471: Sedimentary Basins of the World, v5. Abstract. (Ed. A. D. Miall), Elsevier.

[3] **Britannica,** Triassic Period—Economic significance of Triassic deposits: Encyclopedia Britannica. https://www.britannica.com/science/Triassic-Period/Economic-significance-of-Triassic-deposits

[4] **World Heritage Journeys of Europe,** UNESCO, Underground Europe: https://visitworldheritage.com/en/eu/underground-europe/e72e47f7-90ad-484b-9a6d-3b69e3f5b106

[5] **J. R. Beerbower,** 1968, Search for the Past: 512p, Prentice-Hall.

[6] **ScienceViews.com,** How are dinosaurs classified: https://scienceviews.com/dinosaurs/dinosaurclassification.html

[7] **U.S. Nat. Park Services,** Pterosaurs—Big Bend National Park: https://www.nps.gov/bibe/learn/nature/pterosaur.htm

[8] **S. Ray, S. Bhat, D. Mukherjee, P. M. Dutta,** Feb 2016, Vertebrate fauna from the Late Triassic Tiki Formation of India: New finds and their biostratigraphic implications: Jour. Palaeosciences, v65(1):47-59. https://www.researchgate.net/publication/320353156

[9] **L. M. E. Percival, M. Ruhl, S. P. Hesselbo, H. C. Jenkyns, T. A. Mather, J. H. Whiteside,** July 25, 2017, Mercury evidence for pulsed volcanism during end-Triassic mass extinction: PNAS, USA, v114(30): p7929-7934.

[10] **T. J. Blackburn, P. E. Olsen, S. A. Bowring, N. M. Mclean, D. V. Kent, J. Puffer, G. McHone, E. Troy Rasbury, M. Et-Touhami,** May 2013, Zircon U-Pb geochronology links the end-Triassic extinction with the Central Atlantic Magmatic Province: Science, v340: p941-945. Abstract. https://pubmed.ncbi.nlm.nih.gov/23519213/

[11] **Wikipedia,** Central Atlantic Magmatic Province: https://en.wikipedia.org/wiki/Central_Atlantic_magmatic_province

[12] **Britannica,** End Triassic Extinction: Encyclopedia Britannica. https://www.britannica.com/science/end-Triassic-extinction

[13] **Wikipedia,** Manicouagan Reservoir: https://en.wikipedia.org/wiki/Manicouagan_Reservoir

[14] **C. R. Scotese,** PaleoMap Project, www.scotese.com

[15] **N. Hurley and P. Weimer (editors),** May 1998, Gulf of Mexico Petroleum Systems: Bull. AAPG, v82 (5B): p865-1112.

[16] **W. B. Maze,** 1984, Jurassic La Quinta Formation in the Sierra de Perijá, northwestern Venezuela: Geology and tectonic environment of red beds and volcanic rocks, Mem. GSA, v162: p263-282. The Caribbean-South American Plate boundary and regional tectonics. 421p. (Eds. W. E. Bonini, W. E. Hargraves, R. Shagam), Geol. Soc. America. Publisher.

[17] **P. M. Barrett, R. J. Butler, R. Mundil, T. M. Scheyer, R. B. Irmis, M. R. Sánchez-Villagra,** Sept 2014, A palaeoequatorial ornithischian and new constraints on early dinosaur diversification: Proc. Royal Soc. B 281(1791). http://dx.doi.org/10.1098/rspb.2014.1147

[18] **UC-Berkley,** Tetrapods: https://ucmp.berkeley.edu/vertebrates/tetrapods/tetraintro.html

[19] **N. Rogers,** Feb 03, 2016, 'Jurassic 'butterflies' predated true butterflies by 50 million years: AAAS, Science. https://www.science.org/content/article/jurassic-butterflies-predated-true-butterflies-50-million-years

[20] **The Laur Foundation for Paleontology,** Science and Education, Solenhofen fossils: https://www.lauerfoundationpse.org/solnhofen-fossils

[21] **UCMP Berkeley,** The Solenhofen limestone of Germany: https://ucmp.berkeley.edu/mesozoic/jurassic/solnhofen.html

[22] **J. L. Wilson,**1975, Carbonate facies in geologic history: 471p. Springer-Verlag.

[23] **Wikipedia,** Solenhofen Limestone: https://en.wikipedia.org/wiki/Solnhofen_Limestone

[24] **T. Cook,** June 5, 2018, Travels in Geology: Mesozoic Masterpiece: England's Jurassic Coast: Earth Magazine. https://www.earthmagazine.org/article/travels-geology-mesozoic-masterpiece-englands-jurassic-coast/

[25] **Famous Scientists,** Mary Anning—Biography, Facts and Pictures: https://www.famousscientists.org/mary-anning/

[26] **Wikipedia,** Tongue-twister: https://en.wikipedia.org/wiki/Tongue_twister

[27] **Wikipedia,** Jurassic Coast: https://en.wikipedia.org/wiki/Jurassic_Coast

[28] **C. L. Dybas,** Glass Castle in the Sea: Natural History Magazine. https://www.naturalhistorymag.com/features/282622/glass-castles-in-the-sea

[29] **R. Lopez,** Nov 22, 2019, A Raja, a Jain, a Tagore: Meet desi dinosaurs that once roamed India's plains: Hindustan Times, India.

[30] **Wikipedia,** Carnian Pluvial Episode: https://en.wikipedia.org/wiki/Carnian_pluvial_episode

[31] **J. Davis,** Natural History Museum, The Triassic Period: The rise of the dinosaur: https://www.nhm.ac.uk/discover/the-triassic-period-the-rise-of-the-dinosaurs.html

[32] **G. L.W. Lutrell,** 1989, Stratigraphic nomenclature of the Eastern Newark Supergroup of Eastern North America: Bull. USGS, v1572: 136p.

[33] **M. J. Lockley, C. A. Meyer,** Jan 1994, Trackway evidence of a herd of juvenile sauropods from the Late Jurassic of Portugal: Gaia, v10: p27-35.

[34] **V.F. Santos, P. Caetano, A. Pólvora et al.,** 2016, Sesimbra Geocircuit and dinosaur tracksites—RUN: Univ. NOVA de Lisboa, Comunicaciones geológicas, v103(Especial 1): p153-158.

[35] **C. Thorsberg,** April 19, 2024, 11-year-old uncovers fossils of giant ichthyosaur in England: Smithsonian Magazine. https://www.smithsonianmag.com/smart-news/11-year-old-uncovers-fossils-of-giant-ichthyosaur-in-england-the-largest-ever-marine-reptile-ever-found-scientists-say-180984185/

Chapter 8

[1] **NOAA,** What is the Mid-Ocean ridge? Ocean exploration facts: ocean explorer.https://oceanexplorer.noaa.gov/facts/mid-ocean-ridge.html

[2] **NOAA,** What is a hot spot volcano? Ocean exploration facts: https://oceanexplorer.noaa.gov/facts/volcanic-hotspot.html

[3] **USGS,** The long trail of the Hawaiian hotspot: https://pubs.usgs.gov/gip/dynamic/Hawaiian.html

[4] **S. E. Pratt,** Dec 14, 2015, The question mantle plumes: Earth Magazine. https://www.earthmagazine.org/article/question-mantle-plumes/

[5] **S. K. Acharya (Editor),** Dec 04, 2018, Tectonic Setting and Gondwana basin Architecture in the Indian Shield: Developments in Structural Geology and Tectonics, ISSN Book 4: 153p. Elsevier Publisher.

[6] **Wikipedia,** Ninety East Ridge: https://en.wikipedia.org/wiki/Ninety_East_Ridge

[7] **S. Omondi,** April 1, 2019, What is the Ninety East Ridge: World Atlas. https://www.worldatlas.com/articles/what-is-the-ninety-east-ridge.html

[8] **G. Plint,** 2019, Anatomy of a Late Cretaceous transgressive shelf system: The influence of high frequency eustasy and crustal flexure on stratigraphy and paleogeography, basal Kaskapau Formation, Western Canada Foreland Basin: Bull. Can. Pet. Geol., v67(1):p1-46.

[9] **P. A. Scholle,** 1977, Current oil and gas production from North American Upper Cretaceous: USGS Circular 767, 48p.

[10] **P. H. Heckel,** 1974, Carbonate buildups in the geologic record—A review: SEPM Special Pub 18: p90-152. In Reefs Time and Space. 256p. (Ed. L. F. Laporte). DOI: 10.2110/pec.74.18.

[11] **J. L. Wilson,** 1975, Carbonate facies in geologic history: 471p. Springer-Verlag.

[12] **H. H. Wilson,** Jan 1987, The structural evolution of the Golden Lane, Tampico Embayment, Mexico: Jour. Pet. Geology, v10(1): p5-40. Abstract. https://onlinelibrary.wiley.com/doi/abs/10.1111/j.1747-5457.1987.tb00994.x

[13] **F. Viniegra and C. Castillo-Tejero,** 1968, Golden Lane Fields, Veracruz, Mexico: AAPG Data pages, v52(3): p553-554.

[14] **P. A. Scholle, M. Patsoules,** 2008, Depositional and diagenetic pattern and petroleum potential of Cretaceous carbonate platforms, Ionian Islands, Greece: AAPG Data pages 2019. Search & Discovery article #40271.

[15] **P. Enos,** Nov 1977, Tamabra limestone of the Poza Rica trend, Cretaceous, Mexico: SEPM Sp. Pub. No. 25: p273-314, In Deep-water carbonate environments, 336p. (Eds. H Cook & P Enos).

[16] **T. J. Bralower and M. A. Lorente,** Oct 01, 2003, Paleogeography and Stratigraphy of the La Luna Formation and related Cretaceous anoxic depositional systems: Palaios, v18(4-5): p301-304.

[17] **S. K. Ghosh,** 1984, Late Cretaceous condensed sequence, Venezuelan Andes: Geological Society of America Mem. v162: p317-324, Caribbean-South American Plate boundary and regional tectonics. 421p. (Eds. W. E. Bonini, R. B. Hargraves, R. Shagam).

[18] **J. M. Grajales-Nishimura, G. Murillo-Muñetón, C. Rosales-Domínguez, J. C. Bermúdez-Santana, L. G. Velasquillo-Martínez, J. García-Hernández, J. A. Arz, I. Arenillas,** 2009, The Cretaceous Paleogene boundary Chixulub impact: Its effect on the carbonate sedimentation on the western margin of the Yucatan Platform and nearby areas: AAPG Mem. v90: p315-336, In Petroleum Systems in the southern Gulf of Mexico, 502p. (Eds. C. Bartolini and J. R. R. Ramos).

[19] **J. M. Grajales-Nishimura, G. Murillo-Muñeton, C. Rosales-Domínguez, E. Cedillo-Pardo,** 2003, Heterogeneity of lithoclast composition in the deep-water carbonate breccias of the K/T boundary sedimentary succession, southeastern Mexico, and offshore Mexico: AAPG Mem. v79: p312—329. In Petroleum Systems in the southern Gulf of Mexico, 502p. (Eds. C. Bartolini and J. R. Roman Ramos).

[20] **D. Kring and W. V. Boynton,** 1992, Impact melt rock related to the K/T boundary deposits in Haiti: Nature v358, p141-144.

[21a] **Wikipedia,** Chixulub Crater: https://en.wikipedia.org/wiki/Chicxulub_crater

[21] **D. A. Kring & D. D. Durda,** Aug 30, 2002, Trajectories and distribution of material ejected from the Chixulub impact crater: Implications for postimpact wildfires: Jour. Geophysical Res., v107(E8): p6-1—6-22.

[22] **A. R. Hildebrand, G. Penfield, D. A. Kring, M. Pilkington, Z. Camargo, S. Jacobsen, W. V. Boynton,** Sept 01, 1991, Chixulub Crater: A possible Cretaceous-Tertiary boundary impact crater on the Yucatán Peninsula: Geology, v19(9) p867-871.

[23] **E. Howell,** Sept 11, 2019, The asteroid impact that doomed the dinosaurs also spawned a Mega-Tsunami: SPACE.com>News >Science and Astronomy.

[24] **J. M. Grajales-Nishimura, E. Cedillo-Pardo, C. Rosales-Domínguez, D. J. Morán-Zenteno,** April 2000, Chixulub impact: The origin of reservoir and seal in the southeastern Mexico oil fields: Geology, v28: p307-310.

[25] **S. Weeden,** Jan 05, 2015, Meteoric history of Cantarell Field continues for Pemex: hartenergy.com. https://www.hartenergy.com/ep/exclusives/meteoric-history-cantarell-field-continues-pemex-175129

[26] **P. Glisovic and A. Forte,** 2017, On the deep-mantle origin of the Deccan traps: Science, v355(6325): p613-616.

[27] **R. A. Berner,** Dec 2006, GEOCARBSULF: A combined model for Phanerozoic atmospheric O_2 and CO_2: Geochimica et Cosmochimica Acta, v70(23): p5653-5664.

[28] **A. Lyle,** Mar 22, 2019, UAlberta Paleontologists report world's biggest Tyrannosaurus rex. https://www.ualberta.ca/en/folio/2019/03/paleontologists-identify-biggest-tyrannosaurus-rex-ever-discovered.html

[29] **W. S. Persons 4th, P. J. Currie, G. M. Erickson,** Mar 2019, An older and exceptionally large adult specimen of Tyrannosaurus rex: The Anatomical Record, v303(4): p656-672.

[30] **Nat. History Museum,** Argentinosaurus: https://www.nhm.ac.uk/discover/dino-directory/argentinosaurus.html

[31] **M. B. Griggs,** Sept 4, 2014, Meet Dreadnought, largest dinosaur in the World: Smithsonian Magazine. https://www.smithsonianmag.com/smart-news/meet-dreadnought-largest-dinosaur-world-180952612/

[32] **AKSON,** Dec 07, 2018, Scientists find new giant dinosaur: Phys.org. https://phys.org/news/2018-12-scientists-giant-dinosaur.html

[33] **V. Udurawane,** 2016, Quatzalcoatlus, the largest flying animal of all time: Earth Archives. https://eartharchives.org/articles/quetzalcoatlus-the-largest-flying-animal-of-all-time/index.html

[34] **U. S. Nat. Park Service,** April 28, 2020, Pterosaurs in the Big Bend: https://www.nps.gov/bibe/learn/nature/pterosaur.htm

[35] **M. Butterfield,** Dec 22, 2021, Perfectly preserved, 'extremely rare' dinosaur embryo in egg found in China: Global News. https://globalnews.ca/news/8468563/dinosaur-egg-embryo-fossil-found-china/

[36] **Dinosaur World live,** May 24, 2019, Dinosaur Fossils: Where have the most fossils been found: https://dinosaurworldlive.com/blog/dinosaur-fossils-where-have-the-most-fossils-been-found

[37] **UNESCO World Heritage Centre,** Dinosaur Provincial Park (Alberta): https://whc.unesco.org/en/list/71/

[38] **R. Conniff,** May 2018, The Great Chinese dinosaur boom, Smithsonian Institution: https://www.smithsonianmag.com/science-nature/great-chinese-dino-boom-180968745/

[39] **H. Pu, D. K. Zelenitsky, J. Lü, P. J. Currie, K. Carpenter, L. Xu, E. B. Coppelhus, S. Jia, L. Xiao, H. Chuang, T. Li, M. Kundrát, C. Shen,** May 9, 2017, Perinate and eggs of a giant caenagnathid dinosaur from the Late Cretaceous of central China: Nat. Commun., v8:14952. https://www.nature.com/articles/ncomms14952

[40a] **M. Butterfield,** Dec 22, 2021, Perfectly preserved, 'extremely rare' dinosaur embryo in egg found in China: Global News. https://globalnews.ca/news/8468563/dinosaur-egg-embryo-fossil-found-china/

[40] **P. Lal,** 2016, A deep natural history of the Indian subcontinent: INDICA. 468p. Allen Lane, Penguin Books.

[41a] **J. A. Wilson, D. M. Mohabey, S.E. Peters, J. J. Head,** Mar 02, 2010, Predation upon hatchling dinosaurs by a new snake from the Late Cretaceous of India: PLoS Biology, v8(3): p1-5.

[41] **Am. Geosciences Inst.,** How does a footprint become a fossil? Am. Geosciences. Inst. https://www.americangeosciences.org/static/files/education/files/fossil-footprints.pdf

[42] **W. Mingren,** Jan 23, 2018, Footprints on the wall: Were dinosaurs actually climbing the cliffs in Bolivia? Ancient Origins. https://www.ancient-origins.net/unexplained-phenomena/footprints-wall-were-dinosaurs-actually-climbing-cliffs-bolivia-009483

[43] **K. Lin,** Jan 29, 2017, Bolivian dinosaur footprints wall at Cal Orcko Cretaceous Park: https://www.historicmysteries.com/science/bolivia-dinosaur-footprints-wall/8250/

[44] **Sucre Life,** 2017, Sucre's Dinosaur Park and Footprints: https://www.sucrelife.com/sucres-dinosaur-park/

[45] **R. T. Reid,** Oct 27, 2015, Where Dinosaurs walked: Eight of the Best places to see prehistoric footprints: Smithsonian Magazine. https://www.smithsonianmag.com/travel/where-dinosaurs-walked-8-places-prehistoric-footprints-180956982/

[46] **PBS,** Evolution: Change: Deep Time: www.pbs.org>wgbh>chttps://www.pbs.org/wgbh/evolution/change/deeptime/cretaceous.html

[47] **S. Hide,** Mar 26, 2015, On the Dino trail in Villa de Leyva: The City Paper. https://thecitypaperbogota.com/travel/andes-travel-guide/on-the-dino-trail1503271/

[48] **PBS,** The Rise of Mammals: https://www.pbs.org/wgbh/evolution/library/03/1/l_031_01.html

[49] **K. Littler, S. A. Robinson, P. R. Brown, A. J. Nederbragt, R. D. Pancost,** Feb 20, 2011, High Sea surface temperatures during the Early Cretaceous Epoch: Nature Geoscience, v4: p169-172. Abstract.

[50] **C. R. Scotese, H. Song, B. J. W. Mills, and D. van der Meer,** Jan 2021. Phanerozoic Paleotemperatures: The Earth's Changing Climate during the last 540 million years, Earth Science Reviews, v215(11). https:/doi.org/10.1016/j.earscirev.2021.103503. Abstract.

[51] **J. Shelton,** Oct 21, 2019, Mystery solved: Ocean acidity in the last mass extinction: Yale University. https://news.yale.edu/2019/10/21/mystery-solved-ocean-acidity-last-mass-extinction

[52] **G. Keller,** Sept 2005, Impacts, volcanism, and mass extinction: Random coincidence or cause and effect? Aus. Jour. Earth Sci., v52(4-5) p725-757. [McNeal (1985), Courtillot et al (1986) and Officer et al (1987) cited in this paper].

[53] **G. V. R. Prasad and A. Sahni,** Aug 29, 2014, Vertebrate fauna from the Deccan Volcanic Province: Response to volcanic activity: Geol. Soc. America Sp. Paper, v505: p193-211. Abstract.

[54] **Wikipedia,** Xiphactinus: https://en.wikipedia.org/wiki/Xiphactinus

[55] **B. Schoene, M. P. Eddy, K. M. Samperton, C. B. Keller, G. Keller, T. Adatte, S. F. R. Khadri,** Feb 22, 2019, U-Pb constraints on pulsed eruption of the Deccan Traps across the end-Cretaceous mass extinction: Science, v363(6429): p862-866. https://www.science.org/doi/10.1126/science.aau2422

[56] **L. Fuller-Wright,** Mar 4, 2019, Did volcanoes kill the dinosaurs? New evidence points to 'maybe': Princeton University Office of Comm. https://www.princeton.edu/news/2019/03/04/did-volcanoes-kill-dinosaurs-new-evidence-points-maybe

[57] **G. Keller, P. Bhowmick, H. Upadhyay, A. Dave, A. Reddy, B. Jaiprakash, T. Adatte,** Nov 2011, Deccan volcanism linked to the Cretaceous-Tertiary boundary mass extinction: New evidence from ONGC wells in the Krishna-Godavari Basin, Jour. Geol. Soc. India, v78(5): p399-428.

[58] **G. Keller & K.V.S. Perch-Nielsen,** 1995, Cretaceous -Tertiary (K/T) Mass Extinction: Mass Extinction: Effect of global change on calcareous microplankton: Nat. Acad. of Sci. In 'Effect of past global change on life'. 272p. National Academic Press, U.S.A.

[59] **V. Vajda and A. Bercovici,** Nov 2014, The global vegetation pattern across the Cretaceous-Paleogene mass extinction interval: A template for other extinction events: Global & Planetary Change, v122: p29-49.

[60] **R. Wylie,** July 10, 2014, Earth's magnetic flip may have triggered mass extinctions: Discover Magazine. https://www.discovermagazine.com/planet-earth/earths-magnetic-flips-may-have-triggered-mass-extinctions

[61] **J. A. Jacobs,** Dec 01, 2001, The cause of superchrons: Astronomy & Geophysics, v42(6): p6.30-6.33.

[62] **J. A. Tarduno, R. D. Cottrell, A. V. Smirnov,** 2002, The Cretaceous superchron geodynamo: PNAS, USA, v99(22):14020-14025.

[63] **B. Yirka,** May 13, 2020, A possible explanation for the Earth's North magnetic pole moving toward Russia: Phys.org. https://phys.org/news/2020-05-explanation-earth-north-magnetic-pole.html

[64] **Y. Wei, Z. Pu, Q. Zong, W. Wan, Z. Ren, M. Fraenz, E. Dubinin, F. Tian, Q. Shi, S. Fu, M. Hong,** May 15, 2014, Oxygen escape from the Earth during geomagnetic reversals: Implications to mass extinction: Earth and Planetary Science Letters, v394: p94-98.

[65] **H. Baadsgaard, J. F. Lerbekmo, I. Mcdougall,** Feb 2011, A radiometric age for the Cretaceous-Tertiary boundary based upon K-Ar, Rb-Sr, and U-Pb ages of bentonites from Alberta, Saskatchewan, and Montana: Can. Jour. Earth Sci., v25(7). Abstract.

[66] **L. W. Alvarez, L. Alvarez, F. Asaro, H. V. Michel,** 1980, Extraterrestrial cause for the Cretaceous-Tertiary Extinction: Science, v208 (4448): p1095-1108.

[67] **K. Pope, A. C. Ocampo, A. G. Fischer, W. Alvarez,** July 1999, Chixulub impact ejecta from Albion Island, Belize: Earth and Planetary Sci. Letters, v170(4): p351-364. Abstract.

[68] **V. Vajda, A. C. Ocampo, E. Ferrow, C. B. Koch,** Jan 2014, Following the Chixulub impact—evidence from ejecta deposits in Belize, and Mexico: Gondwana Research. Abstract.

[69] **H. D. Bermúdez, J. García, W. Stinnesbeck, G. Keller, J. V. Rodríguez, M. Hanel, J. Hopp, W. H. Schwarz, M. Trieloff, L. Bolívar, F. J. Vega,** Dec 2015, Cretaceous-Paleogene boundary at Gorgonilla Island, Colombia, South America: Terra Nova, v28(1): p83-90. Abstract.

[70] **J. C. Sanford, J. W. Snedden, S. P. S. Gullick,** Mar 2016, The Cretaceous-Paleogene boundary deposit in Gulf of Mexico: Large-scale oceanic basin response to the Chixulub impact: Jour. Geophysical Res. Solid Earth, v121(3): p1240-1261.

[71] **R. A. De Palma, J. Smit, D. A. Burnham, K. Kuiper, P. L. Manning, A. Oleinik, P. Larson, F. J. Maurrasse, J. Vellekoop, M. A. Richards, L. Gurche, W. Alvarez,** April 23, 2019, A seismically induced onshore surge deposit at the K/Pg. boundary, North Dakota: PNAS, USA, v116(17): p8190-8199.

[72] **Univ. Washington News Staff,** Mar 29, 2019, North Dakota site shows wreckage from same object that killed the dinosaurs: https://www.washington.edu/news/2019/03/29/north-dakota-site-shows-wreckage-from-same-object-that-killed-the-dinosaurs/

[73] **B. Oskin,** Mar 27, 2014, Facts about the 1964 Alaska earthquake: Live Science. https://www.livescience.com/44412-1964-alaska-earthquake-facts.html

[74] **D. W. Jolley,** Dec 14, 2021, The Boltysh meteorite impact, Ukraine: verification of a KT boundary multi-impact hypothesis: Univ Aberdeen. Abstract. https://gtr.ukri.org/projects?ref=NE%2FD005043%2F1

[75] **D. W. Jolley, I. Gilmour, E. Gurov, S. Kelley, J. Watson,** Sept 2010, Two large meteorite impacts at the Cretaceous-Paleogene boundary: Geology, Sept 2010, v38(9): p835-838.

[76] **S. A. Stewart & P. Allen,** Aug 01, 2002, A 20-km diameter multi-ringed impact structure in the North Sea: Nature, v418(6897): p520-523.

[77] **G. Keller & Deccan Team,** 2008, 1. Deccan volcanism—An adventure in Science: Princeton University. https://gkeller.princeton.edu/deccan-volcanism/1-deccan-volcanism-%E2%80%93-adventure-science

[78] **L. Yarris,** Mar 9, 2010, Alvarez theory on Dinosaur die-out upheld: Experts find asteroid guilty of killing the dinosaurs: Berkeley lab. https://newscenter.lbl.gov/2010/03/09/alvarez-theory-on-dinosaur/

[79] **Planetary and Space Science Centre (PASSC),** Earth Impact Database: http://www.passc.net/EarthImpactDatabase/New%20website_05-2018/Index.html

[80] **K. Kaiho & N. Oshima,** Nov 9, 2017, Site of asteroid impact changed the history of life on Earth: the low probability of mass extinction: Scientific Reports 7, Article 14855(2017). https://www.nature.com/articles/s41598-017-14199-x

[81] **G. Keller, T. Adatte, Z. Berner, M. Harting, G. Baum, M. Prauss, A. Tantawy, D. Steuben,** 2007, Chixulub impact predates K/T boundary: New evidence from Brazos, Texas: Earth and Planetary Science Letters, v255: p339-356.

[82] **M. A. Richards, W. Alvarez, S. Self, L. Karlstrom,** Nov 2015, Triggering of the largest Deccan eruptions by the Chixulub impact: Bull. Geol. Soc. America, v127(11-12): p1507-1520. Abstract.

[83] **S. Brussatte,** 2018, The Rise and Fall of Dinosaurs—A new history of a lost world: 416p. Harper Collins.

[84] **R. Lopez,** Nov 22, 2019, A Raja, a Jain, a Tagore: Meet desi dinosaurs that once roamed India's plains: Hindustan Times, India.

[85] **Deutesche Presse-Agentur-DPA,** Dec 31, 2023, Indian villagers worship stones believed to be dinosaur eggs: Daily Sabah. https://www.dailysabah.com/life/indian-villagers-worship-stones-discovered-to-be-dinosaur-eggs/news

[86] **Gandhinagar Online,** Indroda Nature Park: https://www.gandhinagaronline.in/guide/indroda-nature-park

Chapter 9

[1] **L. Wald, K. Scharer, C. Prentice,** June 01, 2017, Back to the Future on the San Andreas Fault: USGS. https://www.usgs.gov/programs/earthquake-hazards/science/back-future-san-andreas-fault

[2] **R. Somoza,** Mar 24, 2007, Eocene paleomagnetic pole for South America: Northward continental motion in the Cenozoic, opening of Drake Passage and Caribbean convergence: Jour. Geophysical Res., Solid Earth. v112(B3).

[3] **J. E. Case, T. L. Holcombe, R. G. Martin,** 1984, Map of Geologic Provinces of the Caribbean region: 1-31, Geol. Soc. America Mem. 162. In Caribbean-South American Plate boundary and Regional Tectonics. 421p. (Eds W. E. Bonini, R. B. Hargraves, R. Shagam).

[4] **A. Escalona & P. Mann,** April 01, 2006, An overview of the petroleum system of Maracaibo Basin: Bull. AAPG, v90(4): p657-678.

[5] **O. Jagoutz, F. A. Macdonald, L. Royden,** May 3, 2016, Low latitude arc-continent collision as a driver for global cooling, PNAS, USA, v113(18): p4935-4940. Abstract.

[6] **J. Chu,** Feb 6, 2013, India joined with Asia 10 million years later than previously thought: MIT News: https://news.mit.edu/2013/india-joined-with-asia-10-million-years-later-than-previously-thought-0206

[7] **P. G. DeCelles, D. M. Robinson, G. Zandt,** 2002, Implications of shortening in the Himalayan fold-thrust belt for uplift of the Tibetan Plateau: Tectonics, v21(6): p12-1 to 12-25.

[8] **USGS,** Mar 09, 2015, The Himalayas: Two continents collide. https://pubs.usgs.gov/gip/dynamic/himalaya.html

9 **A. Zijlstra,** June 07, 2018, Fossils of Mount Everest: Volcano Café. https://www.volcanocafe.org/fossils-of-mount-everest/

[10] **S. Pappas,** Oct 11, 2016, How did a chunk of India and Eurasia just disappear? LiveScience: (Reported work of D. Rowley et al., Univ. Chicago).

[11] **A. Alden,** Oct 27, 2019, Geology of the Tibetan Plateau: ThoughtCo. https://www.thoughtco.com/all-about-the-tibetan-plateau-1441240

[12] **A. Zanchi, F. T. Fürsich, M. Santosh.,** April 2015, Cimmerian Terrane: Preface: Jour. Asian Earth Sci., v102: p1-3. Abstract.

[13] **Wikipedia,** Cimmeria. https://en.wikipedia.org/wiki/Cimmeria

[14] **V. D. Mats and T. I. Perepelova,** July 2011, A new perspective on the evolution of the Baikal rift: Geoscience Frontiers, v2(3): p349-365.

[15] **NASA Earth Observatory,** July 2, 2016, Morocco's High Atlas Mountains: https://earthobservatory.nasa.gov/images/89511/moroccos-high-atlas-mountains

[16] **M. W. Mikesell,** Atlas Mountains: Britannica: https://www.britannica.com/place/Atlas-Mountains

[17] **W. Beauchamp, R. W. Allmendinger, M. Barazangi, A. Demnati, M. E. Alji, M. Dahmani,** April 1999, Inversion tectonics and the evolution of the High Atlas Mountains, Morocco, based on a geological-geophysical transect: Tectonics, v18(2): p163-184.

[18] **Wikipedia,** North Anatolian Fault: https://en.wikipedia.org/wiki/North_Anatolian_Fault

[19] **E. V. Sharkov, V. A. Lebedev, A. G. Rodnikov, A. V. Chugaev, N. A. Sergeeva, L. P. Zabarinskaya,** July 2012, Features of the Caucasus segment of the Alps-Himalayan convergence zone: Geology, volcanological, neotectonical and geophysical data. In 'Tectonic – Recent Advances'. 350p. (Ed. E. Sharkov). https://www.intechopen.com/chapters/37861

[20] **L. L. Haynes and B. Honisch,** Sept 29, 2020, The seawater carbon inventory at the Paleocene-Eocene Thermal Maximum: PNAS, USA, v117(39): p24088-24095.

[21] **USGS,** Climate Res. & Dev. Program, Nov 30, 2019, Geologic records of a warmer Arctic: Earth Science Matters Newsletter—Fall 2019 issue: https://www.usgs.gov/news/geologic-records-a-warmer-arctic

[22] **J. D. Wright and M. F. Schaller,** Oct 1, 2013, Evidence for a rapid release of carbon at the Paleocene-Eocene Thermal Maximum: PNAS, USA, 110(40): p15908 -15913.

[23] **A. Tripaty & D. Darby,** Mar 12, 2018, Evidence for ephemeral middle Eocene to early Oligocene Greenland glacial ice and pan-Arctic Sea ice: Nature Communication, v9(1):1038.

[24] **Z. Liu, M. Pagani, D. Zinniker, R. Deconto,** Feb 27, 2009, Global cooling during the eocene-oligocene climate transition: Science, v323 (5918): p1187-1190. Abstract. https://pubmed.ncbi.nlm.nih.gov/19251622/

[25] **A. Zanazzi, M. J. Kohn, B. J. MacFadden, D. O. Terry,** 2007, Large temperature drop at the Eocene-Oligocene transition in central North America, Nature, v445: p639-642, 2007.

[26] **S. Earle,** 2015, 16.1 Glacial periods in Earth's History—Physical Geology: https://opentextbc.ca/geology/chapter/16-1-glacial-periods-in-earths-history/

[27] **A. Escalona & P. Mann,** Feb 24, 2005, Regional tectonics, stratigraphy, reservoir characterization and petroleum systems in an Eocene Foreland Basin, Maracaibo Basin, Venezuela: AAPG Datapages, Search & Discovery article 10078. https://www.searchanddiscovery.com/pdfz/documents/2005/escalona/images/escalona.pdf.html

[28] **S. C. Talukdar & F. Mercano,** 1994, Petroleum systems of the Maracaibo Basin: AAPG Search & Discovery Article #90986. Abstract. https://www.researchgate.net/publication/279908325_Petroleum_systems_in_the_Maracaibo_Basin_Venezuela

[29] **Wikipedia,** Maracaibo Basin. https://en.wikipedia.org/wiki/Maracaibo_Basin

[30] **GECSC,** USGS, 2010, Laramide Basin evolution: https://www.usgs.gov/media/images/laramide-basin-evolution-task-study-area

[31] **The Cretaceous Atlas,** Geology of the Western Interior Seaway: https://www.cretaceousatlas.org/geology/

[32] **Passing Strangeness,** Mar 2009, Australian Inland Sea: https://passingstrangeness.wordpress.com/2009/03/04/the-australian-inland-sea/

[33] **National Museum Australia,** Primary Source Study—Great Artesian Basin Map: https://www.nma.gov.au/learn/kspace/prehistoric-australia-110-million-years-ago/teacher-resources/primary-source-study

[34] **A. R. Loelich & P. Hallock,** 2017, Nummulites: McGraw Hill Access Science.Abstract: https://www.accessscience.com/content/article/a462000

[35] **S. Sarangi, A. Sarkar, M. M. Sarin, S. K. Bhattacharya, M. Ebihara, A. K. Ray,** Dec 2001, Growth rate and life span of Eocene/Oligocene Nummulite tests: inferences from Sr/Ca ratio: Terra Nova, v13(4): p264-269. Abstract. https://onlinelibrary.wiley.com/doi/abs/10.1046/j.1365-3121.2001.00349.x

[36] **AMNH,** Indricotherium: https://www.amnh.org/exhibitions/extreme-mammals/meet-your-relatives/indricotherium

[37] **Wikipedia,** Ring of Fire: https://en.wikipedia.org/wiki/Ring_of_Fire

[38] **S. Green,** May 18, 2018, The Geology of Mount Everest: TripSavvy. https://www.tripsavvy.com/geology-of-mount-everest-755308

[39] **Metageologist,** June 10, 2012, The Geology of Mount Everest: https://all-geo.org/metageologist/2012/06/the-geology-of-mount-everest/

[40] **H. Sakai, M. Sawada, Y. Takigami, Y. Orihashi, T. Danhara, H. Iwano, Y. Kuwahara, Qi. Dong, H. Cai, J. Li,** Dec 21, 2005, Geology of the summit limestone of Mount Qomalangma (Everest) and cooling history of the Yellow Band under the Qomalangma detachment: The Island Arc, v14: p297-310.

[41] **M. P. Searle, R. L. Simpson, R. D. Law, D. Waters, R. R. Parrish,** 2003, The structural geometry, metamorphic and magmatic evolution of the Everest massif, High Himalaya of Nepal-South Tibet: Jour. Geol. Soc. London, v160: p345-366. https://www.researchgate.net/publication/266672263_The_structural_geometry_metamorphic_and_magmatic_evolution_of_the_Everest_massif_High_Himalaya_of_Nepal_-_South_Tibet

[42] **P. Huangfu, Z-H. Li, T. Gerya, W. Fan, K-J. Zhang, H. Zhang, Y. Shi,** Sept 17, 2018, Multi-terrane structure controls the contrasting lithospheric evolution beneath the western and central-eastern Tibetan Plateau: Nat Communications 9, Article Number 3780(2018). https://doi.org/10.1038/s41467-018-06233-x

[43] **J. F. Dewey, R. M. Shackleton, C. Chenfa, S. Yiyin,** Dec 12, 1988, The tectonic evolution of the Tibetan Plateau: Phil. Trans. Royal Soc. https://royalsocietypublishing.org/doi/10.1098/rsta.1988.0135

[44] **J. Li and X. Fang,** 1999, Uplift of the Tibetan Plateau and environmental changes: Chinese Science Bull. v44, p2117-2124. https://doi.org/10.1007/BF03182692

[45] **Wikipedia,** Alps: https://en.wikipedia.org/wiki/Alps

[46] **K. Jennings,** July 23, 2018, 'Part of the Alps' Matterhorn is in Africa: https://www.cntraveler.com/story/part-of-the-alps-matterhorn-mountain-is-in-africa

[47] **S. Khanolkar and J. Sharma,** 2019, Record of Early and Middle Eocene paleoenvironmental changes from lignite mines, western India: Jour. Micropaleontology, v 38(1): p1-24.

[48] **J. Pickrell,** Oct 2019. Remarkable fossil beds capture mammals' recovery after the dino-killing asteroid: Science News. https://www.sciencenews.org/article/fossils-capture-mammals-recovery-after-dinosaur-killing-asteroid

[49] **AMNH,** The evolution of Horses: https://www.amnh.org/exhibitions/horse/the-evolution-of-horses

[50] **S. Williams,** Nov 20, 2014, Ancient relative of horses, rhinoceros originated in India more than 50 million years ago, fossils show: John Hopkins Univ. https://hub.jhu.edu/2014/11/20/india-fossils-perissodactyla/

[51] **K. D. Rose, L. T. Holbrook, R. S. Rana, Kishor Kumar, K. E. Jones, H. E. Ahrens, P. Missiaen, A. Sahni, T. Smith,** Nov 20, 2014, Early Eocene fossils suggest that the mammalian order Perissodactyla originated in India: Nat. Communication v5(5570).

[52] **D. T. Ksepka, T. A. Stidham, T. E. Williamson.**, 2017, Early Paleocene land bird supports rapid phylogenetic and morphological diversification of crown birds after the K-Pg extinction: PNAS, USA, v114: p8047-8052.

[53] **D. J. Field,** July 25, 2017, Big-time insights from a tiny bird fossil: PNAS, USA, v114(30): p7750-7752.

[54] **V. L. Masaitis,** Mar 1998, Popigai crater: Origin and distribution of diamond-bearing impactites: Meteorites and Planetary Science, v33(2): p349-359. Abstract.

[55] **UCMP—Berkeley,** The Oligocene Epoch: https://ucmp.berkeley.edu/tertiary/oligocene.php

[56] **J. Sun, X. Ni, S. Bi, W. Wu, J. Ye, J. Meng, B. F. Windley,** 2014, Synchronous turnover of flora, fauna, and climate at the Eocene-Oligocene boundary in Asia: Scientific Reports 4, Article 7463(2014). https://pmc.ncbi.nlm.nih.gov/articles/PMC4264005/

[57] **T. Lewis,** May 15, 2013, Oldest fossils reveal when Apes and monkeys first diverged, (Reported research of Nancy Stevens): https://www.livescience.com/32029-oldest-monkey-fossil-found.html

[58] **PBS,** The Oligocene Change: Deep Time: https://www.pbs.org/wgbh/evolution/change/deeptime/oligocene.html

[59] **Wikipedia,** Oligocene: https://en.wikipedia.org/wiki/Oligocene

[60] **Britannica,** Primate—Oligocene, Evolution, Adaptations: https://www.britannica.com/animal/primate-mammal/Oligocene

[61] **USGS,** Geology and Ecology of National Parks, Geology of Rocky Mountain National Park: https://www.usgs.gov/geology-and-ecology-of-national-parks/geology-rocky-mountain-national-park

[62] **NS ENERGY,** Nov 9, 2020, Cerrejón Open-pit Coal Mine, La Guajira, Colombia: https://www.nsenergybusiness.com/projects/cerrejon-coal-mine/

[63] **Jonathan Bloch,** Titanoboa—Rare, Beautiful & Fascinating—Florida Museum of Natural History: https://www.floridamuseum.ufl.edu/100-years/

[64] **K. James,** 2005, Distribution and tectonic significance of Cretaceous-Eocene flysch/wild flysch deposits of Venezuela and Trinidad: Transactions 16th Caribbean Geological Conference, Barbados. Caribbean Journal of Earth Science, v39(2005): p29-46.

[65] **Sciencedirect.com,** The Great Artesian Basin: https://www.sciencedirect.com/topics/engineering/great-artesian-basin

Chapter 10

[1] **Historical Geology,** San Andreas Fault and Basin and Range: https://opengeology.org/historicalgeology/san-andreas-fault-and-basin-and-range/

[2] **G. H. Haug and L. D. Keigwin,** March 22, 2004, How the Isthmus of Panama put ice in the Arctic: Oceanus Magazine, W.H.O.I. https://www.whoi.edu/oceanus/feature/how-the-isthmus-of-panama-put-ice-in-the-arctic/

[3] **S. A. Whattam and R. J. Stern,** Dec14, 2015, Arc magmatic evolution and the construction of continental crust at the Central American Volcanic Arc system: Jour. International Geol. Review, v58(6): p653-686.

[4] **A. O'Dea, H. Lessios, R. I. Eytan, A. G. Coates, +25 authors,** Aug 2016, Formation of the Isthmus of Panama: Science Adv. AAAS, v2(8).

[5] **Cardiff University News,** Feb 7, 2019, Volcanic growth 'critical' to the formation of Panama—Cardiff University News: https://www.cardiff.ac.uk/news/view/1434207-volcanic-growth-critical-to-the-formation-of-panama

[6] **NASA Earth Observatory,** Dec 30, 2003, Panama: Isthmus that changed the World: https://earthobservatory.nasa.gov/images/4073/panama-isthmus-that-changed-the-world

[7] **R. Peckyno,** Dec 12, 2011, What is a hot spot and what is a mantle plume? Volcanoworld. https://volcanoworld.wordpress.com/2011/12/12/what-is-a-hot-spot-and-what-is-a-mantle-plume/

[8] **Wikipedia,** Hawaiian Islands: https://en.wikipedia.org/wiki/Hawaiian_Islands

[8a] **M. Regueiro, M. Stamatakis, K. Laskaridis,** Nov 2014, The geology of Acropolis (Athens, Greece). European Geologist, v38(nov.2014): p45-52. https://www.researchgate.net/publication/269076024

[9] **Univ. of Maryland,** The Messinian Salinity Crisis: https://www2.atmos.umd.edu/~dankd/MessinianWeb/_private/HOME.htm

[10] **C. G. C. van Baak, W. Krijgsman, I. Magyar, O. Sztanó, L. A. Golovina, A. Grothe, T. M. Hoyle, O. Mandic, I. S. Patina, S. V. Popov, E. P. Radionova, M. Stoica, I. Vasiliev,** Sept 2017, Paratethys response to the Messinian salinity crisis: Earth-Science Reviews, v172: Abstract. https//doi.org/10.1016/j. earscirev.2017.07.015

[11] **M. Roveri, R. Flecker, W. Krijgsman, J. Lofi, S. Lugli, V. Manzi, F. J. Sierro, A. Bartini, A. Camerlenghi, G. J. de Lange, R. Govers, F. Hilgen, C. Heubscher, P. Th. Meijer, M. Stoica,** 2014, The Messinian Salinity Crisis: Past and future of a great challenge for marine sciences: Marine Geology, v352: p25-58. https://doi.org/10.1016/j.margeo.2014.02.002

[12] **The Conversation,** Mar 29, 2018, Africa is splitting in two—here is why: https://theconversation.com/africa-is-splitting-in-two-here-is-why-94056

[13] **Wikipedia,** East African Rift: https://en.wikipedia.org/wiki/East_African_Rift

[14] **NASA,** Feb 15, 2020, Earthquake in Eastern Africa: https://earthobservatory.nasa.gov/images/15855/earthquake-in-eastern-africa

[15] **Zambia Tourism,** Lake Tanganyika: https://www.zambiatourism.com/destinations/lakes/lake-tanganyika/

[16] **Britannica,** Andes Mountains Definition, Characteristics, Types & Facts: https://www.britannica.com/place/Andes-Mountains

[17] **Wikipedia,** Chinchorro mummies: https://en.wikipedia.org/wiki/Chinchorro_mummies

[18] **J. Houston & A. J. Hurtley,** Oct 2003, The Central Andes west slope rain shadow and its potential contribution to the origin of HYPER-ARIDITY in the Atacama Desert: Inter. Jour. Climatology, v23(12):p1453-1464. Abstract.

[19] **Beautiful World,** Atacama Desert fact and information, map, skeleton—Chile: https://www.beautifulworld.com/south-america/chile/atacama-desert/

[20] **S. Zielinski,** Sept 17, 2014, The Sahara is millions of years older than thought: Smithsonian Magazine. https://www.smithsonianmag.com/science-nature/sahara-millions-years-older-thought-180952735/

[21] **AstrobiologyWiki,** Present day Mars habitability analogue environments on Earth: https://encyclopediaofastrobiology.org/wiki/Present_day_Mars_habitability_analogue_environments_on_Earth

[22] **Wikipedia,** Atacama Giant: https://en.wikipedia.org/wiki/Atacama_Giant

[23] **Atlas Obscura,** Feb 21, 2018, Nazca lines: https://www.atlasobscura.com/places/nazca-lines-peru

[24] **Wikipedia,** Ape: https://en.wikipedia.org/wiki/Ape

[25] **Britannica,** Pasco Knot | plateau, Peru: https://www.britannica.com/place/Pasco-Knot

[26] **R. Cano,** April 29, 2020, How Machu Picchu successfully eluded the Spaniard for generations: We Are The Mighty: https://www.wearethemighty.com/mighty-history/macchu-pichu-remained-hidden/

[27] **History.com,** Aug 21, 2018, Machu Picchu: https://www.history.com/articles/machu-picchu

[28] **R. Dalton,** Oct 1, 2009, Oldest hominid skeleton revealed: Nature (2009). https://doi.org:10.1038/news. 2009.966.

[29] **GeoPostings,** June 13, 2016, Machu Picchu—The Geological Landscape: https://geopostings.com/machu-picchu-the-geological-landscape/

[30] **Wikipedia,** Ciudad Mitad del Mundo: https://en.wikipedia.org/wiki/Ciudad_Mitad_del_Mundo

[31] **Ingapirca,** Inca ruins in Ecuador (2022): parks-and-tribes. https://www.parks-and-tribes.com/inca/inca-culture-ingapirca-ruins.htm

[32] **J. Bennett & S. Shostak,** 2007, Life in the Universe: 485p. Pearson Addison Wesley Publisher.

[33] **M. Cartwright,** Sept 8, 2014, The Inca Road System: Ancient History Encyclopedia. https://www.worldhistory.org/article/757/the-inca-road-system/

[34] **C. Vinasco and U. Cordani,** Nov. 2012, Reactivation episodes of the Romeral Fault system in the northwestern part of Central Andes Colombia, through 39AR-40AR and K-AR results: Bol.de Ciencia de la Tierra, 32(32): P111-124, Medellin.scielo.org.co

[35] **J. F. Duque-Trujillo, C. Bustamente, L. Solari, A. Gomez-Mafla, G. Toro Villegas, S. Hoyos,** 2019, Reviewing the Antioquia Batholiths and Satellite bodies: A record of late Cretaceous to Eocene syn- to post-collisional arc magmatism in the Central Cordillera of Colombia: Andean Geology, v46(1): p82-10.

[36] **B. K. Horton, M. Parra, J. E. Sailor, J. Nie, A. Mora, V. Torres, D. F. Stockli, M. R. Strecker,** July 2010, Resolving uplift of the northern Andes using detrital zircon age signatures: GSA Today, Article:4-10.

[37] **SoftSchools.com,** Missouri River Facts: https://www.softschools.com/facts/rivers/missouri_river_facts/2417/

[38] **GreekMythology.com,** Amazons: https://www.greekmythology.com/Myths/Figures/Amazons/amazons.html

[39] **Wikipedia,** Amazon River: https://en.wikipedia.org/wiki/Amazon_River

[40] **Univ. van Amsterdam,** July 8, 2009, The Amazon River is 11 million years old: ScienceDaily, July 8, 2009. Abstract. https://www.sciencedaily.com/releases/2009/07/090707155827.htm

[41] **V. Sacek,** Sept 2014, Drainage reversal of the Amazon River due to the coupling of surface and lithospheric processes: Earth & Planetary Science Letters, v401: p301-312. Abstract. https://ui.adsabs.harvard.edu/link_gateway/2014E&PSL.401..301S/doi:10.1016/j.epsl.2014.06.022

[42] **S. Perkins,** July 15, 2014, Why the Amazon flows backward: Science Journal. https://www.science.org/content/article/why-amazon-flows-backward

[43] **J. S. Albert, P. Val, C. Hoorn,** Oct 18, 2018, The changing course of the Amazon River in the Neogene: Center stage for Neotropical Diversification: Neotropical Ichthyology, v16(3):180033(1-23). https://doi.org/10.1590/1982-0224-20180033

[44] **E. T. Pimentel and V. M. Hamza,** 2012, Indications of regional scale ground water flows in the Amazon basins: Jour. South Am. Earth Sci., v37: p214-227.

[45] **L. de Gamero,** 1996, The changing course of the Orinoco River during the Neogene: Palaeogeography, Palaeoclimatology and Palaeoecology, v123(4): p385-402. Abstract. https://www.osti.gov/biblio/6064825

[46] **K. S. Valdiya,** Jan 2016, Indo-Gangetic Plains: Evolution and later development: p723-745. Abstract. In 'Making of India', Nov 26, 2015, 924p. Springer. DOI:10.1007/978-3-319-25029-8_22.

[47] **K. Borah, S. Lal, K. S. Prakasam, N. Kanna**, Mar 2015, Sediment thickness beneath the Indo-Gangetic Plain and Siwalik Himalaya inferred from receiver function modelling: Jour. of Asian Earth Sci., v99: p41-56. https://doi.org/10.1016/j.jseaes.2014.12.010

[48] **JOIDES Resolution,** 2015, Bengal Fan: https://joidesresolution.org/expedition/354/

[49] **Wikipedia,** Bengal Fan: https://en.wikipedia.org/wiki/Bengal_Fan

[50] **Wikipedia,** Catarrhini: https://en.wikipedia.org/wiki/Catarrhini

[50a] **I. Nengo, E. Tafforeau, C. C. Gilbert, J. G. Fleagle, E. R. Miller, C. Feibel, D. L. Fox, J. Feinberg, K. D. Pugh, C. Berruyer, S. Mana, Z. Engle, F. Spoor,** Aug 10, 2017, New infant cranium from the African Miocene shed light on ape evolution: Nature Journal, v548: p169-174.

[51] **Banglapedia,** June 18, 2021, Swatch of No Ground: https://en.banglapedia.org/index.php/Swatch_of_no_Ground

[52a] **C. Faccena, P. Glisovic, A. Forte, T. W. Becker, E. Garzanti, A. Sembroni, J. Gvirtzman,** 2019, Role of dynamic topography in sustaining the Nile River over 30 million years. Nature Geoscience, v12, p1012-1017.

[52] **V. Kolla, A. Bandyopadhyay, P. Gupta, B. Mukherjee,** 2012, Morphology and Internal structure of a recent Upper Bengal Fan-Valley Complex: SEPM Sp. Pub. 99: p347-369.

[53] **G. Musumeci & L. Vaselli,** 2012, Neogene deformation and granite emplacement in the metamorphic units of northern Apennines (Italy): Insights from mylonitic marbles in the Porto Azzurro pluton contact aureole (Elba Island): Geosphere, v8(2): p470-490.

[54] **P. Copeland, T. M. Harrison, P. Le Fort,** Dec 1990, Age, and cooling history of Manaslu granite: implications for Himalayan tectonics: Jour. of Volcanology and Geothermal Research, v44(1): p33-50. Abstract. https://www.sciencedirect.com/science/article/abs/pii/037702739090010D

[55] **H. Ito, R. Yamada, A. Tamura, S. Arai, K. Horie, T. Hokada,** Feb 2013, Earth's youngest exposed granite and its tectonic implications: the 10-0.8 Ma Kurobegawa granite: Scientific Reports—Nature, Article 1306. https://www.nature.com/articles/srep01306

[56] **A. Kumar, M. Ali, R. Kumar, M. Kumar, P. Sagar, R. K. Pandey, V. Akhouri, V. Kumar, G. Anand, P. K. Niraj, R. Rani, S. Kumar, D. Kumar, A. Biswapriya, A. K. Ghosh,** Jan 27, 2021, Arsenic exposure in Indo-Gangetic plains of Bihar causing increased cancer risk: Sci. Report-NIH, v11(1):2376. Abstract.

[57] **NatureScott,** Feb 2022, Paleogene and Neogene rocks: https://www.nature.scot/landforms-and-geology/scotlands-rocks-landforms-and-soils/rocks-and-minerals/rocks-formed-foundations-joined/palaeogene-and-neogene-rocks

[58] **D. Roemnich,** Jan 03, 2011, Voyager—What makes the ocean salty and how much salt is in the ocean? Scripps Inst. of Oceanography—UCSD magazine 'exploration now.' https://scripps.ucsd.edu/news/voyager-what-makes-ocean-salty-and-how-much-salt-ocean

[59] **Wikipedia,** Evaporite: https://en.wikipedia.org/wiki/Evaporite

[60] **J. García-Veigas, F. Orti, L. Rosell, C. Ayora, J. M. Rouchy, S. Lugli,** Jan 1995, Messinian salt of the Mediterranean: geochemical study of the salt from the Central Sicily Basin and comparison with the Lorca basin (Spain): Bul. De la Soc. Geol. de France, v166(6): p699-710. Abstract.

[61] **J. Davis,** Megalodon: The truth about the largest shark that ever lived: Natural History Museum. https://www.nhm.ac.uk/discover/megalodon--the-truth-about-the-largest-shark-that-ever-lived.html

[62] **Ktb.gov.tr,** Kaymakli Underground City: https://www.ktb.gov.tr/EN-113739/kaymakli-underground-city.html

[63] **Wikipedia,** Kaymakli Underground City: https://en.wikipedia.org/wiki/Kaymakli_underground_city

[64] **A. Voiland,** Jan 7, 2019, Lithium harvesting at Salar de Uyuni: Earth Observatory-NASA. https://earthobservatory.nasa.gov/images/144976/lithium-harvesting-at-salar-de-uyuni

[65] **R. Wooster and C. M. Sanders,** April 2019, TSHA|Spindletop Oilfield: Handbook of Texas Online. https://www.tshaonline.org/handbook/entries/spindletop-oilfield

[66] **J. D. Beckman and A. K. Williamson,** 1990, Salt-dome locations in the gulf coastal plain, south-central United States: USGS Report 90-4060. https://www.usgs.gov/publications/salt-dome-locations-gulf-coastal-plain-south-central-united-states

[67] **P. T. Gómez-Cabrera and M. P. A. Jackson,** 2009, Regional Neogene salt tectonics in the offshore Salina del Istmo basin, southeastern Mexico: AAPG Mem., v90:p1-28. In Petroleum Systems in the southern Gulf of Mexico, AAPG Mem. 90, 502p. (Eds. C. Bartolini and J. R. Román Ramos).

[68] **S. Ortuño Arzate, S. M. Esteban, R. M. Cuauhtémoc, R. R. Joel,** 2009, Two-dimensional and three-dimensional numerical simulation of petroleum systems approaching the deep-water Gulf of Mexico (Kayab area, Campeche Sound, Mexico) Definition of thermally mature and prospective areas: AAPG Mem., v90:p117-136. In Petroleum Systems in the southern Gulf of Mexico. AAPG Mem. 90, 502p. (Eds. C. Bartolini & J. R. Román Ramos).

[69] **K. Milam,** April 2021, Evolving understanding of Salt diapirs is critical to oil exploration: AAPG Explorer. https://explorer.aapg.org/story/articleid/59985/evolving-understanding-of-salt-diapirs-is-critical-to-oil-exploration

[70] **C. Kahn,** Sept 3, 2009, BP drills deepest-ever oil well in the Gulf of Mexico: The Ledger. https://www.theledger.com/story/news/2009/09/03/bp-drills-deepest-ever-oil-well-in-gulf-of-mexico/26640424007/

[71] **AAPG,** May 1998, Gulf of Mexico Petroleum Systems: AAPG Bull., v82/5B: p865-1112.

[72] **V. M. Chavez Valois, M. de Lourdes C. Valdés, J. I, Juárez Placencia, I. Alor Ortiz, M. Mata Jurado, R. V. Yánez, M. Guerrero Tristán, and S. Ghosh,** 2009, A new multidisciplinary focus in the study of the Tertiary plays in the Sureste basin, Mexico: AAPG Mem., v90:p155-190. In Petroleum Systems in the southern Gulf of Mexico: AAPG Mem. 90, 502p. (Eds. C. Bartolini and J. R. Román Ramos).

[73] **PBS Nature,** Mar 21, 2012, Going Aquatic: Cetacean Evolution: https://www.pbs.org/wnet/nature/ocean-giants-going-aquatic-cetacean-evolution/7577/

[74] **R. Lloyd,** Sept 08, 2008, Theory of flightless birds shot down: LiveScience. https://www livescience. com/2846-theory-flightless-birds-shot.html

[75] **Wikipedia,** Great American Interchange: https://en.wikipedia.org/wiki/Great_American_Interchange

[76] **NASA,** Mar 23, 2008, Panama: Isthmus that changed the world: https://earthobservatory.nasa.gov/images/4073/panama-isthmus-that-changed-the-world

[77] **K. McDougall,** August 1996, Benthic foraminiferal response to the emergence of the Isthmus of Panama and coincident paleooceanographic changes: Marine Micropaleontology, v28(2): p133-169. Abstract. https://doi.org/10.1016/0377-8398(95)00081-X

[78] **Nicaragua.com,** The Bull Sharks of Lake Nicaragua: https://www.nicaragua.com/blog/the-bull-sharks-of-lake-nicaragua/

[79] **D. F. Su,** 2013, The Earliest Hominins: Sahelanthropus, Orrorin, Ardipithecus: Nature Education Knowledge v4(4):11. https://www.nature.com/scitable/knowledge/library/the-earliest-hominins-sahelanthropus-orrorin-and-ardipithecus-67648286/

[80] **C. Attila, E. Aydar, M. Akif Sarikaya,** April 27, 2020, Volcanism, and evolution of the landscapes in Cappadocia: OpenEdition Books. https://books.openedition.org/ifeagd/3212

[81] **C. Heller,** Sep 21, 2015, Turkey's 'Fairy Chimneys' are millions ofWW years in the making: Smithsonian Magazine. https://www.smithsonianmag.com/travel/fairy-chimneys-turkey-180956654/

[82] **F. Dorey,** Jan 07, 2019, Sahelanthropus tchadensis: Australian Museum. https://australian.museum/learn/science/human-evolution/sahelanthropus-tchadensis/

Chapter 11

[1] **Smithsonian Inst.,** Pleistocene volcano list: https://volcano.si.edu/volcanolist_pleistocene.cfm

[2] **The Geol. Soc. UK.,** Plate Tectonics Continental-continental: The Himalayas—The geological: https://www.geolsoc.org.uk/Plate-Tectonics/Chap3-Plate-Margins/Convergent/Continental-Collision

[3] **K. M. Gregory-Wodzicki,** 2000, Uplift history of the central and Northern Andes: A review: GSA Bull., v112(7): p1091-1105. http://geomorphology.sese.asu.edu/Papers/gregory-wodzicki_andes_uplift_gsab00.pdf

[3a] **NOAA (Wikimedia),** Oct 4, 2019, File: Sea level temp 140ky.gif. https://commons.wikimedia.org/wiki/File:Sea_level_temp_140ky.gif

[4] **V. Gomitz,** Jan 2007, NASA, Science Briefs: Sea level rise, after the ice melted and today.

[5] **H. Lu, S. Yi, Z. Xu, Y. Zhou, L. Zeng, F. Zhu, H. Feng, L. Dong, H. Zhuo, K. Yu, J. Mason, X. Wang, Y. Chen, Q. Lu, B. Wu, Z. Dong, J. Qu, Xunming Wang, Z. Guo,** 2013, Chinese deserts and sand fields in the Last Glacial Maximum and Holocene Optimum: Abstract. Chinese Science Bull., v58(23): p1-9. https://link.springer.com/article/10.1007/s11434-013-5919-7

[6] **P. La Jeunesse and G. St-Onge,** Feb 2008, Subglacial origin of the Lake Agassiz-Ojibway: Nature Geoscience, v1(3): p184-188. https://www.researchgate.net/publication/31969704

[7] **J. Daley,** June 28, 2018, How the Belize Barrier Reef beats the endangered list: Smart News. https://www.smithsonianmag.com/smart-news/belize-reef-beats-endangered-list-180969485/

[7a] **P. Lasalle and R.G. Rogerson,** Feb 7, 2006, Champlain Sea: The Canadian Encyclopedia, https://www.thecanadianencyclopedia.ca/en/article/champlain-sea

[8] **C. Zazo, J. L. Goy, C. Hillaire-Marcel, P-Y Gillot, V. Soler, J. A. González-Delgado, C. J. Dabrio, B. Galheb,** Oct 2002, Raised marine sequences of Lanzarote and Fuerte revisited—A reappraisal of relative sea-level changes and vertical movements in the eastern Canary Islands during the Quaternary: Quat. Sci. Reviews, v21(18-19): p2019-2046. https://www.researchgate.net/publication/223851239

[9] **R. C. Speed and H. Cheng,** Jan 2004, Evolution of marine terraces and sea level in the last interglacial, Cave Hill, Barbados: Abstract. Bull. G.S.A., v116(1): p219-232. https://website.whoi.edu/gfd/wp-content/uploads/sites/14/2018/10/Speed_2004_49558.pdf

[10] **NPS.gov.,** Mar 31, 2020, Bering Land Bridge, National Preserve Alaska: https://www.nps.gov/bela/index.htm

[11] **NPS.gov.,** Feb 2022, Fossilized Footprints: White Sands National Park, USA: https://www.nps.gov/whsa/learn/nature/fossilized-footprints.htm

[12] **Wikipedia,** Adam's Peak: https://en.wikipedia.org/wiki/Adam's_Peak

[13] **HeritageDaily,** May 15, 2020, Doggerland-Europe's lost land: https://www.heritagedaily.com/2020/05/doggerland-europes-lost-land/117925

[14] **S. Natarajan,** Sept 30, 2021, Ram Setu Expedition: What to expect from this shallow sea mission? https://www.thequint.com/explainers/explained-the-ram-setu-adam-bridge-under-water-expedition-in-2021

[15] **Profilpelajar.com,** Adam's Bridge: https://profilpelajar.com/en/Adam's_Bridge

[16] **NASA MODIS Web,** July 27, 2018—Palk Strait: https://modis.gsfc.nasa.gov/gallery/individual.php?db_date=2018-07-27

[17] **A. Kumar,** Sept 14, 2007, Photos no proof of Ram Setu: NASA: https://tech.hindustantimes.com/tech/news/photos-no-proof-of-ram-setu-nasa-story-iSqcGrPS96TCb25hqjxF3M.html

[18] **N. Kilvert,** Oct 27, 2020, 'Detached' reef bigger than Empire State Building discovered in 500 meters off Queensland: ABC Science. https://www.abc.net.au/news/science/2020-10-27/reef-detached-discovered-cape-york/12816760

[19] **B. J. Winney, R. L. Hammond, W. Macasero, B. Flores, A. Boug, V. Biquand, S. Biquand, M. W. Bruford,** Sept 2004, Crossing the Red Sea: Phylogeography of the Hamadryas baboon, Papio Hamadryas hamadryas: Mol Ecology, v13(9):2819-2827. https://pubmed.ncbi.nlm.nih.gov/15315692/

[20] **Wikipedia,** Red Sea: https://en.wikipedia.org/wiki/Red_Sea

[21] **M. B. Pitrelli,** April 9, 2020, It's one of the harshest places on Earth—and travelers love it: CNBC. https://www.cnbc.com/2020/04/09/danakil-depression-hot-and-harsh-but-travelers-love-it.html

[22] **Brilliant Ethiopia,** Danakil Depression: https://www.brilliant-ethiopia.com/regions/danakil-depression

[23] **NASA,** Aug 26, 2014, Curiosities of the Danakil Depression: NASA Earth Observatory. https://earthobservatory.nasa.gov/images/84239/curiosities-of-the-danakil-depression

[24] **K. Norman,** April 02, 2018, How did the first people get to Australia? RealClearScience. https://www.realclearscience.com/articles/2018/04/02/how_did_the_first_people_get_to_australia_110595.html

[25] **MLH Cheltenham College,** Dec 16, 2021, Case study of one coastal landscape that is being managed: storymaps.arcgis.com/stories/843601cebe404ca7901500ebf7a75366

[26] **N. Eyles, H. T. Mullins, A. C. Hine,** 1991, The seismic stratigraphy of Okanagan Lake, British Colombia: a record of rapid deglaciation in a deep "fiord-lake" basin: Sedimentary Geology, v73(1-2):13-41. Abstract.

[27] **The Peterborough Examiner,** Oct 18, 2012, 'Drumlins': https://www.pressreader.com/canada/the-peterborough-examiner/20121018/281651072346668

[28] **Wikipedia,** Finger Lakes: https://en.wikipedia.org/wiki/Finger_Lakes

[29] **Oak Ridges Trail Assoc.,** Oak Ridges Moraine: https://www.oakridgestrail.org/moraine/

[30] **Soil Sci. Soc. America,** June 15, 2018, Why do the Loess Hills of Iowa need to be farmed in terraces: https://soilsmatter.wordpress.com/2018/06/15/why-do-the-loess-hills-of-iowa-need-to-be-farmed-in-terraces/

[31] **J. Bradburn,** July 2, 2015, "Toronto Feature" Scarborough Bluffs: The Canadian Encyclopedia. https://www.thecanadianencyclopedia.ca/en/article/toronto-feature-scarborough-bluffs

[32] **N. Eyles,** Aug 28, 2020, Scarborough Bluffs is on the move, a little bit at a time, Univ. of Toronto Scarborough: https://utsc.utoronto.ca/news-events/our-community/scarborough-bluffs-move-little-bit-time

[33] **Alberta.ca,** 2022, Okotoks Erratic—'Big Rock': https://www.alberta.ca/okotoks-erratic-big-rock

[34] **NASA Earth Observatory,** Aug 05, 2012, Perito Moreno Glacier, Argentina: https://earthobservatory.nasa.gov/images/78754/perito-moreno-glacier-argentina

[35] **L. Lipsett,** Nov 12, 2018, How do corals build their skeletons? Oceanus Magazine. https://www.whoi.edu/oceanus/feature/how-do-corals-build-their-skeletons/

[36] **Wikipedia,** Niagara Falls: https://en.wikipedia.org/wiki/Niagara_Falls

[37] **G. Larson and R. Schaetzl,** 2001, Origin and evolution of the Great Lakes: Jour. Great Lakes Research, v27(4): p518-546. Abstract. https://www.researchgate.net/publication/252246174

[38] **Niagara Parks,** Niagara Falls Geology Facts & Figures: https://www.niagaraparks.com/visit-niagara-parks/plan-your-visit/niagara-falls-geology-facts-figures/

[39] **Wikipedia,** Salar de Uyuni: https://en.wikipedia.org/wiki/Salar_de_Uyuni

[40a] **Wikipedia,** Older Dryus, https://en.wikipedia.org/wiki/Older_Dryas

[40] **K. Andrews,** June 14, 2016, What causes an ice age and what would happen if the earth endured another one? ABC Science. https://www.abc.net.au/news/science/2016-06-15/what-is-an-ice-age-explainer/7185002

[41] **National Center for Environmental Information,** Glacial-Interglacial cycles: https://www.ncei.noaa.gov/sites/default/files/2021-11/1%20Glacial-Interglacial%20Cycles-Final-OCT%202021.pdf

[42] **A. Buis,** Feb 27, 2020, Milankovitch (Orbital) cycles and their role in Earth's climate: NASA. https://science.nasa.gov/science-research/earth-science/milankovitch-orbital-cycles-and-their-role-in-earths-climate/

[43] **Volcano Hazards,** USGS, Oct 10, 1969, Dome fountain of episode 10, Oct 10-13, 1969: https://www.usgs.gov/media/images/dome-fountain-episode-10-oct-10-13-1969

[44] **S. Breslin,** Feb 13, 2019, Fiery Hawaii Lava Bubble Photo goes viral 49 years later: https://weather.com/news/news/2018-04-05-fiery-bubble-lava-dome-photo-usgs

[45] **Wikipedia,** Lower Puna eruption: https://en.wikipedia.org/wiki/2018_lower_Puna_eruption

[46] **R. G. Andrews,** Dec 12, 2018, America's most hazardous volcano erupted this year. Then it erupted and erupted: https://www.nytimes.com/2018/12/12/science/kilauea-hawaii-volcano-eruption.html

[47] **USGS,** Kilauea Volcano Erupts: https://www.usgs.gov/news/featured-story/kilauea-volcano-erupts

[48] **C. Leblang,** Oct 02, 2015, Zooxanthellae, and their symbiotic relationship with marine corals: Microbe Wiki. https://microbewiki.kenyon.edu/index.php/Zooxanthellae_and_their_Symbiotic_Relationship_with_Marine_Corals

[49] **PBS,** June 18, 2018, Laze, vog and other volcano vocabulary inspired by Kilauea: https://www.pbs.org/newshour/science/laze-vog-and-other-volcano-vocabulary-inspired-by-kilauea

[50] **T. Puiu,** April 26, 2019, Yellowstone supervolcano erupted twice 630,000 years ago: https://www.zmescience.com/science/geology/yellowstone-supervolcano-erupted-twice-04323/

[51] **USGS,** What is a supervolcano? What is a super-eruption? https://www.usgs.gov/faqs/what-a-supervolcano-what-a-supereruption

[52] **USGS,** How big is the magma chamber under Yellowstone? https://www.usgs.gov/faqs/how-big-magma-chamber-under-yellowstone

[53] **C. J. Southworth,** 2012, The Bishop Tuff: An overview of the world's roughest and toughest volcanic landform: https://sierra.sitehost.iu.edu/papers/2012/southworth.html

[54] **E. Ting,** Aug 14, 2018, California supervolcano discovered to be hiding 240miles^3 of magma: SFGATE. https://www.sfgate.com/news/article/California-supervolcano-magma-Long-Valley-Caldera-13153657.php

[55] **National Ocean Service,** NOAA, Zooxanthellae … What's that? Corals Tutorial: https://oceanservice.noaa.gov/education/tutorial_corals/coral02_zooxanthellae.html

[56] **C. A. Chesner,** 1998, Petrogenesis of the Toba Tuffs, Sumatra Indonesia: Jour. of Petrology, v39(3): p397-438. https://academic.oup.com/petrology/article-abstract/39/3/397/1483413

[57] **R. Raj,** 2008, Occurrence of volcanic ash in the Quaternary alluvium deposits, Lower Narmada basin, western India: Jour. of Earth System Dynamics, v117(1): p41-48.

[58] **M. R. Rampino and S Seif,** 1992, Volcanic winter and accelerated glaciation following the Toba super-eruption: Nature, v359: p50-52. Abstract. https://doi.org/10.1038/359050a0

[59] **J. A. Bayona Viveros, G. Suárez Reynoso, M. G. Ordaz-Schroeder,** Jan-Mar 2017, A probabilistic seismic hazard assessment of the Trans-Mexican Volcanic Belt, Mexico, based on historical and instrumentally recorded seismicity: Geofisica Internacional, v56 (1). https://doi.org/10.19155/geofint.2017.056.1.7

[60] **D. M-Zenteno,** 1994, Geology of the Mexican Republic: AAPG Studies in Geology #39,160p.

[61] **J. L. Arce, P. W. Layer, J. L. Macías, E. Morales-Casique, A. G. Palomo, F. J. Jiménez Dominguez, J. Benowitz, A. Vasquez,** 2019, Geology and stratigraphy of the Mexico Basin (Mexico City)—Central Trans-Mexican Volcanic Belt: Jour. of Maps, v15(2): p1-13.

[62] **J. Bradbury,** 1971, Paleolimnology of Lake Texcoco, Mexico, Evidence from diatoms: Limnology and Oceanography, v16(2): p180-200. https://doi.org/10.4319/lo.1971.16.2.0180

[63] **K. K. Landes,** 1973, Mother Nature as an Oil Polluter: A.A.P.G. Bull., v57(4): p637-641. pubs. https://doi.org/10.1306/819A430E-16C5-11D7-8645000102C1865D

[64] **Bible Gateway,** Genesis 14:10 NIV: https://www.biblegateway.com/passage/?search=Genesis%20 14%3A10&version=NIV

[65] **Bible Ref,** What does Exodus 2:3 mean? https://www.bibleref.com/Exodus/2/Exodus-2-3.html

[66] **Biblia,** Genesis 6:13-14: https://www.biblegateway.com/passage/?search=Genesis%206%3A13-14&version=NIV

[67] **Smithsonian Inst.,** Oct 2010, The whale fish mystery: Smithsonian Ocean Team. https://ocean.si.edu/ocean-life/fish/whalefish-mystery

[68] **Wikipedia,** Lake Bermudez: https://en.wikipedia.org/wiki/Lake_Bermudez

[69] **A. I. Levorsen,** 1954, Geology of Petroleum: A.A.P.G. Foundation, USA, 2001, 724p.

[70] **UNESCO World Heritage Centre,** Aug 17, 2011, La Brea Pitch Lake: https://whc.unesco.org/en/tentativelists/5645/

[71] **P. Bissessar,** Sept 9, 2021, Legends of the Pitch Lake: Angelo Bissessarsingh's Virtual Museum of Trinidad and Tobago. https://www.facebook.com/groups/191766699268/posts/10159610571069269/

[72] **T. Daley,** 2019, La Brea Pitch Lake: The largest tar pit in the world: GEOExPRo, v16(4). https://geoexpro.com/la-brea-pitch-lake-the-largest-tar-pit-in-the-world/

[73] **Wikipedia,** Pitch Lake: https://en.wikipedia.org/wiki/Pitch_Lake

[74] **D. Thrift-Viveros,** NOAA.Gov. June 15, 2015, Who thinks crude oil is delicious? These ocean microbes do: https://response.restoration.noaa.gov/about/media/who-thinks-crude-oil-delicious-these-ocean-microbes-do.html

[75] **Univ. of Texas,** May 09, 2016, Genetic Potential of oil-eating bacteria from the BP oil spill decoded: https://news.utexas.edu/2016/05/09/potential-of-oil-eating-bacteria-from-bp-oil-spill-decoded/

[76] **I. M. Head, D. Martin Jones, S. R. Larter,** Nov 20, 2003, Biological activity in the deep subsurface and the origin of heavy oil: Nature, v426 (6964): p344-352. Abstract. https://doi.org/10.1038/nature02134

[77] **K. Nooren, W. Z. Hoek, T. Winkels, A. Huizinga,** 2017, Usumacinta-Grijalva beach-ridge plain in southern Mexico: A high-resolution archive of river discharge and precipitation: Earth Surf Dynamics, v5(3): p529-556. http://dx.doi.org/10.5194/esurf-5-529-2017

[78] **A. J. Kopf,** 2002, Significance of mud volcanism: Reviews of Geophysics, v40(2):2-1:p2-52. https://doi.org/10.1029/2000RG000093

[79] **O. G. Gutierrez,** 2013, Chronologie et dynamique de la formación des terrasses fluviales dans des chaines des montagnes avec une surrection modere: l'example de Venezuela et de l'Albanie: https://theses.hal.science/tel-00934484v1

[80] **A. Norris,** Wikipedia, Dead Sea: https://en.wikipedia.org/wiki/Dead_Sea

[81] **The Editors of Encyclopedia Britannica,** Altiplano | Andes Plateau, Bolivia-Peru-Chile Border: https://www.britannica.com/place/Altiplano

[82] **Jessica S,** Aug 22, 2016, Floating gardens of Xochimilco: https://www.journeymexico.com/blog/floating-gardens-xochimilco

[83] **Arqueología Mexicana,** Danza del Volador: https://arqueologiamexicana.mx/mexico-antiguo/danza-del-volador

[84] **J. Bhattacharya and R. G. Walker,** 1992, (9) Deltas: Geol. Assoc. of Canada, Geotext 1: p157-177. In Facies Models Response to Sea Level Change. 454p. (Eds. R.G. Walker and N. P. James).

[85] **R. P. Agarwal and R. Bhoj,** 1992, Evolution of Koshi River fan, India: Structural implications and geomorphic significance: Int. Jour. of Remote Sensing, v13(10): p1891-1901. Abstract.

[86] **AMNH,** Earthquake risk in Bangladesh: Museum of Natural History. https://www.amnh.org/learn-teach/curriculum-collections/earthquake-risk-in-bangladesh

[87] **R. G. Walker and G. Plint,** 1992, (12) Waves and storm dominated shallow marine systems: Geol. Assoc. of Canada, GeoText1: p219-238. In Facies Models Response to Sea Level Change. 454p. (Eds. R.G. Walker and N. P. James).

[88] **S. E. Pratt,** NASA Earth Observatory, Turkey's Cotton Castle: https://earthobservatory.nasa.gov/images/149127/turkeys-cotton-castle

[89] **Geology Science,** Aug 19, 2023, Pamukkale Travertines, Turkey | Geology, Formation: https://geologyscience.com/gallery/geological-wonders/pamukkale-turkey/

[90] **H. Kumsar, Ö. Aydan, C. Simsek, F. D'Andria,** Sep 2015, Historical earthquakes that damaged Hierapolis and Laodikeia antique cities and their implications for earthquake potential of Denizli basin: Bull. Engineering Geology and the Environment, v75(2):p519-536. http://dx.doi.org/10.1007/s10064-015-0791-0

[91] **E. C. Rankey and B. Reigl,** Dec 2006, Form, function, and feedbacks in a tidally dominated ooid shoal, Bahamas: Sedimentology, v53(6): p1191-1210.

[92] **J. B. Svendsen,** Jan 2003, Parabolic halite dunes on the Salar de Uyuni, Bolivia: Sed. Geology, v155(1): p147-156. Abstract.

[93] **F. Risacher and B. Fritz,** June 2000, Bromine geochemistry of Salar de Uyuni and deeper salt crusts, Central Altiplano, Bolivia: Chemical Geology, v167(3): p373-392. https://doi.org/10.1016/S0009-2541(99)00251-X

Chapter 12

[1] **C. Sagan,** 1977, The Dragons of Eden: 263p., Random House.

[2] **D. F. Su,** 2013, The Earliest Hominins: Sahelanthropus, Orrorin, Ardipithecus: Nature Education Knowledge v4(4):11. https://www.nature.com/scitable/knowledge/library/the-earliest-hominins-sahelanthropus-orrorin-and-ardipithecus-67648286

[3] **B. Hilton Barber and L. Berger,** 2004, Field Guide to Cradle of Humankind, 212p, Struik Publisher.

[4] **Wikipedia,** Ape: Wikipedia: https://en.wikipedia.org/wiki/Ape

[5] **Editors of Encyclopedia Britannica,** Peking Man—anthropology: https://www.britannica.com/topic/Peking-man

[6] **E. Callaway,** Aug 24, 2022, Seven-million-year-old femur suggests ancient human relative walked upright: Nature News. https://www.nature.com/articles/d41586-022-02313-7

[7] **Arizona State Univ.,** Lucy's Story, Institute of human origins: https://iho.asu.edu/about/lucys-story

[8] **Smithsonian Inst.,** Laetoli Footprint Trails: https://humanorigins.si.edu/evidence/behavior/footprints/laetoli-footprint-trails

[9] **E. Wayman,** April 9, 2012, Mrs. Ples: A hominid with an identity crisis: https://www.smithsonianmag.com/science-nature/mrs-ples-a-hominid-with-an-identity-crisis-59680909/

[10] **Brand of South Africa,** Jan 13, 2006, Mystery of the "Taung Child" solved: https://brandsouthafrica.com/108824/community/mystery-of-the-taung-child-solved/

[11] **R. Picheta,** Jan 19, 2019, 'Missing Link' in human history confirmed after long debate – CNN: https://www.cnn.com/2019/01/19/health/australopithecus-sediba-human-history-scli-intl/index.html

[12] **S. Knapton,** Feb 20, 2018, Language started 1.5m years earlier than previously thought as scientists say Homo Erectus were first to talk: The Telegraph. https://www.telegraph.co.uk/news/2018/02/20/language-started-15m-years-earlier-previously-thought-scientists/

[13] **L. R. Berger, D. J. de Ruiter, S. E. Churchill, P. Schmid, K. J. Karlson, P. H. G. M. Dirks, J. M. Kibli,** April 2010, Australopithecus sediba: A new species of Homo-like Australcpithecus from South Africa: Science, v328(5975): p195-204. https://doi.org/10.1126/science.1184944

[14] **B. M. Fagan,** 1998, World Prehistory—A brief Introduction: 348p. Longman Pub.

[15] **R. Dunbar,** June 1992, Neocortex size as a constraint on group size in primates: Jour. Human Evolution, v22(6): p469-493. Abstract. https://doi.org/10.1016/0047-2484(92)90081-J

[16] **Wikipedia,** 4.2-kiloyear event: https://en.wikipedia.org/wiki/4.2-kiloyear_event

[17] **C. V. Ward and A. S. Hammond,** 2016, Australopithecus and Kin: Nature-Education, v7(3):1. https://www.nature.com/scitable/knowledge/library/australopithecus-and-kin-145077614/

[18] **Britannica,** Human Evolution—Increasing brain size: https://www.britannica.com/science/human-evolution/Increasing-brain-size

[19] **B. Pobiner,** Mar-April 2016, Meat eating among the earliest humans: American Scientist, v104(2): p110.

[20] **Sang Hee Lee with Shin Young Yoon,** 2018, Close Encounters with Humankind: A paleoanthropologist investigates our evolving species: 352p. W.W. Norton & Co.

[21] **R. Schuster,** Dec 5, 2016, Archeologists find 780,000-year-old remains of prehistoric man's meal. https://www.haaretz.com/archaeology/2016-12-05/ty-article-magazine/780-000-year-old-remains-of-plant-diet-found-in-israel/0000017f-f87e-d887-a7ff-f8fe05f00000

[22] **Britannica,** Human Evolution—recuction in tooth size: https://www.britannica.com/science/human-evolution

[23] **H. Osborne,** Dec 18, 2019, Last of the Homo Erectus killed in 'Mass Death' event as climate change drove species to extinction: Newsweek Magazine.

[24] **G. Forth,** May 3, 2022, Between Ape and Man: 336p. Pegasus Books.

[25] **K. Kris Hirst,** May 3, 2019, The discovery of fire in the early Stone age: ThoughtCo. https://www.thoughtco.com/the-discovery-of-fire-169517

[26] **J. A. J. Gowlett,** June 5, 2016, The discovery of fire by humans: a long and convoluted process: Phil. Trans. Royal Soc. B Biol. Sci., v371(1696). https://doi.org/10.1098/rstb.2015.0164

[27] **M. B. Loayza Careaga,** Aug 1, 2023, Why don't humans have fur? The Scientist. Issue 1. https://www.the-scientist.com/why-don-t-humans-have-fur-71201

[28] **J. F. O'Connell, J. Allen, M. A. J. Williams, A. N. Williams, C. S. M. Turney, N. A. Spooner, J. Kamminga, G. Brown, A. Cooper,** Aug 21, 2018, When did H. sapiens first reach Southeast Asia and Sahul? PNAS, USA, v115(34): p8482-8490.

[29] **DBpedia,** About: Torralba and Ambrona (archaeological site): https://dbpedia.org/page/Torralba_and_Ambrona_(archaeological_site)

[30] **Smithsonian's Human Origins,** Jan 3, 2024, Homo habilis: https://humanorigins.si.edu/evidence/human-fossils/species/homo-habilis

[31] **R. Gutman-Wei,** Dec 12, 2019, A 'Mic Drop' on a Theory of Language evolution: The Atlantic Magazine. https://www.theatlantic.com/science/archive/2019/12/when-did-ancient-humans-start-speak/603484/

[32] **Wikipedia,** Archaic Homo sapiens: https://en.wikipedia.org/wiki/Archaic_Homo_sapiens

[33] **F. Dorey,** Oct 16, 2020, Homo sapiens—modern humans: Australian Museum. https://australian.museum/learn/science/human-evolution/homo-sapiens-modern-humans/

[34] **Q. Ji, W. Wu, Y. Ji, Q. Li, X. Ni,** Aug 28, 2021, Late Middle Pleistocene Harbin cranium represents a new Homo species: The Innovation, v2(3):100132. https://doi.org/10.1016/j.xinn.2021.100132

[35] **R. Khan,** June 30, 2021, The Human family tree, it turns out, is complicated: Nautilus Magazine, July 01, 2021.

[36] **J. Rios-Garaizar, O. Lopez-Bulto, E. Iriarte, C. Perez-Garrido, R. Pique, A. Aranburu, M. J. Iriarte-Chiapusso, I. Ortega-Cordellat, L. Bourguignon, D. Garate, I. Libano,** Mar 28, 2018, A Middle paleolithic wooden digging stick from Aranbaltza III, Spain: Jour. PLoS ONE, v13(3): e0195044. https://doi.org/10.1371/journal.pone.0195044

[37] **B. Latza Nadeau,** May 11, 2021, Gnawed bones of 9 Neanderthals found in Italian cave an 'extraordinary' discovery: CNN May 16, 2021. https://www.cnn.com/2021/05/11/world/neanderthal-remains-italy-scn/index.html

[38a] **S. Landow,** Feb 19, 2021, Ancient relic points to a turning point in Earth's history 42,000 years ago: Univ. New South Wales, https://www.unsw.edu.au/newsroom/news/2021/02/ancient-relic-points-to-a-turning-point-in-earth-s-history-42-00

[38] **J. Timperley,** Mar 10, 2023, Why don't humans have fur? BBC. https://www.bbc.com/future/article/20230310-why-dont-humans-have-fur

[39] **A. Piore,** April 2020, How we learned to love Neanderthals .... and a lot of other hominids, too: Nautilus Magazine. https://nautil.us/how-we-learned-to-love-neanderthalsand-a-lot-of-other-hominids-too-237774/

[40] **Dictionary.com,** Archaic Homo sapiens Definition and Meaning: https://www.dictionary.com/browse/archaic-homo-sapiens

[41] **M. Martinon-Torres,** Feb 15, 2018, The Past, Present and Future of human evolution: Nature, v554, p.296-297.

[42] **X. Wu,** 2004, On the origin of modern humans in China: Quaternary International, v117(1): p131-140. Abstract. https://doi.org/10.1016/S1040-6182(03)00123-X

[43] **Bible Gateway,** KJV, Genesis 1, Quote 26: https://www.biblegateway.com/passage/?search= Genesis 1 and version=KJV

[44] **L. Chutel,** Sep 16, 2018, What the oldest drawing found in South Africa tells us about our human ancestors: Quartz Africa. https://qz.com/africa/1389000/what-the-oldest-drawing-found-in-south-africa-tells-us-about-our-human-ancestors

[45] **M. Walker, M. J. Head, J. Lowe, M. Berkelhammer, S. BjÖrck, H. Cheng, L. C. Cwynar, D. Fisher, V. Gkinis, A. Long, R. Newnham, S. O. Rasmussen, H. Weiss,** April 23, 2019, Subdividing the Holocene Series/Epoch: Formalization of stages/ages and subseries/subepochs, and designation of GSSPs and auxiliary stratotypes: Jour. Quaternary Science, v34(3): p173-186. Abstract. https://doi.org/10.1002/jqs.3097

[46] **B. Handwerk,** Jan 13, 2021, 45,000-Year-Old pig painting in Indonesia may be oldest known animal art: Smithsonian Mag. https://www.smithsonianmag.com/science-nature/45000-year-old-pig-painting-indonesia-may-be-oldest-known-animal-art-180976748/

[47] **T. Puiu,** June 27, 2019, Ancient Neanderthal genomes reveal surprising twist in their settling of Europe: ZME Science. https://www.zmescience.com/science/anthropology/neanderthal-genome-europe-settling-0423/

[48] **Lowenmensch.de,** The Lion Man - Der Löwenmensch: http://www.loewenmensch.de/lion_man.html

[49] **S. Florek,** 2022, Dolni Vestonice—Archeological Site: Australian Museum. https://australian.museum/learn/cultures/international-collection/dolni-vstonice-archaeological-site/

[50] **B. Haughton,** May 04, 2011, Göbekli Tepe—the World's first Temple? Ancient History Encyclopedia. https://www.worldhistory.org/article/234/gobekli-tepe---the-worlds-first-temple/

[51] **The News Lens,** April 02, 2017, What remains of Asia's Traditional Sky Burial Sites: https://international.thenewslens.com/article/65071

[52] **UNESCO,** Cueva de la Manos, Rio Pinturas: https://whc.unesco.org/en/list/936/

[53] **Wikipedia,** Megafauna: https://en.wikipedia.org/wiki/Megafauna

[54] **U. College of London,** Nov 4, 2019, Deep Sea vents had ideal conditions for origin of life: https://www.ucl.ac.uk/news/2019/nov/deep-sea-vents-had-ideal-conditions-origin-life

[55] **Wikipedia,** Lascaux: https://en.wikipedia.org/wiki/Lascaux

[56] **L. Sheriff,** Sept 17, 2021, 'Last Journey' offers clues to an ancient civilization: CNN Travel. https://www.cnn.com/travel/article/rock-art-serrania-la-lindosa-colombia-last-journey-scn/index.html

[57] **A. Buffenstein,** June 23, 2015, 20,000-year-old rock paintings discovered in remote Colombian jungle: Artnet. https://news.artnet.com/art-world/rock-paintings-discovered-colombian-jungle-310006

[58] **UCMP,** Berkeley, The La Brea Tar pits: https://ucmp.berkeley.edu/quaternary/labrea.html

[59] **S. Onary, A. D. Rincón, A.S. Hsiou,** Aug 14, 2018, Fossil snakes (Squamata, Serpentes) from the tar pits of Venezuela: taxonomic, paleoenvironmental and paleobio-geographical implications for the North of South America during the Cenozoic/Quaternary boundary: PeerJ, v6:e5402. https://doi.org/10.7717/peerj.5402

[60] **Australian Museum,** Nov 13, 2018, Megafauna: https://australian.museum/learn/australia-over-time/megafauna/

[61] **BBC News,** Mar 24, 2012, 'Humans killed off Australia's giant beasts: https://www.bbc.com/news/science-environment-17488447

[62] **ABC,** Mar 3, 2016, Fact check: Does Australia have one of the 'highest loss of species in the world'? https://www.abc.net.au/news/2015-08-20/fact-check-does-australia-have-one-of-the-highest-extinction/6691026

[63] **D. F. Belknap,** Hominin Evolution: Britannica. https://www.britannica.com/science/Quaternary

[64] **R. Aridi,** Oct 20, 2020, Droughts and human interference wiped out Madagascar's gigantic wildlife 1,500 years ago: Smithsonian Mag. https://www.smithsonianmag.com/smart-news/droughts-and-human-interference-wiped-out-madagascars-gigantic-wildlife-1500-years-ago-180976094/

[65] **C. Haggi, C. M. Chiessi, U. Merkel, S. Mulitza, M. Prange, M. Schulz, E. Schefub,** Oct 1, 2017, Response of Amazon rainforest to late Pleistocene climatic variability: Earth and Planetary Science Letters, v479: p50-59. https://ui.adsabs.harvard.edu/link_gateway/2017E&PSL.479...50H/doi:10.1016/j.epsl.2017.09.013

[66] **L. W. Alvarez, W. Alvarez, F. Asaro, H. V. Michel,** 1980, Extraterrestrial cause for the Cretaceous-Tertiary Extinction: Science, v208(4448): p1095-1108. https://doi.org/10.1126/science.208.4448.1095

[67] **T. Ghose,** Dec 7, 2012, Volcanoes, not meteorite, killed the Dinosaurs, Scientist argues: https://www.livescience.com/25324-volcanoes-killed-dinosaurs.html

[68] **Wikipedia,** Hippopotamus in Colombia: https://en.wikipedia.org/wiki/Hippopotamuses_in_Colombia

[69] **V. Richter,** July 6, 2015, The big five mass extinctions: Cosmos Magazine. https://cosmosmagazine.com/history/palaeontology/the-big-five-mass-extinctions/

[70] **A. M. C. Sengör, S. Atayman, S. Ozeren,** Sept 16, 2008, A scale of greatness and causal classification of mass extinctions: Implications for mechanisms: PNAS, USA, v105(37): p13736-13740. doi:10.1073/pnas. 0805482105.

[71] **M. Conte,** April 29, 2020, Pentagon officially releases UFO videos: CNN. https://www.cnn.com/2020/04/27/politics/pentagon-ufo-videos/index.html

[71a] **A. Hurt,** Mar 22, 2021, Life elsewhere in the Universe: When did we first consider the possibility? https://www.discovermagazine.com/the-sciences/life-elsewhere-in-the-universe-when-did-we-first-consider-the-possibility

[72] **NASA/JPL-Caltech,** Mar 3, 2021, Infographic: Profile of Planet 51 Pegasi b—Exoplanet Exploration: https://science.nasa.gov/resource/infographic-profile-of-planet-51-pegasi-b/

[73] **D. Cook,** Nov 2020, Perseverance Rover cruising to the Red Planet: AAPG Explorer, p6,8.

[74] **M. Gillon, E. Jehin, S. M. Lederer, L. Delrez, J. de Wit, A. Burdanov, V. Van Grootel, A. Burgasser, A. H. M. J. Triaud, C. Opitom, B-O. Demory, D. K. Sahu, D. B. Gagliuffi, P. Magain, D. Queloz,** May 12, 2016, Temperate Earth-sized planets transiting a nearby ultracool dwarf star: Nature, v533(7602): p221-224. https://www.science.org/doi/10.1126/science.aay3253

[75] **Wikipedia,** TRAPPIST-1: https://en.wikipedia.org/wiki/TRAPPIST-1

[76] **A. Strickland,** June 10, 2021, Newly discovered planet could have water clouds: https://www.cnn.com/2021/06/10/world/exoplanet-atmosphere-study-scn

[77] **Population Matters,** Welcome to Anthropocene: https://populationmatters.org/welcome-to-the-anthropocene/

[78] **Environmental News Network,** Oct 31, 2018, WWF report reveals staggering extent of human impact on Planet: https://www.enn.com/articles/55902

[79] **Y. Malhi,** Aug 26, 2017, The concept of Anthropocene. Annual Review of Environment & Resources, v42: p25.1-25.28. https://doi.org/10.1146/annurev-environ-102016-060854

[80] **M. Williams, J. Zalasiewicz, C. N. Waters, M. Edgeworth, C. Bennett, A. D. Barnosky, E. C. Ellis, M. A. Ellis, A. Cearreta, P. Haff, J. A. Ivar do Sul, R. Leinfelder, J. Mcneill, E. Odada, N. Oreskes, A. Revkin, D. D. Richter, W. Steffen, C. P. Summerhayes, J. Syvitski, D. Vidas, M. Wagreich, S. L. Wing, A. P. Wolfe, A. Zhisheng,** Feb 03, 2016, The Anthropocene: a conspicuous stratigraphical signal of anthropomorphic changes in production and consumption across the Biosphere: Earth's Future, v4(3): p34-53. https://doi.org/10.1002/2015EF000339

[81] **Europlanet Society,** Sept 20, 2019, Could Venus have been habitable? https://www.europlanet.org/could-venus-have-been-habitable/

[82] **P. Kelley,** Nov 20, 2018, Study brings new climate model of small star TRAPPIST 1's seven intriguing worlds: UW News. https://www.washington.edu/news/2018/11/20/study-brings-new-climate-models-of-small-star-trappist-1s-seven-intriguing-worlds/

[83] **P. J. Crutzen and E. F Stoermer,** 2000, The "Anthropocene": IGBP Global Newsletter, No. 41: p17-18.

[84] **Wikipedia,** Anthropocene: https://en.wikipedia.org/wiki/Anthropocene

[85] **E. Chung,** Mar 5, 2024 "Goodbye, Anthropocene? Scientists vote against new epoch": CBC News. https://www.cbc.ca/news/science/anthropocene-epoch-scientists-vote-1.7134449

Chapter 13

[1] **G. Forth,** May 03, 2022, Between Ape and Man: Pegasus Books, 336p.

[2] **New World Encyclopedia,** Red Sea: https://www.newworldencyclopedia.org/entry/Red_Sea

[3] **Niagara Parks,** Niagara Falls facts | Geology. Facts and Figures: https://www.niagaraparks.com/visit-niagara-parks/plan-your-visit/niagara-falls-geology-facts-figures/

[4] **K. Andrews,** June 14, 2016, What causes an ice age and what would happen if the Earth endured another one? https://www.abc.net.au/news/science/2016-06-15/what-is-an-ice-age-explainer/7185002

[5] **R. A. Berner,** Dec 2006, A combined model for atmospheric O_2 and CO_2: Geochimica & Cosmochimica Acta, v70(23): p5653-5664. https://doi.org/10.1016/j.gca.2005.11.032

[6] **NASA,** Nov 15, 2011, Supervolcanoes: Not a threat for 2012: https://phys.org/news/2011-11-supervolcanoes-threat.html

[7] **S. Landow,** Feb 19, 2021, Ancient relic points to a turning point in Earth's history 42,000 years ago: Univ. New South Wales, https://www.unsw.edu.au/newsroom/news/2021/02/ancient-relic-points-to-a-turning-point-in-earth-s-history-42-00

[8] **NASA,** Apophis: https://science.nasa.gov/solar-system/asteroids/apophis/

[9] **NASA,** Asteroid fast facts: https://www.nasa.gov/solar-system/asteroids/asteroid-fast-facts/

[10] **P. Smith,** April 4, 2022, An icy mystery deep in Arctic Canada: BBC. https://www.bbc.com/travel/article/20220403-an-icy-mystery-deep-in-arctic-canada

[11] **A. E. Snyder-Beattie, T. Ord, M. B. Bonsall,** July 30, 2019, An upper bound for the background rate of human extinction: Nature, Sci. reports 9, Article 11054. https://doi.org/10.1038/s41598-019-47540-7

[12] **J. R. Gott,** May 27, 1993, Implications of the Copernican principle for our future prospects: Nature v363: p315-319.

[13] **National Science Foundation,** April 7, 2004, How long does it take for earth's magnetic field to reverse? Science Daily: https://www.sciencedaily.com/releases/2004/04/040408085209.htm

[14] **C. Leary,** Oct 21, 2021, 8 Ancient civilizations destroyed by climate change: https://www.treehugger.com/ancient-civilizations-were-destroyed-climate-change-4869712

[15] **ETH Zurich,** Jan 14, 2022, "Earth's interior is cooling faster than expected": Science Daily. https://www.sciencedaily.com/releases/2022/01/220114115640.htm

[16] **CNRS,** Dec 16, 2013, When will Earth lose its oceans?: ScienceDaily. https://www.sciencedaily.com/releases/2013/12/131216142310.htm

[17] **S. Frazier,** Sept 26, 2019, Why the Sun won't become a Black Hole—NASA.gov: https://www.nasa.gov/image-article/why-sun-wont-become-black-hole/

[18] **NASA Earth Observatory,** July18, 2009, Mount Tambora Volcano, Sumbawa Island, Indonesia: https://earthobservatory.nasa.gov/images/39412/mount-tambora-volcano-sumbawa-island-indonesia

[19] **A. Farnsworth, Y. T. E. Lo, P. J. Valdes, J. R. Buzan, B. J. W. Mills, A. S. Meredith, C. R. Scotese, H. R. Wakeford,** Sept 25, 2023, Climate extremes likely to drive land mammal extinction during next supercontinent assembly: Nature Geoscience, v16: p901-908. http://dx.doi.org/10.1038/s41561-023-01259-3

INDEX

About the Author

SANTOSH GHOSH is a widely recognized geologist, academician, industry practitioner and speaker with direct experience in the energy industry, research, academia, and public sector mineral exploration. He has published and presented his work extensively in international geological journals and conferences. After his B.Sc. (Honors) and M.Sc. in geology from the Presidency College, Calcutta University, India, he earned his Ph.D. in geology (1977) from the Binghamton University, New York, as a U.S. Government Fulbright Scholar. A certified petroleum geologist, he is emeritus member of American Association of Petroleum Geologists (AAPG), USA, and Society for Sedimentary Geology (SEPM), USA. This book, a testimony of 5-decades-long international experience, reflects his life-long passion for geology and how it unravels the mystique of the universe and the planet we inhabit. He highlights climate change through time, including his concern about its human impact. He now lives in the Greater Toronto Area, Canada.

Made in United States
Cleveland, OH
18 October 2026

46341973R00291